Mapping Strategies
in Chemical Oceanography

ADVANCES IN CHEMISTRY SERIES **209**

Mapping Strategies in Chemical Oceanography

Alberto Zirino, EDITOR
Naval Ocean Systems Center

Based on a symposium sponsored by
the Division of Analytical Chemistry
at the 185th Meeting
of the American Chemical Society,
Seattle, Washington,
March 20–25, 1983

American Chemical Society, Washington, D.C. 1985

Library of Congress Cataloging in Publication Data
Mapping strategies in chemical oceanography.
  (Advances in chemistry series, ISSN 0065-2393; 209)

  "Papers originally presented at the Analytical
Chemistry Division Symposium 'Analytics for Mesoscale
and Macroscale Processes'"—Pref.

  Bibliography: p.
  Includes index.

  1. Chemical oceanography—Congresses.

  I. Zirino, Alberto, 1941-   . II. American Chemical
Society. Division of Analytical Chemistry. III. American
Chemical Society. Meeting (185th: 1983: Seattle, Wash.)
IV. Analytical Chemistry Division Symposium "Analytics
for Mesoscale and Macroscale Processes" (1983: Seattle,
Wash.) V. Series.

QD1.A355   no. 209   [GC110]
540 s   [551.46'01]   85-20265
ISBN 0-8412-0862-X

Advances in Chemistry Series

M. Joan Comstock, *Series Editor*

Advisory Board

FOREWORD

The ADVANCES IN CHEMISTRY SERIES was founded in 1949 by the American Chemical Society as an outlet for symposia and collections of data in special areas of topical interest that could not be accommodated in the Society's journals. It provides a medium for symposia that would otherwise be fragmented because their papers would be distributed among several journals or not published at all. Papers are reviewed critically according to ACS editorial standards and receive the careful attention and processing characteristic of ACS publications. Volumes in the ADVANCES IN CHEMISTRY SERIES maintain the integrity of the symposia on which they are based; however, verbatim reproductions of previously published papers are not accepted. Papers may include reports of research as well as reviews, because symposia may embrace both types of presentation.

ABOUT THE EDITOR

ALBERTO ZIRINO was first introduced to "chemical" mapping in 1964 while working as a technician at the Scripps Institution of Oceanography (Food Chain Group) and simultaneously pursuing an M.S. degree in spectroscopy at San Diego State College. At that time, bubble-segmented colorimetric analysis was being adapted to ocean mapping by F. A. J. Armstrong and J. D. H. Strickland. Zirino became interested in the electrometric measurement of trace metals in seawater while completing a Ph.D. thesis on the measurement of zinc in seawater by anodic stripping voltammetry at the University of Washington. There he was able to carry out voltammetric determinations at sea on several major oceanic mapping expeditions led by R. C. Dugdale, F. A. Richards, and M. L. Healy. In 1970, he joined the Naval Ocean Systems Center as a National Research Council Postdoctoral Fellow. He became Head of the Center's Marine Environment Branch in 1979. Since 1978, he has been teaching Introductory Oceanography and Chemical Oceanography as an adjunct professor at San Diego State University. Zirino has presented numerous invited papers at national and international meetings on such diverse subjects as chemical speciation of trace metals in seawater, voltammetric techniques, real-time methods in chemical oceanography, and, recently, satellite chemical oceanography. In June 1984, he was a visiting professor at the Institute of Biophysics (C.N.R.) at the University of Pisa. He has authored or coauthored over 25 journal articles and research papers, and he is a member of the American Chemical Society and the American Geophysical Union.

This book is dedicated to the memory of Prof. F. A. Richards (1917–1984), mentor, friend, and colleague, who clearly understood, loved, and encouraged interdisciplinary oceanic exploration.

CONTENTS

ix

PREFACE

THIS BOOK IS A COLLECTION of invited papers originally presented at a symposium entitled "Analytics for Mesoscale and Macroscale Processes." This symposium focussed on current approaches to chemical and biological mapping of the oceans, and it brought together many of the scientists currently engaged in this task. By presenting the collective experiences of seasoned investigators rather than just descriptions of methods, this book attempts to give a practical overview of "chemical" mapping, including discussions on method development, sampling, data collection, and analysis. No single chapter covers all of these topics in great detail; however, all of the chapters stress environmental applications as well as methodology.

The general approach to mapping is to develop high-resolution analytical techniques for the measurement of chemical, biological, and physical parameters so that their changes in time and space can be followed. Chemical mapping, therefore, is considered throughout this book to be an interdisciplinary analytical problem, and no artificial boundaries are created to separate traditional scientific fields. Indeed, the authors consist largely of physicists, chemists, and engineers who use methods originally developed in their respective fields and who have joined biologists to study biological questions. This consequence is natural when working in an environment in which chemistry, biology, and physics are inextricably related, and in which changes over smaller spatial and temporal scales are either controlled by or manifested in biological processes.

I hope that by bringing forth many novel or very recent observations made with new and diverse methodologies the book will encourage traditional analytical scientists to adapt their own methods to marine environmental monitoring and perhaps join the authors in their chosen task of ocean exploration.

I gratefully acknowledge the Analytical Division of the American Chemical Society, the Office of Naval Research, and the Naval Facilities Command for sharing the registration and travel costs associated with the symposium. Many thanks are also due to Florence Edwards and Deborah Corson, the American Chemical Society's in-house editors who did much of the work associated with the production of the book and who deserve most of the credit for getting it into print. I also wish to thank my colleagues and sponsors at the Naval Ocean Systems Center, San Diego, who have supported the concept of real-time ocean mapping over the past 15 years.

Finally, I wish to thank Antonio, Herta, Barbara, Laura, Aline, and Marco Zirino for contributing liberally, though indirectly, "their" time to this book and to marine chemistry.

ALBERTO ZIRINO
Naval Ocean Systems Center
San Diego, CA 92152
August 1984

New Problems for Chemical Oceanographers

EDWARD J. GREEN
Office of Naval Research, Arlington, VA 22217

The development of new methodologies in chemical ocean-ography—techniques such as underway sampling, fiber-op-tic technology, laser-induced fluorescence, flow-injection analysis, and other new autoanalyzer methods—promises to initiate a revolution in the scope of problems that can be ad-dressed by marine chemists. The new problems are to be found in chemical variations occurring on short spatial scales and short time scales in upwelling and other frontal regions where chemical patchiness and biological patchiness are strongly interactive.

WILLIAM DITTMAR PUBLISHED HIS RESULTS FROM THE FIRST 2 years of the H.M.S. *Challenger* expedition almost 100 years ago (*1*). His results demon-strated that, to a very good first approximation, seawater could be consid-ered a two-component system (water and sea salt). Consequently, 99.5% (by weight) of all the chemical components of seawater became relatively uninteresting to water-column chemists. However, 80 of the elements of the periodic table could be found in the remaining 0.5%. Consequently, marine chemists have tended to concentrate on the minor and trace compo-nents of seawater because the interesting variations occur in them; that is, they reveal information about the processes and mechanisms that influence the chemistry of seawater. Chemical oceanographers, as all scientists do when their field of study matures, naturally progress from the survey mode to process-oriented investigations, from correlations of observations to hy-potheses of mechanisms, and from the general understanding of the chem-istry of seawater to highly specialized and focused interests.

Evolution of Chemical Oceanography

Chemical oceanography has evolved from the study of the major elements to the investigation of minor components—the nutrients, silica, and dis-solved oxygen—then on to radiotracers, trace metals, organics, and bio-

0065-2393/85/0209-0001$06.00/0

genic gases (Figure 1). Simultaneously, analytical methods and techniques have also evolved. Initially our basic tool was based on the hydrographic wire and the Nansen bottle, a technology rooted in the 19th century. Despite limited vertical and horizontal resolution, the data obtained still represented the most valuable base for discussion of the general circulation of the oceans.

In the past 20 years the Nansen-bottle technology has been replaced with the conductivity, temperature, and depth (CTD)–rosette system. Vertical resolution of samples is now within centimeters, temperature measurements are within the accuracy of the best reversing thermometers, and data are transmitted directly to computer, which not only eliminates tedium but also safeguards against transcription and calculation errors. The advent of the CTD–rosette has, however, some disadvantages. Nansen bottles are purely mechanical, and, therefore, can be deployed, maintained, and repaired by technicians with little specialized training. CTD systems are sophisticated electronic instruments that require personnel trained especially in their maintenance. Moreover, they are linked to complex computers and software that also need highly skilled and trained operators. Therefore, data acquisition has not only become much more expensive, it is now limited, at least in the modern sense, to large institutions. The net result of these changes is that much more coordination and planning are

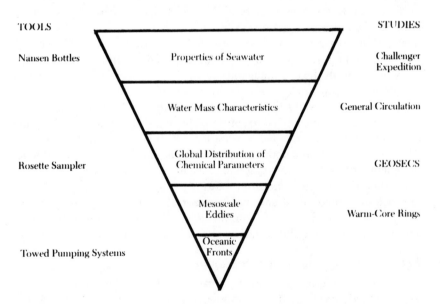

Figure 1. The evolution of chemical oceanography, 1884–1984. Oceanic fronts are characterized by areas of enhanced biological activity—the major process by which the chemistry of the upper ocean varies on a short time scale.

required of scientists who wish to justify and mount major field efforts in oceanography.

These trends will continue and the specialization and sophistication of data gathering will intensify. A major driving force in this direction is the increasing interest in problems characterized by shorter time and space scales. For many years the equipment and techniques available to physical oceanographers made it almost impossible to study any phenomenon other than the general circulation in the ocean. These very large scales of motion, the general circulation, are set by the large-scale wind stress pattern, the large-scale solar heating and cooling, and the broad bathymetry of the ocean basins.

However, as advanced techniques and equipment allowed measurement on finer grids and shorter time scales, the mesoscale fluctuations were found to be much more energetic than the general circulation. The great bulk of the energy supplied to the ocean does *not* drive the general circulation, but rather drives the finer scale fluctuations in the ocean, with horizontal space scales of tens to several hundreds of kilometers, and time scales not of decades or years, but of many days to several months.

Possible New Tools

We foresee a parallel development in chemical oceanography. In the past 7 years, considerable attention has been paid to the chemistry of Gulf Stream rings, features with spatial scales of tens of kilometers and lifetimes of months. To study still smaller phenomena—ocean fronts that may have spatial scales of tens of meters and evolution periods of days—the old hydrographic cast method of observation is inadequate. One possible direction for the future is the deployment of underway towed pumping systems. Some investigators have already made significant progress toward this goal.

Advances in plankton sampling have been made by Alex W. Herman.[1] He deploys his Batfish vehicle to measure chlorophyll a fluorescence, conductivity, temperature, and depth with a CTD device, and zooplankton numbers with a conductivity sensor. Eugene Traganza[2] has been surface sampling with an on-deck pump in his studies of the California current upwelling system. In this region, where strong chemical variability extends all the way to the surface, significant work can be done with surface sampling. In most cases, however, sampling over a range of depths within the photic zone will be necessary. David R. Schink[3] and his colleagues have employed a modified version of a commercially available towed pump to study biogenic dissolved gases in the ocean. Richard J. Mousseau[4] and his

---

[1] Bedford Institute of Oceanography.
[2] Naval Postgraduate School.
[3] Texas A&M University.
[4] Gulf Science and Technology Company.

coworkers have had a similar hydrocarbon "sniffer" operational in petroleum exploration for several years. The National Oceanic and Atmospheric Administration has put some effort into developing a six-port pumping system for the study of dispersion of wastes in the sea.

Other Needed Instrumentation

Underway water sampling is only part of the solution. Underway chemical analyses are essential to the investigation of small-scale, rapidly evolving features. Continuous, in-situ specific electrodes for all the chemical parameters of interest would be ideal, but currently we have only the pH electrode and the dissolved oxygen sensor, and very few ideas on how to proceed with in-situ sensors for other chemical species at trace concentrations. The next best thing is to mate underway water sampling with continuous or very rapidly repetitive, on-board chemical analyses. Considerable progress is being made in this area also.

The autoanalyzer represented a substantial advance in the ability to make repetitive on-board chemical measurements. The development of flow-injection analysis promises to enlarge enormously the scope and speed of such methods. The flow-injection methods were pioneered in clinical chemistry, and are being developed for oceanography by Kenneth S. Johnson[5] and Robert L. Petty.[5] Dana Kester[6] and Richard W. Zuehlke[6] are also developing new autoanalyzer techniques for the rapid analysis of certain trace metals in seawater. One of the most technologically exciting prospects for the future involves the use of fiber optics to transmit a spectroscopic signal from an in-situ sensor to the ship's deck.

Just as with underway water sampling, these analytical systems are becoming complex, expensive, and require highly trained technical specialists to maintain and operate them. This effort and expense can only be justified if significant oceanographic problems are being tackled. Understanding the biology and chemistry of frontal processes is such a significant problem.

The Importance of Fronts

Ocean fronts are regions where vertical advection and the exchange of momentum and other properties are intense. In addition, two-sided convergences are very effective in collecting and concentrating suspended particulate matter. For these reasons, in part, fronts are regions of enhanced chemical variability. Moreover, an understanding of the role of biological fronts is needed in both diagnostic and prognostic biological productivity models.

A number of intensely interesting questions can be posed about the

---

[5] University of California, Santa Barbara.
[6] University of Rhode Island Graduate School of Oceanography.

biology and chemistry of fronts: Do distributions of chemical properties at ocean fronts follow general patterns; what mechanisms—physical, biological, or chemical—are important for these patterns in chemical properties; can predictive chemical models be developed based on physical distributions and remote sensing imagery; what diffusional processes occur across ocean fronts and how can understanding of these processes be used for ocean prediction purposes? Many of the contributors to this volume are in the forefront in answering these questions.

Recent Progress in Upper Ocean Chemistry

Traganza and his students have studied the eddies of the California upwelling (2–3). His system allows continuous chlorophyll a and temperature measurements. Nutrients are processed by autoanalyzer every 2 min (~ 500-m spacing at 9 knots); adenosine triphosphate (ATP) and guanosine triphosphate (GTP) are processed every 10 min (~ 2.5-km spacing). These densely spaced data show the expected close correlation of temperature and nutrient chemistry; the cold water is high in nutrients, the warm water is low. Most interesting, however, is the location of biomass concentrations as determined by ATP and chlorophyll. Traganza's data show strongly that the blooms are occurring not in the cold, nutrient-rich water, but rather on the thermochemical gradients. This occurrence is particularly true for the equatorial side of cyclonic cold water spirals.

This observation was supported by the agreement between in-situ data and color imagery from the Nimbus-7 coastal zone color scanner. Because of the location of the phytoplankton blooms in the sharp nutrient gradients, and because these features persist, Traganza has suggested that the gradients may act as "natural chemostats" in which the specific growth rate is equal to the dilution rate, and the total biomass is proportional to the concentration of the nutrient supply. His study provides evidence that, in certain situations, the biological patchiness may be set initially by the chemical patchiness that is a consequence of the physical oceanography.

The reverse situation may be true with respect to the distribution of dissolved molecular hydrogen in the ocean. Schink's observations (4) of diel variation of hydrogen concentrations in the controlled ecosystem population experiment containers and in the Marine Ecosystem Research Laboratory tanks have been confirmed in the open ocean by Herr et al. (5). Hydrogen is probably produced either in the guts of zooplankton or fish, or by cyanobacteria. Dissolved hydrogen typically decreases with depth below the photic zone; apparently, the gas is consumed by microflora in the deep ocean. Such consumption occurs in freshwater (6) and anaerobic (7) systems.

The studies in this book (by Zuehlke and Kester, Herman, Schink, Traganza, and others) indicate that biological activity is the major process changing the chemistry of the upper ocean on a short time scale. Unraveling the biological–chemical relationships will provide some of the most challenging problems for chemical oceanographers in the years ahead.

Literature Cited

1. Dittmar, W. in "Report on the Scientific Results of the Exploring Voyage of H.M.S. *Challenger*. Physics and Chemistry"; Murray, J., Ed.; H.M. Stationery Office: London, 1884; vol. 1, pp. 1–251.
2. Traganza, E. D.; Nestor, D. A.; McDonald, A. K. *J. Geophys. Res.* 1980, 85, 4101–6.
3. Traganza, E. D.; Conrad, J. C.; Breaker, L. C. in "Coastal Upwelling," Richards, F. A., Ed.; American Geophysical Union: Washington, D.C., 1981; pp. 228–41.
4. Bullister, J. L.; Guinasso, N. L., Jr.; Schink, D. R. *J. Geophys. Res.* 1982, 87, 2022–34.
5. Herr, F. L.; Frank, E.; Leone, G. M.; Kennicutt, M. C. *Deep-Sea Res.* 1984, *31*, 13–20.
6. Conrad, R.; Aragno, N.; Seiler, W. *Appl. Environ. Microbiol.* 1983, 45, 502–10.
7. Sorensen, J.; Christensen, D.; Jorgensen, D. D. *Appl. Environ. Microbiol.* 1981, *42*, 5–11.

RECEIVED for review July 15, 1983. ACCEPTED January 19, 1984.

Flow-Injection Analysis for Seawater Micronutrients

KENNETH S. JOHNSON, ROBERT L. PETTY, and JENS THOMSEN

Marine Science Institute, University of California, Santa Barbara, CA 93106

Flow-injection analysis (FIA) is a technique for automating chemical analyses. The principles of FIA are reviewed here. Methods for applying FIA to the anayses of nitrate, nitrite, phosphate, silicate, and total amino acids in seawater are examined. Analyses of other nutrients, metals, and carbonate system components are also discussed. Various techniques to eliminate the refractive index effect are reviewed. Finally, several examples of the application of FIA to oceanographic problems are presented.

T HE BENEFITS OF AUTOMATING NUTRIENT ANALYSES IN SEAWATER have been recognized and utilized for several decades. Automated analyses allow samples to be processed faster and generally with better precision and accuracy than is possible with most manual methods. The only technique that was available at a reasonable cost for the automated analysis of seawater nutrients, until recently, was segmented continuous-flow analysis (CFA). Segmented CFA is characterized by the use of air bubbles to segment the liquid in the reaction tube so that dispersion of the sample is limited.

Segmented CFA has been used widely at sea, and it has proven to be an extremely valuable tool. The analyses of all major nutrients in seawater have been automated by segmented CFA, including the determination of nitrate and nitrite (*1, 2*), silicate (*2, 3*), phosphate (*2, 4*), and ammonia (*5*). The applications of segmented CFA to analyses in seawater were summarized previously (*6–8*).

A second technique that is capable of automating chemical analyses was developed by Ruzicka and Hansen (*9*), and Beecher, Stewart, and Hare (*10*). This technique, called flow-injection analysis (FIA), is also a form of continuous-flow analysis. However, it differs from segmented CFA in several respects. The main distinction between segmented CFA and FIA is that the continuous mixing of sample and reagents in a turbulent stream segmented by bubbles is replaced by periodic mixing in an unsegmented, laminar stream. Periodic mixing is achieved by injecting the sample, or in some cases the reagent, into a carrier stream flowing to the detector. The analyte species forms in the reaction zone that contains both sample and reagent. The reaction does not have to go to completion. The residence time of the sample is very reproducible because the flow is incompressible; therefore, the extent of reaction will be similar in all of the samples. This similarity results in several advantages, relative to segmented CFA, including faster analyses and less complex equipment.

FIA is readily adaptable to the determination of dissolved nutrients in seawater. Methods for the determination of nitrate (11, 12), phosphate (13), and silicate (14) in seawater have been developed. We have used FIA methods at sea and they have proven to be quite reliable. Because high sampling rates are possible with FIA (15), it is well-suited for oceanographic investigations of small-scale and mesoscale features in the ocean. It is particularly well-suited to applications where it is interfaced with the effluent of a submersible pumping system.

This chapter examines the use of FIA in oceanographic investigations and begins by reviewing the principles of FIA. The methods that can be used for the analysis of micronutrients (nitrate, nitrite, phosphate, silicate, and total amino acids) in seawater will then be discussed. Finally, the application of FIA to the determination of the nutrient structure across an ocean thermal front will be presented.

Principles of Flow-Injection Analysis

The principles of FIA were reviewed in several articles (16–20) and a book (15). Therefore, only a summary is given here.

A simple FIA system consists of a length of small bore tubing [< 1-mm inside diameter (ID), ~ 1 m long] through which a carrier stream is pumped, a valve that can introduce the sample (typically 20–200 μL) into the stream as a discrete plug, and a flow-through detector at the end of the tubing (Figure 1). The successful operation of this system requires reproducible sample injection, timing, and analyte dispersion. Dispersion must be reproducible so that the sample is diluted with the carrier stream by the same amount during each analysis. The residence time of the sample in the reaction tube must also be constant so that the extent of reaction will be the same in all samples and standards. The sample volume injected must be reproducible because it affects the dispersion reproducibility.

Although this seeems to be a difficult combination of conditions to achieve, it is in fact simple. Inexpensive rotary injection valves, such as those used for low-pressure liquid chromatography, will inject highly reproducible volumes. Dispersion and timing are also reproducible if a constant flow rate ($\pm$ 1%) is maintained in the reaction tube. Constant flow rates can be produced with peristaltic pumps. A simple FIA system can, therefore, be assembled from common components.

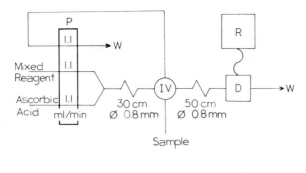

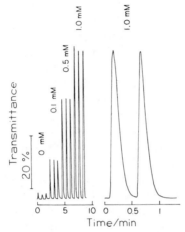

Figure 1. The manifold (top) for the determination of phosphate by FIA. Key: P, the pump (Gilson Mini Puls); IV, injection valve (Rheodyne Model 50); D, LED photometer (24) with an IR LED (880-nm maximum emission); and R, recorder. Typical results for a series of standards are shown (bottom).

A variety of operations can be performed on the sample as it flows through the reaction tube. In the simplest case, the sample merely disperses into an inert carrier stream. It is then measured directly at the detector, as in the determination of seawater pH. The contact time between the sample and the pH electrode is short so that only a fraction of the steady-state response is obtained. However, this fraction is very reproducible [± 0.002 pH, (15)], which is the key to the success of FIA.

The carrier may also contain a reagent that reacts with the sample to form an analyte species. For example, the carrier stream used for the analysis of phosphate contains molybdate and ascorbic acid, which combine with the injected phosphate to form the reduced phosphomolybdate dye (Figure 1). Reagents may also be pumped into the carrier stream at T-fittings if the sequential addition of several reagents is necessary. This technique is used in the determinations of silicate and nitrate. More complex operations, such as gas diffusion through membranes, solvent extractions, or column operations may also be used (15–20). Any type of detector that has a flow-through cell can be interfaced to an FIA system.

The periodic injection of the sample into the flowing carrier stream creates a heterogeneous composition in the carrier stream. The detector will normally see the undiluted carrier stream flowing past it. The sample bolus will, however, pass the detector at intervals and cause a response to the analyte species. The output of the detector is a normally flat baseline interspersed with peak-shaped responses shortly following each sample injection (Figure 1).

Air bubbles are necessary in segmented CFA to prevent excessive broadening of the sample bolus as it flows through the reaction manifold. However, segmentation of the flow is not necessary in FIA to limit tailing of the sample. The most important factor limiting sample tailing is the use of narrow (< 1 mm) tubes. The flow in a narrow tube will be laminar (Reynolds number < 20) at the flow rates (< 10 mL/min) used in FIA (17). Radial diffusion can proceed at a rate comparable with flow along the tube axis under these conditions (21). Any portion of the sample that lags behind in the low-velocity region along the tube wall will rapidly diffuse back into the high-velocity region in the center of the tube where it will "catch up" with the rest of the sample (15, 17).

These processes were studied in detail by Vanderslice et al. (22) with numerical models. Their results were confirmed experimentally (22, 23). Their numerical results (22) show that dispersion will depend on the diffusion coefficient of the sample compound. Gerhardt and Adams (23) used these results to determine diffusion coefficients of inorganic and organic compounds with an accuracy of a few percent by injecting the compounds in an FIA system and measuring their dispersion.

A reaction need not go to completion before the sample enters the detector in FIA (15–20). The extent of reaction will be the same in all samples and standards if constant flow rates and sample volumes are maintained. Successful FIA systems have been used in which the extent of reaction was less than 10%. However, the extent of reaction is surprisingly high for many reactions with residence times less than 30 s. The reaction used for the determination of phosphate is greater than 90% complete in less than 15 s (13) even though the manual method calls for at least 5 min for full color development (2). This extent of reaction was accomplished by heating a portion of the manifold to 50 °C. Greater than 90% of the nitrate in seawater is reduced to nitrite in a cadmium reductor in less than 2 s (11, 12). The reaction of nitrite to form an azo dye is complete in less than 15 s (15).

A sample injected into a carrier stream must mix with the reagents before a reaction can proceed. A minimum dispersion factor of about 3 is necessary to produce uniform mixing of the reagent and sample so that smooth analyte peaks are formed (Figure 1). The dispersion factor is defined as the ratio between the sample concentration in the injector, before mixing has occurred, and the maximum sample concentration at the detector in the absence of any reaction. The dispersion factor will increase as the residence time increases (22). If the sample must mix with several reagents added sequentially, or if the reaction kinetics are slow, then the dispersion factor may exceed 10.

The dispersion of the sample decreases the analyte concentration and reduces the sensitivity of an analysis. If two or more reagents must be added to the carrier stream, or if the initial sample concentration is low, then dispersion may lower the analyte concentration below detection limits. However, by reversing the role of the sample and the carrier–reagent stream from that used in normal FIA, the relationship between dispersion and sensitivity is also reversed (13). In this procedure, called reverse flow-injection analysis (rFIA), the sample is used as the carrier stream and the reagents are injected. The sample concentration in the zone of the injected reagent will increase as time and dispersion also increase. This increase results in a greater sensitivity compared to a similar analysis by FIA in which the sample is injected and its concentration decreases as time and dispersion increase. The manifold for the determination of phosphate by rFIA is shown in Figure 2.

The continually increasing sample concentration in the reagent bolus in rFIA is the reason for the increase in sensitivity of rFIA compared to conventional FIA. If the analyte is produced by a reaction of the sample and re-

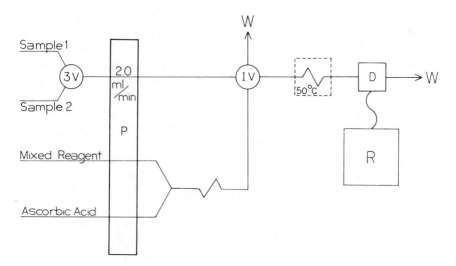

Figure   2. The manifold for the determination of phosphate in seawater (13). All symbols are as in Figure 1. In addition, 3V is a three-way valve (Rheodyne Model 5301) used to switch between samples and standards.

agent, however, the analyte peak will still tend to be diluted with increasing dispersion. Thus, there is some trade-off between increasing dispersion in rFIA and analytical sensitivity.

The sensitivity of FIA and rFIA, using a photometric detector with 1- or 2-cm path-length cells, can approach that obtained by manual methods using 5- and 10-cm cells as long as the reaction kinetics are not unusually slow. The analyte concentrations approach the values that would be found if the sample and reagent were mixed in batch, particularly in rFIA. The detector baseline is also quite stable in FIA because the cell is stationary and few air bubbles are present. Electronic noise and refractive-index interferences are the main factors that affect the detection limits of an analysis. The effects of refractive index errors, which are discussed later, can usually be eliminated by matching the refractive indices of the carrier stream and the injected solution. The electronic noise in the simple light-emitting diode (LED) photometers (24) that we used for most colorimetric analyses is about ± 0.0005 absorbance units, which is about an order of magnitude better than the reproducibility obtained with long path-length cells when they are removed from the spectrophotometer in each manual analysis. Thus, the signal-to-noise ratios for analyses performed by FIA and rFIA are about the same as those obtained with manual methods if the extent of reaction is 50% or greater. The detection limits for the analyses of phosphate, silicate, and nitrate confirm these results (Table I).

The question of whether to use FIA or rFIA is not just one of greater sensitivity, however. There are additional trade-offs. The sample size necessary for a high-sensitivity analysis by FIA is about 500 μL (15); a 5-mL sample is required for rFIA (13). If the sample volume is limited, as in the analysis of sediment-pore waters, then FIA might be the method of choice. In addition, interfacing an FIA analysis to an autosampler is easier. An additional three-way valve is necessary in an rFIA setup (Figure 2) to avoid drawing air when changing samples. Discrete samples can also be analyzed at a higher rate by using FIA because the length of tubing from the sample inlet to the injector can be made shorter. The mixing zone between adjacent samples will, therefore, pass much more quickly. The consumption of reagent solution is lower in rFIA because the sample is used as the carrier stream rather than the reagent. This consideration is important if reagents are expensive, or if data are to be taken at sea for extended periods.

The behavior of blank peaks is also different in FIA and rFIA. Turbidity in the samples will add to the peak height in FIA; however, sample turbidity has little effect in rFIA because the carrier stream has nearly the same turbidity as the carrier stream–reagent bolus. However, if there is a significant reagent blank, it will automatically be compensated for in FIA because the carrier stream contains all of the reagents.

In summary, rFIA seems best suited for analysis of near-surface samples where concentrations are low. In situ pumps can be used to sample the

Table I. Analytical Figures of Merit for the Determination of Seawater Micronutrient Elements

Species	Method	Detection Limit (µM)	Sample Rate (per hour)	
			Discrete	Continuous
NO_3	Reduction to NO_2 and determination, as an azo dye, using rFIA	0.1	30	75
NO_2	Determination, as an azo dye, using rFIA	0.1	30	75
PO_4	Determination, as a reduced phosphomolybdate dye, using rFIA	0.08	30	60
SiO_4	Determination, as a reduced silicomolybdate dye, using rFIA	0.5	30	80
Total amino acids	Fluorimetric determination with 1,2-benzenedicarbaldehyde using FIA	0.01	150	150

NOTE: Discrete samples were analyzed in duplicate; continuous samples refer to analysis of a stream of seawater.

water and deliver it to the rFIA system where semicontinuous measurements of concentration are made. FIA would, perhaps, be better suited to the analysis of deep samples obtained from a hydrocast. Nutrient concentrations in these samples are larger, and FIA allows the discrete samples to be processed faster. The same apparatus is used in both applications, and only a simple rearrangement of the manifold is required to convert from one application to the other.

Apparatus

A flow-injection analyzer can be assembled from readily available components (15, 25). The minimum requirements are an injection valve, pump, detector, recorder, and manifold tubing and connectors. Peristaltic pumps are most often used in FIA to propel the liquids. The flow produced by the pump should be steady and have low pulsation. A peristaltic pump should have 8–10 rollers on the pump head. A recently developed (25) low-cost FIA system uses air pressure to propel the liquid streams. This pressure produces a virtually pulse-free flow, but it may be prone to motion-related problems at sea.

Injection valves designed for FIA are produced by a number of manufacturers. These valves are often microprocessor controlled (26) and are quite versatile. However, for all of the analyses described later, a simple low-pressure rotary injection valve designed for liquid chromatography is adequate (e.g., Rheodyne Model 50). Pneumatic and electric actuators are available for automated operation. An injection valve that can be machined quite easily from poly(tetrafluoroethylene) (PTFE, Teflon) and Plexiglas is described elsewhere (25, 27).

Any type of detector with a flow-through cell can be used for FIA. Photometric detectors are most often used in FIA (15–18, 25). However, many other analyses using fluorimeters (28, 29), refractometers (24), atomic absorption (30, 31), and inductively coupled plasma emission spectrometers (32) have been described. Electrochemical detectors based on potentiometry with ion-selective electrodes (15, 33), anodic stripping voltammetry (15, 34), potentiometric stripping (35), and amperometry (36) have also been used.

Betteridge et al. (24), designed an extremely simple and inexpensive LED photometer for use in FIA. A block diagram of the detector and its electronics is shown in Figure 3. An LED is used as the light source, and a phototransistor is the detector. These components are glued into the block of Plexiglas that forms the flow cell, and they act as the windows of the cell. These detectors are ideal for use at sea. They are insensitive to motion because the light source and detector are rigidly mounted in the flow cell. The maximum emission wavelengths of the LEDs range from 560 to 950 nm.

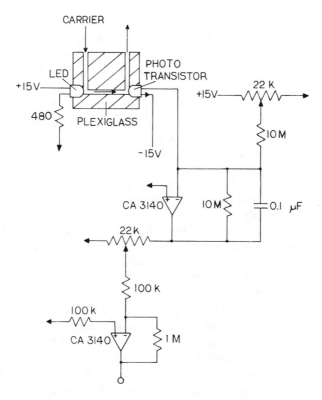

Figure 3. Partial schematic of the LED photometer electonics and flow cell designed by Betteridge et al. (24).

The LED photometer can, therefore, be used for the determination of nitrate, silicate, phosphate, and ammonia with the analyses commonly used for these groups (2, 7).

The manifold can be constructed from any appropriate small-diameter tubing. We used PTFE tubing because it is inert and can be obtained in various diameters. Altex-type tubing connectors and T-fittings provide a convenient means of joining the tubing components. A low-cost tool for flaring PTFE tubing for these fittings was described elsewhere (37).

Future directions for FIA include improved methods for determining concentrations in FIA (19, 20), an advanced modular system using fiber optics or ion-selective field-effect transistors in the detectors (20), and a system of solid microconduits for handling the solutions (20). These developments suggest even greater reductions in the size and cost of automated chemical analyses.

Analytical Methods

Most of the work done with FIA has tended toward developing high analysis rates, rather than high-sensitivity analyses (15–20). However, sensitivity is the most important factor when developing analyses for seawater micronutrients. The concentrations of the essential micronutrients such as nitrate, nitrite, ammonia, phosphate, and silicate are all regularly found to be less than 1 μM in surface seawater (38). Nitrate concentrations as low as 0.01 μM have been found (39) with a chemiluminescence detector (40). Analytical methods for these compounds must have detection limits that are considerably less than 1 μM if the factors that limit primary production in the euphotic zone of the ocean are to be understood.

Anderson (11) was the first to report on the use of FIA for the analysis of seawater micronutrients. He developed a method for the simultaneous determination of nitrate and nitrite. The chemical reactions for the analysis of nitrate were based on the reduction of nitrate to nitrite by a copperized cadmium column placed in the flow path. The nitrite was then analyzed as an azo dye (11). This reaction sequence is conventionally used in both segmented CFA and manual analyses of nitrate and nitrite in seawater (2, 6, 7). The detection limits are 0.1 μM for nitrate and 0.05 μM for nitrite.

The high sensitivity and precision of the analyses (relative standard deviation less than 1%) demonstrated the potential of FIA in marine research. However, the sampling rate obtained by Anderson (11) was rather low. Only 30 samples could be analyzed per hour, although both nitrate and nitrite were determined on each sample. Long residence times were necessary to obtain sufficient extent of reaction.

Following Anderson's work, we investigated the application of FIA to the determination of dissolved phosphate in seawater (13). Concentrations of phosphate above 4 μM are rare in seawater, and concentrations near the surface are typically less than 0.5 μM. The determination of phosphate requires an analysis with a very high sensitivity. Initial experiments indicated that an FIA procedure lacked sufficient sensitivity for surface water samples, and this finding led us to the development of an rFIA method using the manifold shown in Figure 2. The detection limit obtained with this analysis is 0.05 μM with a 1-cm path length flow cell. This result is comparable to that obtained by manual methods with 10-cm path length cells (2, 7). The sampling rate is 90 per hour when a continuous seawater stream is analyzed.

After developing this method, however, we learned that the Bausch and Lomb flow cell used in that work (13) has a design that acts to minimize refractive index errors (see following discussion). As a result, the manifold shown in Figure 2 does not work well with a simple flow cell such as that used in the LED photometer (Figure 3). We have, therefore, redesigned the manifold to minimize the refractive index error (Figure 4). This redesigned manifold works well and has performance characteristics similar to those of the original analysis, that is, a detection limit of 0.08 μM and a sample

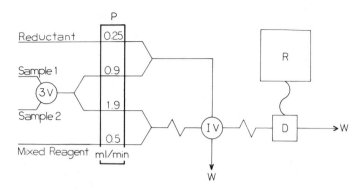

Figure 4. Modified manifold for the determination of phosphate.

All symbols are as in Figures 1 and 2. The LED photometer flow cell has a 2-cm path length and 1.5-mm inside diameter (ID). The LED and phototransistor are both Radio Shack (276–143 and 276–145, respectively). The mixed reagent is the same as that used by Johnson and Petty (13), but 250 mL is diluted to 1L before use. The reductant contained 4 mL of stannous chloride stock solution (11 g of $SnCl_2$ in 100 mL of concentrated HCl) and 2.5 mL of concentrated sulfuric acid in 100 mL of water. The preinjector mixing column is 35 cm of 0.8-mm ID PTFE tubing, and postinjector column is 60 cm of 0.8-mm ID PTFE tubing.

throughput of 75 per hour. The use of stannous chloride as a reductant, rather than ascorbic acid, allows the analysis to be performed at room temperature. The detector output for a series of seawater standards using this manifold is shown in Figure 5. The calibration curve is linear and has a slope similar to that obtained in our initial phosphate analysis $\{A = 0.0156 + 0.0169 [PO_4], (\mu M), R^2 = 0.999\}$. A 2-cm path-length flow cell was used, however, so the extent of reaction with stannous chloride as the reductant at room temperature is about 50% of that obtained with ascorbic acid at 50 °C.

rFIA procedures for the determination of nitrate, nitrite, and silicate in seawater have also been developed (12, 14). The manifolds for these analyses are shown in Figures 6 and 7. The nitrate and nitrite analysis is based on the same reactions used by Anderson (11). The rFIA technique allows about 75 determinations per hour with a detection limit of 0.1 μM, a rate 2.5 times greater than obtained by Anderson using FIA.

Silicate is determined as a reduced silicomolybdate dye (14). The method is sensitive to the reductant used, either stannous chloride or ascorbic acid. The reaction kinetics with stannous chloride are much faster than with ascorbic acid. The analysis with stannous chloride is more sensitive, therefore, and has a detection limit of 0.5 μM compared to 1.0 μM when ascorbic acid is used. Eighty determinations can be made per hour at these detection limits. A detection limit of 0.1 μM can be obtained if the sampling rate is decreased to 50 per hour (14). The analysis with stannous chloride has a smaller salt error than with ascorbic acid, but the interference due to

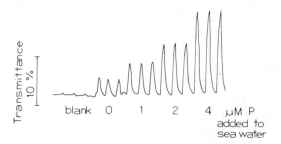

Figure 5. Recorder signals obtained with the manifold shown in Figure 4. The blank solution is artificial seawater (7). The remaining solutions were natural seawater containing ~0.9 μM PO_4^{3-} to which phosphate standard solution was added.

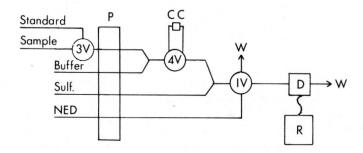

Figure 6. Manifold for the determination of nitrate and nitrite. A four-way valve (Rheodyne Model 50, 4V) is used to switch the cadmium column out of line for the determination of nitrite only. (Reproduced with permission from Ref. 12. Copyright 1983, American Society of Limnology and Oceanography.)

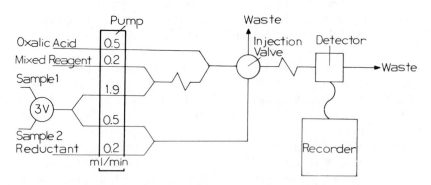

Figure 7. Manifold for the determination of silicate (14).

phosphate is larger. The interference should be insignificant in most marine applications, however.

Petty et al. (29) developed an FIA procedure for the determination of total amino acids in seawater. Total dissolved amino acids can be determined with a detection limit of 0.01 μM and a sampling rate of 150 per hour with the manifold shown in Figure 8. The amino acids are determined as fluorescent isoindoles (41), which are formed by reaction of primary amines with 1,2-benzenedicarbaldehyde (o-phthaldialdehyde) and mercaptoethanol. The fluorescence quantum yields of the isoindoles formed from most amino acids are similar (41, 42), and the sensitivity of the FIA analysis is similar for most amino acids (29). Ammonia also forms a fluorescent product with 1,2-benzenedicarbaldehyde, but its fluorescence quantum yield is about a factor of 15 lower (29, 42). The ammonia interference is low, therefore, in most areas of the ocean.

Other analyses of oceanographic interest have been adapted to FIA. The detection limits reported for these analyses are often too high to be of use in oceanographic investigations. However, great improvements in detection limits can be made with the techniques just discussed.

Ammonia has been measured by using phenol and hypochlorite to form 4-(4-oxocyclohexadienylimino)phenol (indophenol) (15). Ruzicka and Hansen (15) also discussed a sensitive method that uses a gas-diffusion cell to transfer ammonia from the sample stream to an acid–base indicator stream. Urea can be determined as ammonia after reaction with urease

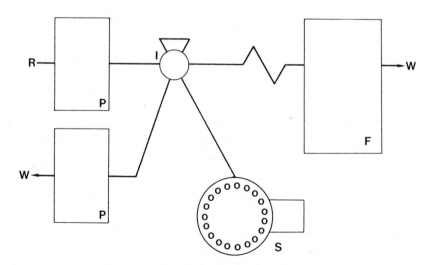

Figure 8. Manifold for the determination of total amino acids by FIA (29). Key: S, an automatic sample changer; F, a filter fluorimeter (Turner Model 111); and R, the 1,2-benzenedicarbaldehyde reagent used as the carrier stream.

(15). Leggett et al. (43) have adapted the determination of sulfide as methylene blue to FIA. Nanomolar quantities of hydrogen peroxide can be determined by fluorescence by using horseradish peroxidase to catalyze the reaction of hydrogen peroxide with leucodiacetyldichlorofluorescein (28).

Components of the inorganic carbon system can also be determined by FIA. The measurement of pH by FIA (15, 33) with a glass electrode appears to be a practical alternative to static measurements. The pH can be measured with a precision of ±0.002 pH units on 240 samples per hour. This method of measurement would eliminate the relatively long period of time for glass electrodes to reach equilibrium. However, the effects of liquid junction potentials in the heterogeneous carrier stream and the effects of solution composition on the degree of electrode response need to be considered further. A total dissolved-CO_2 analysis, which might be particularly well-suited for pore water samples, is discussed elsewhere (44). Acid–base pseudotitrations are also possible in FIA (15, 19). Pseudotitrations differ from conventional titrations because the endpoints occur when the equivalent concentrations of the sample and titrant are the same, rather than when their equivalent amounts are equal as in a conventional titration (19). Ramsing et al. (45) have developed the techniques for a rapid pseudotitration that could be suitable for a moderate precision (0.5%) determination of alkalinity.

Dissolved metal determinations with a large range of sensitivities are also possible. Manganese can be determined colorimetrically with dihydroxyiminomethane (formaldioxime) (46). An iron determination with an amperometric detector that can resolve iron(II) and iron(III) has been reported (36). Dissolved iron can also be determined with ferrozine (47). The detection limits obtained with these analyses are on the order of 1 μM. This level is not low enough for work in open ocean water, but the analyses would be of use in anoxic water and the interstitial water of sediments.

Flow-injection analysis is also well-suited for the automation of anodic stripping voltammetry. Metals can be plated from the sample solution as it passes over the electrode. Stripping is then carried out in the deoxygenated carrier stream (15, 34). The sample itself does not have to be deoxygenated. Detection limits of 3 nM have been reported for lead by this technique (34).

Olsen et al. (48, 20) have described an interesting method for the determination of lead in polluted seawater using FIA and flame atomic absorption spectroscopy. The system incorporates a Chelex-100 column for on-line preconcentration of the sample. The preconcentration and elution step improves the detection limit for lead by a factor of four (50 nM). Further increases in sensitivity are easily possible. The combination of this preconcentration step with a more sensitive detector, such as anodic stripping voltammetry, may make possible the determination of trace metals in seawater on a routine basis.

Refractive Index Effect

When designing a high-sensitivity FIA or rFIA analysis with a photometric detector, the influence of refractive index must be considered (*24, 49*). The heterogeneous composition of the carrier stream, which is caused by injection of the reagent or sample, can create a refractive index gradient along the reaction-tube axis. The parabolic velocity profile across the radial axis, characteristic of laminar flow, will cause the isolines of refractive index to form a parabolic profile (*24*). These parabolic isolines of refractive index will act as a liquid lens and will alternately focus light on the detector and refract it away as they pass through the flow cell (*24*). The refractive index signal is easily noted because of its resulting sinusoidal shape (Figure 9). This signal will occur whether or not the sample has a different molar absorptivity than the carrier if there is a large refractive index difference between the carrier stream and the injected solution; therefore, eliminating this interference is essential.

The simplest way to remove the signal due to refractive index differences is to match the refractive indices of the carrier stream and the injected solution. For example, the reagent *N*-(1-naphthyl)ethylenediamine (NED) is injected into the seawater stream in the analysis of nitrate and nitrite in seawater by rFIA (*12*). The concentration of NED in the reagent solution is low (1 g/L) compared to the seawater carrier stream. A refractive index signal equivalent to 5 μM of nitrate will be detected when seawater is analyzed. However, the refractive index signal is completely suppressed by adding 10 g of NH_4Cl to each liter of NED reagent solution. The refractive index signal can be suppressed by a similar technique in the analyses of silicate and phosphate. This procedure works well for the analysis of oceanic water where changes in salinity are small. If salinity changes larger than 30% of the total salinity are expected, as in estuarine samples, then a different procedure should be used to eliminate refractive index errors. Bergamin et al. (*49*) suggested that refractive index signals can be suppressed by injecting the sample into an inert carrier stream that then merges with the reagent streams. This method leads to an unacceptable decrease in sensitivity for most oceanographic measurements.

Automatic matching of the refractive indices of the injected reagent and the sample stream is possible in rFIA (*14*). The reagent to be injected in the silicate analysis is combined with a portion of the sample stream before the injection valve (Figure 7). The resulting solution has the proper reagent concentration in a matrix that has the same refractive index as the sample stream. The blank peaks produced by this method are shown for seawater of various salinities in Figure 10. Also shown for comparison are the blank peaks in the same seawater samples when the refractive index of the injected reagent (ascorbic acid) is adjusted to match only that of undiluted seawater. Large refractive index peaks are produced in the latter case when

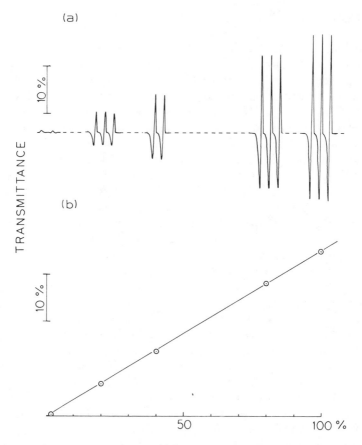

Figure 9. *Refractive index signals obtained when 150 μL of distilled water is injected into a seawater carrier stream (a). The postinjector column was 90 cm of 0.8-mm ID tubing. On the abscissa, 100% corresponds to seawater with a salinity of 33.3 ppt and 0% corresponds to distilled water. The seawater was diluted by volume to produce the intermediate solutions. The peak-to-peak signal is also plotted (b).*

diluted seawater is analyzed; they are almost completely suppressed when automatic refractive index matching is used.

The silicate in the injected reagent cannot react in the sample stream because the oxalic acid added upstream of the injector destroys the excess molybdate. However, the quenching of the silicate reaction before injection of the reagent is not essential. Automatic refractive index matching also works well in the phosphate analysis (Figures 4 and 5). Any phosphate in the injected reductant is reactive in this case. The additional phosphate will only make the peaks proportionately larger. Standards must be analyzed by the same procedure used for the samples.

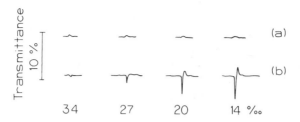

Figure 10. *The blank signals obtained in seawater of various salinities in the silicate analysis (14). Key: top, automatic refractive index matching using the manifold shown in Figure 7; and bottom, the refractive index of the injected reagent was adjusted with sulfuric acid to minimize the blank signal when 34 ppt of seawater was analyzed. Automatic refractive index matching was not used.*

In addition to the procedures just discussed, the refractive index signals can be reduced by proper design of the flow cell. Increasing the diameter of the flow cell can result in a large decrease in the refractive index signal. The increased diameter and lower flow rate reduce the curvature of the liquid lens and result in less refraction of the light. The Bausch and Lomb flow cell (Model 33–30–01) used in our initial phosphate analysis (13) has a 3.5-mm diameter and is designed so that the sample flows into the cell at six ports arranged radially around one end of the cell, rather than the usual single port. These features help reduce the refractive index signal by about a factor of 20 compared to a flow cell based on the design shown in Figure 3. Flow cells with a tapered flow path can also act to reduce the refractive index signal (50). Ham (51) found that long path-length cells also tend to reduce the refractive index effect.

The refractive index effect does have one redeeming feature. The signal is directly proportional to the salt content of the sample (24). Injecting deionized water into a seawater carrier stream is, therefore, a simple means of measuring the salinity of the carrier stream (Figure 9). The relative precision of replicate analyses is about 0.5% when 100 μL of deionized water is injected into seawater of 33 ppt* salinity (unpublished work). Cooling the seawater to 10 °C did not cause a significant change in the signal. Although the precision of the salinity analysis is not sufficient for physical oceanographic measurements, it would be useful for estuarine research.

Field Applications

On two cruises we used rFIA for the determination of nitrate in the Santa Barbara Channel. The results obtained on these cruises are examined briefly here to illustrate the capabilities of rFIA for use in oceanographic investigations.

*Parts per thousand

Oceanographic (52, 53) and satellite thermal measurements (54) show indirect evidence of a thermal front in the eastern end of the Santa Barbara Channel. This front is the boundary between warm southern water carried by the California Counter Current into the channel and cold water flowing south. The source of the cold water may be upwelling in the channel and Point Conception regions, or it may represent a branch of the California Current. The purpose of the two cruises was to investigate the biological and chemical structure across this front. The front was located by measuring the temperature at 2 m until large gradients were found. These gradients were followed into the 85-m isobath, where the ship was anchored for the remainder of each cruise.

Vertical profiles were made to 70-m depths with a submersible pump at the anchor station. A thermistor, fluorometer, and the inlet for the rFIA system were placed in line with the effluent from the pump. The temperature and chlorophyll fluorescence were recorded continuously. The concentration of nitrate was determined 75 times per hour. The residence time of water in the pump was about 4 min. A typical vertical profile is shown in Figure 11. The cadmium column was switched out of line at each depth to

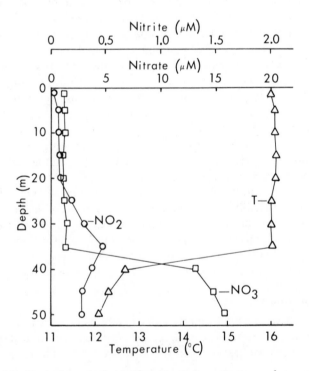

Figure 11. Typical vertical profiles of nitrite, nitrate, and temperature. (Reproduced with permission from Ref. 12. Copyright 1983, American Society of Limnology and Oceanography.)

allow nitrite to be determined. The primary nitrite maximum of 0.5 μM is readily detectable.

The temperature, nitrate, and chlorophyll fluorescence data obtained on each cruise are plotted in Figures 12 and 13. Large oscillations appear in the data. These oscillations have the same period as the semidiurnal tidal

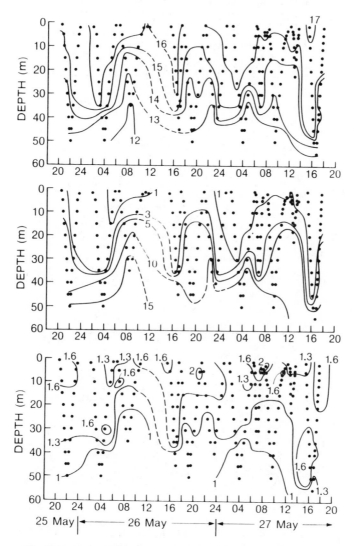

Figure 12. Temperature (top), nitrate (middle), and chlorophyll fluorescence (bottom) at an anchor station in the Santa Barbara Channel, May 25–27, 1982. Dots represent the depths sampled. Station location is 34°01.6 N, 119°19.3 W.

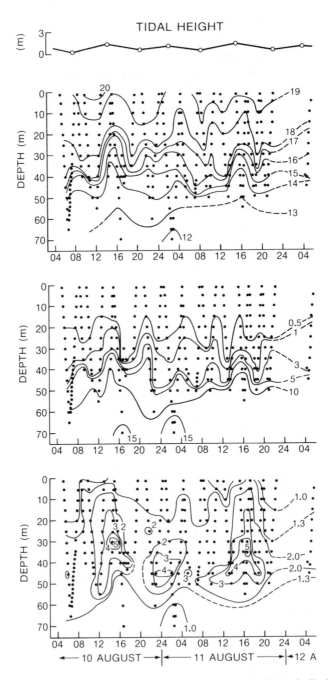

Figure 13. Temperature (top), nitrate (middle), and chlorophyll fluores-cence (bottom) at an anchor station in the Santa Barbara Channel, August 10–12, 1982. Station location is 34°04.0 N, 119°19.5 W.

cycle in the channel. A cursory inspection of Figures 12 and 13 indicates the presence of internal tides [internal waves of tidal period (55)] with a 40-m amplitude. However, closer examination of the temperature and nitrate data indicates that these are not the vertical oscillations of an internal tide, but are horizontal motion of the front as it is advected back and forth past the anchored ship by tidal currents.

To illustrate this point, nitrate is plotted versus temperature for a series of three vertical profiles with the pump beginning at 10:00 A.M., August 10, 1982 (Figure 14). Distinctively different temperature–nitrate plots are obtained from the first and last profiles. This finding indicates the presence of two water types at the anchor station in a short period of time. The intermediate profile shows both water types and the frontal boundary separating them. The results shown in Figure 14 could only be a result of the horizontal oscillation of two water types, separated by a sharp boundary. An idealized picture of the position of the front is shown in the inset in Figure 14. Solar heating and uptake of nitrate could not be responsible because these results are repeated throughout the two data sets. Vertical

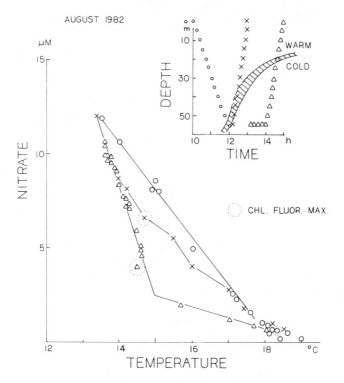

Figure 14. Nitrate vs. temperature for three vertical profiles beginnning at 10:00 A.M. August 10, 1982. The inset shows the sample depths and the position of the front. Symbols correspond to those in the inset.

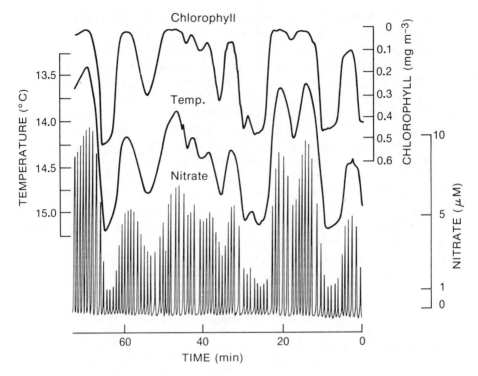

Figure 15. Nitrate (top), temperature (middle), and chlorophyll (bottom) measured as a function of time at 15 m in the Santa Barbara Channel. (Reproduced with permission from Ref. 12. Copyright 1983, American Society of Limnology and Oceanography.)

oscillations induced by internal tides would require that whole water masses disappear during part of an oscillation, and then reappear.

The sharpness of the boundary in the frontal region is shown in Figure 15. The submersible pump was lowered to a depth of 15 m and held there while the ship drifted in the frontal region before it was anchored. The extreme fluctuations in temperature, nitrate, and chlorophyll must be due to the interleaving of the two water masses in the front. An upper limit for the horizontal span of the frontal boundary can be calculated as follows: the transition from one water type to another occurs in 3–4 min in Figure 15; if the current shear from the surface to 15 m is 1 knot (0.5 m/s) then the water is only advected 100 m in the time it takes for the transition from one water mass to another to occur. The oscillations in Figure 15 suggest that rapid changes in current direction occur. If so, then the horizontal span of the front may also be smaller.

Acknowledgments

This work was supported by Office of Naval Research Contract N000 14-82-0740.

Literature Cited

1. Brewer, P. B.; Riley, J. P. *Deep-Sea Res.* 1965, 12, 765–72.
2. Strickland, J. D. H.; Parsons, T. R. "A Practical Handbook of Seawater Analysis"; 2nd ed., 310 p.; *Bull. Fish. Res. Bd. Can.* 1972, 167.
3. Brewer, P. G.; Riley, J. P. *Anal. Chim. Acta* 1966, 35, 514–19.
4. Chan, K. M.; Riley, J. P. *Deep-Sea Res.* 1966, 13, 467–71.
5. Slawyk, G.; MacIsaac, J. J. *Deep-Sea Res.* 1972, 19. 521–24.
6. Riley, J. P. In "Chemical Oceanography, Vol. 3"; 2nd ed.; Riley, J. P.; Skirrow, G., Eds; Academic: New York, 1975; pp. 193–514.
7. Grasshoff, K. "Methods of Seawater Analysis"; Verlag Chemie GmbH; Weinheim, FRG, 1976; pp. 1–317.
8. Whitledge, T. E.; Malloy, S. C.; Patton, C. J.; Wirick, C. D. "Automated Nutrient Analyses in Seawater"; Brookhaven National Laboratory BNL 51398: Upton, N.Y., 1981; pp. 1–216.
9. Ruzicka, J.; Hansen, E. H. *Anal. Chim. Acta* 1975, 78, 145–57.
10. Beecher, G. R.; Stewart, K. K.; Hare, P. E. *Clin. Nutr.* 1975, 1, 411.
11. Anderson, L. *Anal. Chim. Acta* 1979, 110, 123–28.
12. Johnson, K. S.; Petty, R. L. *Limnol. Oceanogr.* 1983, 28, 1260–66.
13. Johnson, K. S.; Petty, R. L. *Anal. Chem.* 1982, 54, 1185–87.
14. Thomsen, J.; Johnson, K. S.; Petty, R. L. *Anal. Chem.* 1983, 55, 2378–82.
15. Ruzicka, J.; Hansen, E. H. "Flow Injection Analysis"; Wiley: New York, 1981; pp. 1–207.
16. Betteridge, D. *Anal. Chem.* 1978, 50, 832A–846A.
17. Ruzicka, J.; Hansen, E. H. *Anal. Chim. Acta* 1978, 99, 37–76.
18. Ranger, C. *Anal. Chem.* 1981, 53, 20A–32A.
19. Stewart, K. K. *Anal. Chem.* 1983, 55, 931A–940A.
20. Ruzicka, J. *Anal. Chem.* 1983, 55, 1040A–1053A.
21. Taylor, G. *Proc. R. Soc. London Ser. A.* 1953, 219, 186–203.
22. Vanderslice, J. T.; Stewart, K. K.; Rosenfeld, A. G.; Higgs, D. J. *Talanta* 1981, 28, 11–18.
23. Gerhardt, G.; Adams, R. N. *Anal. Chem.* 1983, 54, 2618–20.
24. Betteridge, D.; Dagless, E. L.; Fields, B.; Graves, N. F. *Analyst* 1978, 103, 897–908.
25. Ruzicka, J.; Hansen, E. H.; Ramsing, A. U. *Anal. Chim. Acta* 1982, 134, 55–71.
26. Karlberg, B. I. *Am. Lab.* 1983, 73–77.
27. Hansen, E. H.; Ruzicka, J. *J. Chem. Ed.* 1979, 56, 677–79.
28. Kelley, T. A.; Christian, G. D. *Anal. Chem.* 1981, 53, 2110–14.
29. Petty, R. L.; Michel, W. C.; Snow, J. P.; Johnson, K. S. *Anal. Chim. Acta* 1982, 142, 299–304.
30. Wolf, W. R.; Stewart, K. K. *Anal. Chem.* 1979, 51, 1201–5.
31. Tyson, J. F.; Appleton, J. M. H.; Idris, A. B. *Analyst* 1983, 108, 153–58.
32. Broekart, J. A. C.; Leis, F. *Anal. Chim. Acta* 1979, 109, 73–83.
33. Ruzicka, J.; Hansen, E. H.; Ghose, A. K.; Mottola, H. A. *Anal. Chem.* 1979, 51, 199–203.
34. Wang, J; Dewald, H. D.; Greene, B. *Anal. Chim. Acta* 1983, 146, 45–50.
35. Hu, A.; Dessy, R. E.; Graneli, A. *Anal. Chem.* 1983, 55, 320–28.
36. Dieker, J. W.; van der Linden, W. E. *Anal. Chim. Acta* 1979, 114, 267–74.
37. Baron, R. E. *Anal. Chem.* 1982, 54, 339–40.
38. Riley, J. P.; Chester, R. "Introduction to Marine Chemistry"; Academic: New York, 1971; pp. 152–78.

39. Garside, C. *Trans. Am. Geophys. Union* 1982, *63*, 995.
40. Garside, C. *Mar. Chem.* 1982, *11*, 159–67.
41. Chen. R. F.; Scott, C.; Trepman, E. *Biochim. Biophys. Acta* 1979, *576*, 440–55.
42. Lindroth, P.; Mopper, K. *Anal. Chem.* 1979, *51*, 1667–74.
43. Leggett, D. J.; Chen, N. H.; Mahadevappa, D. S. *Anal. Chim. Acta* 1981, *128*, 163–68.
44. Baadenhuijsen, H., Seuren-Jacobs, H. E. H. *Clin. Chem.* 1979, *25*, 443–45.
45. Ramsing, A. U.; Ruzicka, J.; Hansen, E. H. *Anal. Chim. Acta* 1981, *129*, 1–17.
46. Gine, M. F.; Zagatto, E. A. G.; Bergamin Filho, H. *Analyst* 1979, *104*, 371–75.
47. Dutt, V. V. S. E.; Eskander-Hana, A.; Mottola, H. *Anal. Chem.* 1976, *48*, 1207–11.
48. Olsen, S.; Pessenda, L. C. R.; Ruzicka, J.; Hansen, E. H. *Analyst* 1983, *108*, 905–17.
49. Bergamin Filho, H.; Reis, B. F.; Zagatto, E. A. G. *Anal. Chim. Acta* 1978, 427–31.
50. Little, J. N.; Fallick, G. J. *J. Chromatogr.* 1975, *112*, 389–97.
51. Ham, G. *Anal. Proc. (London)* 1981, *18*, 69–70.
52. Dorman, C. E.; Palmer, D. P. In "*Coastal Upwelling*"; Richards, F. A., Ed; American Geophysical Union: Washington, D.C., 1981; pp. 44–56.
53. Cox, J. L.; Willason, S.; Harding, L. *Bull. Mar. Sci.*, 1983, *33*, 213–26.
54. Bascom, W. *Environ. Sci. Technol.* 1982, *16*, 226A–236A.
55. Roberts, J. "Internal Gravity Waves in the Ocean"; Dekker: New York, 1975; pp. 1–274.

RECEIVED for review July 5, 1983. ACCEPTED December 21, 1983.

Determining Trace Gases in Air and Seawater

S. D. HOYT and R. A. RASMUSSEN

Department of Environmental Science, Oregon Graduate Center, Beaverton, OR 97006

A systems approach using stainless steel sample cans, vacuum extraction flasks, and a field purge–cryogenic trap apparatus was used to collect air and seawater samples for the determination of 30–50 trace gases by gas chromatography (GC)–electron capture–flame ionization–flame photometric detection and GC–mass spectrometry (MS). The field sampling equipment was operated on board the National Oceanic and Atmospheric Administration's Discoverer in the central Pacific. This equipment provided an economical method for the collection of virtually contaminant-free samples for studying the global mass balance of trace gases. Analytical procedures are described for the GC–MS technique, which can simultaneously measure and identify 23 trace gases at 1–1000-parts per trillion by volume with a precision of between 5 and 20 %. The study methods were successful in determining the concentrations and relative saturations of 15 trace gases and showed the ocean to be a source of OCS, CH₃Cl, CHCl₃, and, potentially, a source of organobromine compounds.

THE MEASUREMENT OF TRACE GASES in the atmosphere and the ocean is necessary to understand their global budgets. The ocean can act as a source or a sink for gases produced in it by biological activity; it is the major source for OCS (1, 2), CH_3I (3), CH_3Cl (4, 5), $(CH_3)_2S$ (6–8), and $CHCl_3$ (9).

Many of these naturally produced gases play important roles in atmospheric chemistry. For example, OCS may maintain the stratospheric sulfate layer (10, 11). Changes in the concentration of this aerosol layer could alter the global temperature. Dimethyl sulfide is produced in the ocean and is released to the atmosphere where it probably is rapidly oxidized to SO_2, which contributes substantially to the background acidity of rainwater (12). Methyl chloride, which is produced in the ocean, is the dominant source of natural chlorine in the atmosphere, and plays a key role in the stratospheric ozone cycle (13, 14).

The recognition of the importance of oceanic sources of trace gases has led to increased research on the measurement and modeling of these gases. Early efforts concentrated on measuring one or two gases in samples obtained from limited areas. These results showed that concentration distributions in the ocean are variable, and that accurate estimates of the global fluxes for biogenic gases require large numbers of samples to be analyzed from many locations (2). To do in situ analyses on large numbers of samples is costly and usually results in the measurement of a limited number of compounds. Accordingly, we developed procedures to obtain the number and diversity of samples required and to routinely measure 30–50 trace gases in each sample. The analytical method developed uses liquid oxygen to concentrate the gases in two steps on freezeout loops and a fused-silica capillary column to separate them (15–21). A quadrupole mass spectrometer plus several specific gas chromatographic (GC) detectors are used to measure the many different compounds.

In this chapter we describe an easy to use and versatile method for contaminant-free sampling, storage, and field processing of large volumes of analyte for multiple gas chromatography (GC)–electron capture (EC)–flame ionization (FI)–flame photometric detection (FPD) and GC–mass spectrometry (MS). The procedures developed enable a wide variety of trace gases (30–50 species) to be measured at trace levels [parts per trillion by volume (pptv)] in air and water samples. The components of the system are (1) small volume (800 mL), internally passivated, stainless steel cans for gas sampling and storage; (2) vacuum extraction flasks for sampling, extraction, equilibration, and storage of water–gas samples; (3) a purge–cryogenic trap apparatus for multiple and composited extractions of water samples, providing large volumes of purged trace gases; and (4) fused-silica DB-1 capillary columns with Nafion driers (Du Pont) for routine high-resolution analyses on GC–EC–FI–FPD and GC–MS–data storage (DS) instruments. The large volumes of trace gases obtained also provide enough sample for replicate analyses to determine precision, reproducibility, special analyses, and exchange of material for interlaboratory comparisons. The total approach makes it possible to do global studies of 30–50 trace gases in both the sea and the air. This information could not be obtained previously at an acceptable cost.

The sampling systems are robust and can be operated anywhere because they require no electrical lines. Processing samples outdoors has virtually eliminated the trace contaminants invariably encountered from processing samples indoors. Combining the gases purged from many samples into fewer containers via the cryogenic transfer step gives the needed efficiency of sampling required for global studies. The approach is preferable to, and less costly than, operating sophisticated GC instruments in the field because it provides much more data on many different trace gases than can be obtained by any one instrument operated in situ. The trace gases studied with these procedures are principal contributors to the global mass balance and atmospheric cycles of C, N, S, Cl, Br, and I; as such, the trace gases can affect the earth's climate and its perturbation.

Collection of Air and Seawater Samples

Air Sample Collection. Air samples are collected in stainless steel (SS-314) bottles that are internally electropolished by using the Summa process. Bottles with internal volumes of 800 mL, 1.6 L, and 35 L are used. The bottles are fitted with a purge T and Nupro SS-4H4 bellows valves. A typical sample bottle and purge T are shown in Figure 1. The bottles are cleaned by heating to 150 °C under high vacuum (20 mtorr) for several hours and then filled with "zero air" (AADCO 737A) to prevent contamination until used in the field.

Air is collected cryogenically to prevent pump contamination and to obtain a large volume of air under high pressure (about 400 psi). The high volume-to-surface ratio in these samples (300–600 psi) provides excellent stability for many trace gases (22) compared to similar samples stored at low pressure (< 5 psi). The bottles are first connected to the vacuum side of a pump (Metal Bellows, Inc., MB-158), Valves A and B (Figure 1) are opened, and the bottle is flushed with air. The valves are closed, and the bottle is immersed several inches in liquid nitrogen. When the base of the bottle reaches equilibrium with liquid N_2 (-196 °C), the air in the bottle condenses. The vacuum thus created is sufficient to draw in a prescribed air volume (50 L) in 15–20 min through Valve B.

Collection of Seawater by Vacuum Extraction Flask Method. For measurement of background concentrations of biogenic gases present at the parts per trillion level, special precautions against contamination must be taken in collecting and handling the samples. To avoid this problem a vacuum flask was modified to provide a single container to conveniently collect, store, and equilibrate the trace gases in seawater. Concentrations in the equilibrated headspace are quantitatively determined by a modified procedure (23).

A 1-L Erlenmeyer flask is fitted with a graded glass-to-Kovar metal seal welded to a high-vacuum valve (Nupro SS or B-4H4) as shown in Figure 2. Swagelok pipe fittings are used at the other end of the valve for an airtight connection to a funnel and the instrument inlet sampling system. The flask can withstand pressures of 30 psi and vacuum to 10 mtorr. The outside of the glass flask is coated with a resilient polymer to minimize breakage and provide safety in handling. These flasks are sent under vacuum (10 mtorr) to the field.

Seawater samples are collected in a stainless steel bucket when the ship slows as it comes to station. On-board analysis of the bucket samples for

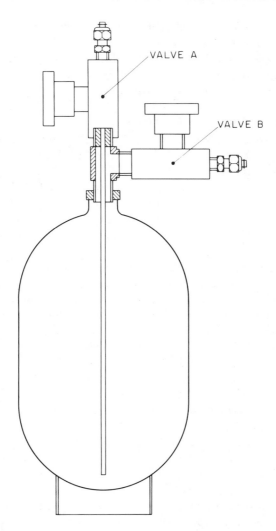

*Figure 1.   Stainless steel bottle (1.6 L) internally electropolished by Summa
process, fitted with purge T and two SS-4H4 Nupro valves, and used for
cryogenic air sample collections.*

CCl_3F (F-11) and CCl_2F_2 (F-12) shows the bucket samples to be free of contamination. The extraction flask is filled by attaching a stainless steel funnel (1 L) to the flask and filling the funnel with seawater from the bucket.
About 650 mL of water is aspirated into the flask without allowing any air
to enter. The flask is then pressurized with 15 psi of zero air. The results are
equal volumes of air and water equilibrated at 25 °C and 1-atm pressure.
Containing the sample under positive pressure reduces contamination

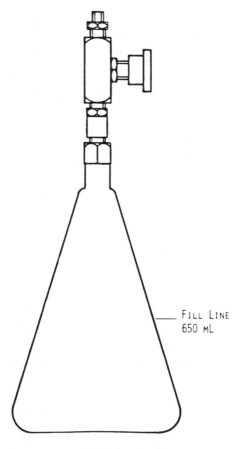

Figure 2. Vacuum extraction flask for headspace seawater collection and analysis. Seawater is added to the fill line, and the flask is pressurized with zero air to 15 psi.

leaking in from the outside and facilitates the transfer of the headspace air to the respective GC–EC–FID–MS instruments.

Seawater samples processed in the manner just stated are placed in a freezer and returned to the laboratory frozen until analyzed. Biological and chemical activity between the time samples are collected and analyzed are greatly reduced. For analysis the flasks are taken from the freezer, placed in a water-bath shaker, and agitated for 30 min until the samples reach 25 °C. The bottle is connected directly to the sample intake line on the instrument via $\frac{1}{16}$-in. outside diameter (OD) stainless steel tubing.

The trace gases in the headspace are related to the concentration of each gas in the water by Henry's law. After being aspirated into the flask, the gases partition between the seawater and the headspace so that the total concentration of a specific gas in the flask is $C_l^{(o)} V_l$, where $C_l^{(o)}$ is the initial concentration in the seawater, and V_l is the volume of water. Upon pressurization and equilibration the gases will partition according to Henry's law so that the headspace concentration C_g and the liquid concentration C_l are in equilibrium. The initial seawater concentration can be found from the mass balance, with $V_g = (V_T - V_l)P$, according to the equation

$$C_l^{(o)} = C_g V_g + C_l V_l \tag{1}$$

From Henry's law $C_l = C_g/H$, so $C_l^{(o)}$ is equal to

$$C_l^{(o)} = C_g \left(\alpha + \frac{1}{H} \right) \tag{2}$$

In these equations P is the headspace pressure in atmospheres, α is the gas-to-seawater ratio (V_g/V_l), and V_T is the total volume of the flask (here 1 L). By this procedure the initial concentration of a dissolved gas $(C_l^{(o)})$ in seawater can be determined. Normally single extractions are done, but subsequent extractions can be used to check the amount of gas remaining in the seawater. The uncertainties associated with the headspace method can be evaluated by referring to Equation 2. In addition to the headspace gas concentration, C_g, the uncertainty in $C_l^{(o)}$ is dependent on the uncertainties in H and the ratio α. By using a propagation of error procedure on Equation 2, the uncertainties associated with the vacuum extraction flask can be estimated for different gases (2). The results show that for a gas that is not very soluble $(H \stackrel{\sim}{>} 3)$, the total uncertainty is $\pm 5\%$. For a more soluble gas $(H \stackrel{\sim}{<} 0.5)$ whose Henry's constant is not accurately known in seawater, the uncertainty is $\pm 30\%$.

Purge and Cryogenic Trapping.   An alternative method to the vacuum extraction flask is to purge the gases from the seawater in the field and bring only the recovered gases back to the laboratory. The advantages of this procedure are (1) large volumes of seawater can be purged in the field to get the needed sample volume for analysis, (2) storage of the extracted gases without seawater is less subject to biological degradation or accentuation, and (3) the major sample preparation is done in the field, and time in the laboratory is thus saved.

The system is designed to purge seawater with zero air and cryogenically collect the purged gases in 800-mL stainless steel cans equipped with single Nupro SS-4H4 valves. Because the samples are analyzed on several instruments in addition to the GC–MS, a sample volume of 2.7 L (~ 50 psi) is desired. To ensure that the concentrations of the gases are measurable, a 500-mL seawater sample is purged. The design of the field purge and cryogenic trap system is shown in Figures 3 and 4. Figure 3 shows the arrange-

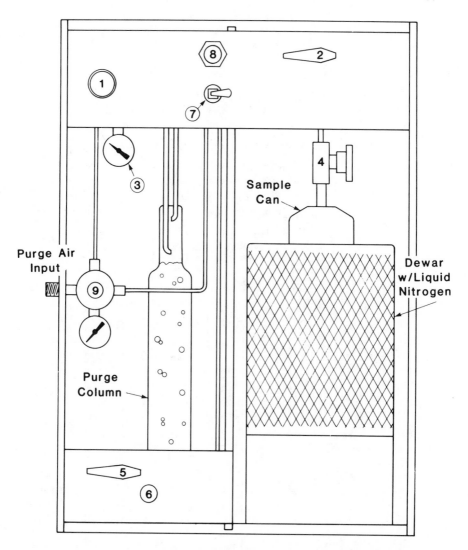

Figure 3. Diagram of shipboard purge–cryogenic trap module. Seawater samples are purged of their trace gases, which are collected and stored in the sample can.

ment of the system, and Figure 4 shows the purge box used on-board ship for sample collection.

The zero-air purge gas is prepared with an AADCO 737A zero-air generator and is stored in 35-L stainless steel containers at 400 psi. Zero air was chosen because helium will not cryocondense in the sample cans at liquid N_2 temperatures, and commercial prepurified N_2 contains too much con-

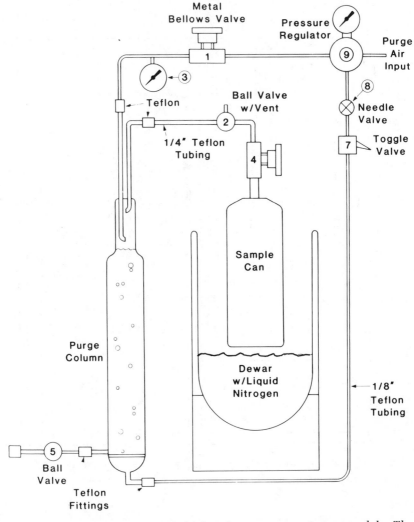

Figure 4. Drawing of actual shipboard purge–cryogenic trap module. The numbers correspond to those shown in Figure 3. (See text for complete discussion.)

tamination. Before filling the purge chamber with seawater, the zero air in the sample cans is used to flush the purge column and the inlet port. This flushing cleans the lines and the purge cylinder.

The seawater samples are collected in 250-mL glass syringes that have been flushed with fresh seawater and used to fill the purge chamber through the fill port. After the purge cylinder is filled with seawater to the 500-mL mark, Valve 2 (Figure 4) is placed in the vent mode, and Valve 1 is opened to flush the headspace above the seawater. The design of the purge column permits this procedure to be done without disturbing the water. Limiting the pressure to 10 psi prevents the glass column from being over-pressurized. When the flushing cycle is completed, Valves 2 and 1 are closed. The bottom of the sample can is immersed in about 2 in. of liquid nitrogen and cooled, and Valves 4 and 7 are opened. Opening Valves 4 and 7 starts the purging of the seawater. The purge rate is adjusted to 50 mL/min by using needle Valve 8. Equilibrated in liquid N_2, the can maintains ~ 10 in. of vacuum, measured on Gauge 3. The vacuum assists in the degassing process and increases the purging efficiency for soluble compounds. At the end of the purge, Valves 4 and 7 are shut. The sample can is removed from the liquid N_2 and disconnected from the system. Keeping the system under positive pressure prevents room air from entering the system and is essential to prevent contamination of the next sample.

Comparison of the Vacuum Extraction Flask and Purge–Cryogenic Trapping Methods.   The extraction efficiencies of these methods were determined by purging a seawater sample for 60 min, replacing the sample can, and purging for another 60 min. This process was repeated four times; the calculated extraction efficiencies are listed in Table I. The results show the first extraction efficiency to be 90 % or better for most compounds. For this reason a 1-h purge of each 500-mL sample was chosen. Longer purge times would result in dilution of the components that had already been completely (>90 %) extracted. Therefore, the uncertainty of the purge–trap method would be ≤ 10 % for most compounds and ≤ 30 % for some of the more difficult to extract bromine compounds. This uncertainty can be reduced to about 10 % by adjusting the results for the extraction efficiency. Those compounds that had lower efficiencies were chloroform ($CHCl_3$) and bromoform ($CHBr_3$). The results agree well with conventional purge–trap devices in which $CHCl_3$ and $CHBr_3$ are also difficult to purge completely.

Comparisons of the vacuum extraction flask method versus the purge–cryogenic trap method were made on duplicate seawater samples. The results are shown in Table II. The compounds measured represent a wide range of Henry's constants (solubilities) and provide a representative evaluation of the two methods. In Run 1 of Table II the $CHCl_3$ concentration was not measured. In Runs 1–4 the CF_2Cl_2 concentrations were below the detection limit of the GC–ECD used for the analysis. The comparisons are for single samples and agree reasonably well except for 1,1,1-trichloroeth-

Table I. Extraction Efficiencies of Purge and Trap Module

	% Removal		
Compound	1st Extraction	2nd Extraction	3rd Extraction
CH_3Cl	100	—	—
CH_3Br	100	—	—
CCl_3F	100	—	—
CH_3I	91	100	—
C_2H_5Br	100	—	—
CH_2ClBr	97	100	—
$CHCl_3$	81	99	100
CH_3CCl_3	87	98	100
CCl_4	95	99	100
$CHCl_2Br$	83	100	—
$CHClBr_2$	78	100	—
CH_2Br_2	71	100	—
C_2Cl_4	82	100	—
$CHBr_3$	58	100	—

NOTE: Each extraction was carried out for 60 min. Data were collected on a GC–ECD with a capillary column. Although a fourth extraction was done, the components had already been extracted.

ane (CH_3CCl_3). The CH_3CCl_3 results for Runs 1–3 from the purge–cryogenic trap method were high. The purge column was found to be contaminated internally with CH_3CCl_3. When the unit was thoroughly cleaned, the CH_3CCl_3 results for Run 4 were in very good agreement with the vacuum extraction flask data. The results for $CHCl_3$ also agree remarkably well considering the high solubility of this compound. The lower $CHCl_3$ values obtained with the purge–cryogenic trap method may be a result of the lower extraction efficiency (81%) of $CHCl_3$ (see Table I).

The tests demonstrate the ability of the purge–cryogenic trap method to recover quantitatively the dissolved gases in seawater, which can then be

Table II. Comparison of Vacuum Extraction Headspace (HS) Samples with Purge–Cryogenic Trap Samples (P–T)

	H^a at 25 °C	Run 1		Run 2		Run 3		Run 4	
Compound		HS	P–T	HS	P–T	HS	P–T	HS	P–T
N_2O	2.0	207	196	326	296	194	240	338	359
CCl_2F_2	20	33	—	42	—	30	—	32	—
CCl_3F	6	145	144	92	79	61	110	72	95
CH_3CCl_3	0.94	277	392	326	444	229	562	252	250
CCl_4	1.5	128	133	197	192	129	163	156	182
$CHCl_3$	0.24	—	—	29,000	21,000	3800	3500	12,000	12,000

[a] See Reference 2.

sent back to the laboratory for analysis by GC–MS or by other instruments. The approach is simple, rugged, and easily operated on-board ship or in other field situations. Although the uncertainties in the method are about 10%, this level is not significant compared to the concentration variations for duplicate seawater samples from a given location.

Analysis of Trace Gases

Two types of standards were prepared for the analysis of trace gases in air and seawater. The first type is background air samples collected cryogenically in 35-L stainless steel tanks at Cape Meares, Oregon. The stability of trace levels of halocarbons in these containers is excellent over periods as long as 3–5 years (22). The second type of standard is zero air spiked with the sulfur, iodinated, and brominated compounds present in seawater at concentrations higher than those found in the background marine air. These standards with low concentrations [parts per billion by volume (ppbv)] are more convenient for routine calibration purposes.

Spiked standards are prepared by adding known amounts of the pure compounds to clean 35-L stainless steel containers under vacuum. They are then filled cryogenically with contaminant-free zero air (AADCO 737A) to a pressure of about 300 psi. The zero air is tested for trace pptv contaminants before use with capillary columns via GC–ECD for halogenated compounds and by GC–FPD for trace sulfur gases. The final volume of air in the tank is determined with a Kurz Model 545-1 digital flow calibrator during the fill procedure and from the static volume–final pressure relationship. The two methods usually agree within ± 2%.

The analyses were made on various chromatographs, generally with liquid N_2 to cryogenically focus the analytes at the head of the column before initiating the temperature program. Carbonyl sulfide, CS_2, and $(CH_3)_2S$ were measured on an HP5840A with an FPD with an SE-30 (J&W) fused-silica, 30-m capillary column. The column was run into the detector to ~ 3 mm below the flame base, the H_2 and O_2–air connections were reversed, and the detector temperature was 150 °C. The gas settings were as follows: He carrier gas, 12 psi; H_2, 200 mL/min; O_2, 16 mL/min; N_2 makeup gas, 10 mL/min; and air, 35 mL/min. The oven temperature program was begun at − 65 °C with a 1.5-min hold before programming it at 12 °C/min to 100 °C. A 6-in. length of the column was formed into a loop outside the oven to cryogenically focus, in liquid N_2, the contents of the packed freezeout loop before making the final transfer of the analytes to the head of the column.

Halocarbons were measured on a PE3920B with an ECD with a DB-1 (J&W) fused-silica 30-m capillary column. Gas settings were as follows: He, 18 psi, and the Ar(95%) + CH_4 (5%) makeup gas, 50 mL/min. The detector temperature was 375 °C, and the temperature program was started at − 60 °C, held for 2 min, then increased at 8° C/min to 100 °C. Integration of both the sulfur gases (HP5840A) and the halocarbons (PE3920B) was made with HP3388A systems.

The C_2–C_4 hydrocarbons were measured isothermally (45 °C) with a flame ionization detector (FID) on a 10-ft by ⅛-in. OD Teflon column packed with phenyl isocyanate–Porasil C (PIC) in an HP5790A and integrated with an HP3390A integrator. The C_3–C_{10} hydrocarbon speciation was done on HP5790A and PE3920B instruments with DB-1 capillary columns programmed from −50 to 100 °C at 4 °C/min. Conventional freezeout loops, 10-in. by ⅛-in. stainless steel tubing packed with 60-mesh glass beads, were used to transfer the analytes to the head of the capillary columns. Either HP3390A or HP3388A integrators were used to record and process the chromatograms.

Nafion dryers (24) were used to remove water vapor from the samples prior to the liquid oxygen freezeout loop. If a dryer is not used, the accumulation of cryofrost eventually plugs the freezeout trap of the capillary column. Nafion tubing dryers were chosen rather than desiccants because they are clean at the low pptv level, do not cause decomposition of gases such as OCS, $(CH_3)_2S$, and CH_3Cl, and are readily available. The 5-Å–13× molecular sieve in the dryers is periodically regenerated to remove water. Occasionally a dryer becomes contaminated with C_2Cl_4. The C_2Cl_4 can be removed by applying heat while flushing with zero air, or the Nafion tubing can be replaced. Each day the dryers are tested by running a zero air blank to check for contamination.

Gas Chromatography–Mass Spectrometry Instrumentation. A Finnigan 4021 GC–MS–DS modified to use a Hewlett-Packard 5790A gas chromatograph was used to measure trace halocarbon and sulfur gases. In marine air and seawater samples their concentrations are in the ppbv–pptv range. Again, one or two freezeout steps are used to concentrate the samples (25) and to obtain sharp and reproducible peak shapes for low boiling compounds such as OCS, CH_3Cl, and CF_2Cl_2. The complete system is shown in Figure 5 with three input methods.

The syringe system is used for air samples because it is convenient and accurately measures sample volumes between 25 and 100 mL. The pressure in the sample bottle is used to flush and fill the 100-mL syringe through a flow restrictor when toggle Valve 1 (Figure 5) is open. When the syringe has been filled, Valve 1 is closed, the six-port valve is turned to "load," Valve 2 is opened, and Valve 4 is placed in vent. Gravity forces the air at 100 mL/min through the freezeout loop. When the syringe is empty, Valves 2 and 4 are closed, the sample valve is turned to "inject," the freezeout loop is placed in 85 °C water, and the sample is transferred to the capillary trap.

The vacuum system is used to transfer larger sample sizes (100–500 mL) onto the freezeout loop with high precision. This system has Teflon lines between the sample container and the Nafion dryer to reduce possible decomposition of the sulfur gases. A GAST Model DOA-P-101-AA pressure

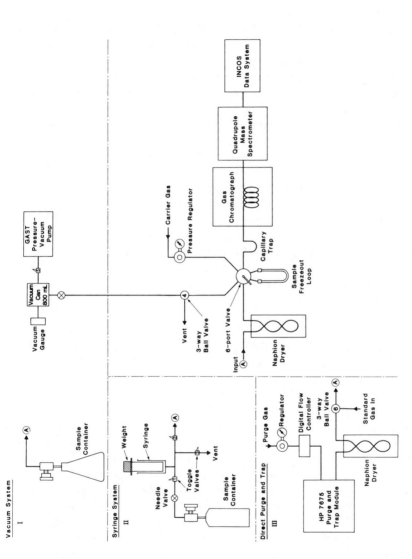

Figure 5. Schematic diagram of capillary column GC–MS–DS system, showing the three means of sample introduction, Nafion dryer, liquid oxygen freezeout loop, and the capillary freezeout trap.

vacuum, oilless diaphragm pump is used to evacuate the 800-mL stainless steel can (Figure 5). The vacuum is measured with a Validyne Model CD23 digital pressure-gauge system. When the can has been evacuated, toggle Valve 5 is closed, isolating the can from the vacuum pump. The three-way ball Valve 4 is turned to the "vacuum" position, and sample is pulled through the freezeout loop into the can. A needle valve is used to control the flow rate to about 50 mL/min. The volume of the vacuum can-gauge system has been calibrated, so that the volume of sample loaded onto the freezeout loop can be determined from the pressure difference. When the desired pressure difference is reached, Valve 4 and the valve on the sample bottle are closed.

A direct purge to the freezeout loop instead of a Tenax or other polymer trap can be done as shown in Figure 5. A Hewlett-Packard Model 7675A purge–trap module was used, and the Tenax trap was bypassed so the purged compounds would flow through Nafion dryers into the freezeout trap. A pressure gauge and digital flow controller on the purge gas line regulated the flow through the freezeout trap.

In the systems just discussed, the analytes are transferred to a capillary freezeout loop using a six-port Valco Model 6P stainless steel valve. This trap is fashioned from a 6-in. loop of the fused-silica column and is attached directly to the valve with Vespel-graphite ferrules. The loop is placed in a short Dewar flask, which can hold either liquid N_2 or hot water. For analysis a 30-M, DB-1 (Durabond) with a 1.0-μm phase loading is used. The oven temperature cycle and the mass spectrometer program are coordinated to start together. The starting temperature is -70 °C. After the contents of the capillary trap are transferred to the column, the oven temperature is held for 1.5 min at -70 °C before initiating the temperature program at 20 °C/min to the final temperature of 125 °C. Helium carrier gas is used, and the column is operated at a constant pressure of 12 psi with a Porter Model 8286 stainless steel pressure regulator.

The Finnigan 4021 quadrupole analyzer is equipped with a pulsed positive-ion, negative-ion chemical ionization (CI) unit and is operated in the multiple ion detection (MID) mode. The filament, CI box, and electron multiplier (EM) are turned on manually when the source pressure drops to its normal value after the injection surge. The CI dynode voltage is -3 kV, and the electron multiplier voltage is -1.85 kV. The electron energy is 70 eV with an emission current of 0.5 mA. The manifold temperature is 100 °C, the source temperature is 250 °C, and the GC transfer line temperature is about 150 °C. Because the peaks are narrow on the capillary run, typically about 2 s, a scan time of 0.3 s was chosen for the MID descriptor to get multiple scans over each peak for good quantitation. Each MID descriptor was set to measure two to five ions per scan, depending on the number of compounds eluting in a particular region of the chromatogram. A list of the MID descriptors used by the computer during the analyses is shown in Table III, which includes the MID descriptors, retention time intervals, and the compounds measured in a particular region. The actual retention times of the compounds on the DB-1 column along with the quantitation ion are shown in Table IV.

Computer Programs. The Finnigan 4021 computer system controls the data acquisition by changing the MID descriptors during the chromatographic run and provides the automatic data processing needed for qualitative and quantitative analysis.

The data acquisition program outlined in Table III changes the descriptors at the retention times shown and plots the reconstructed ion chromatogram (RIC). To measure other compounds, their retention times must first be determined and their ions inserted into the appropriate descriptor. This data acquisition program is convenient because a different program can be optimized for each type of sample. For example, one program is used for determining trace gases from midoceanic samples, and another program is used for analyzing coastal seawater for volatile priority pollutants.

Data processing for each of the data acquisition programs prints out the data for each compound with its characteristic ions (Table IV). The computer integrates each peak by the peak height and peak area. The actual concentration is calculated from standards run with the samples.

Identification of Compounds in Air and Seawater. Many of the biogenic compounds in seawater are at concentrations too low to be measured in the scanning mode. Accordingly, the compounds must be selected in advance, and an MID descriptor must be inserted at the approximate retention time. To determine the retention time, standards are run in the scanning mode, and the data are tabulated in a form similar to Table III. This information is used to write the MID descriptors and the computer program. Compounds such as OCS, CH_3Cl, and CCl_3F were easily identified by retention time and one mass peak. Brominated compounds, which were found in seawater at low concentrations, were identified by retention time and two mass peaks. The RICs of a background marine air sample, a seawater headspace sample, and a purge–cryogenic trap sample with their identified peaks are shown in Figures 6 and 7.

Quantitation of Trace Gases and Instrument Performance. Concentrations are determined by using external standards prepared by static dilution. Precision of analysis is determined by replicate analysis of five standard runs over 2 h. The results are shown in Table V. Peak areas are used for OCS, CCl_2F_2, and CH_3Cl because of variable peak heights associated with the injection procedure for the low boiling materials and the "tailing" of CH_3Cl on the DB-1 column. Generally, peak height and peak area for OCS and CCl_2F_2 give almost equivalent precision. The remainder of the compounds are quantified by using peak heights. Table V shows that for compounds present at low concentration, such as CH_3Br and C_2Cl_4, the peak height measurement is better than the peak area. This result occurs

Table III. Computer Data System and MID Parameters

MID Descriptor	Retention Time^a (min)	MID m/e	MID time (s)	Target Compound	Other Compounds
AA	0	60	0.105	OCS	
	↓	85	0.105	CCl_2F_2	
	4.3				
AB	4.3	50	0.105	CH_3Cl	
	↓	94	0.105	CH_3Br	
		96	0.052	CH_3Br	
	7.3				
AC	7.3	62	0.026	$(CH_3)_2S$	
		101	0.052	CCl_3F	
	↓	108	0.052	C_2H_5Br	
		142	0.052	CH_3I	
	7.6				
AD	7.6	62	0.026	$(CH_3)_2S$	$trans$-$C_2H_2Cl_2$
		84	0.052	CH_2Cl_2	
		96	0.052	$trans$-$C_2H_2Cl_2$	
	↓	108	0.052	C_2H_5Br	
		110	0.052	C_2H_5Br	
	9.0				
AE	9.0	83	0.052	$CHCl_3$	
		85	0.052	$CHCl_3$	
	↓	128	0.052	CH_2BrCl	
		130	0.105	CH_2BrCl	
	9.8				
AF	9.8	97	0.105	CH_3CCl_3	
	↓	117	0.105	CCl_4	CH_3CCl_3
	10.4				
AG	10.4	83	0.052	$CHCl_2Br$	
		85	0.052	$CHCl_2Br$	
		95	0.052	C_2HCl_3	CH_2Br_2
	↓	172	0.052	CH_2Br_2	
		63	0.052	1,2-$C_3H_3Cl_2$	
	11.5				
AH	11.5	79	0.052	$(CH_3)_2S_2$	$CHBr_2Cl$
	↓	94	0.105	$(CH_2)_2S_2$	
		127	0.052	$CHBr_2Cl$	
	12.8	164	0.052	C_2Cl_4	
AI	12.8	107	0.052	$C_2H_4Br_2$	
	↓	109	0.052	$C_2H_4Br_2$	
		171	0.052	$CHBr_3$	
	14.0	173	0.105	$CHBr_3$	

^{a}Retention time measured from time of injection from the capillary column.

Table IV. Retention Times, MID Ions, Parent Ions, and
Quantitation Ions of Sulfur, Bromine, and Chlorine Compounds

Compound	Retention[a] Time (min)	MID Ions		Parent Ion	Quantitation Ion
OCS	3.1	60	62[b]	60	60
CCl_2F_2 (F-12)	3.8	85	87	85	85
CH_3Cl	4.6	50	52	50	50
CH_3Br	6.1	94	96	94	94
CCl_3F (F-11)	7.3	101	103	101	101
CH_3I	7.6	142	127	142	142
C_2H_5Br	7.7	108	110	108	108
$(CH_3)_2S$ (DMS)	7.7	62	64[b]	62	62
CH_2Cl_2	7.9	84	86	49	84
trans-$C_2H_2Cl_2$	8.5	62	96	61	62
CH_2BrCl	9.2	128	130	49	130
$CHCl_3$	9.3	83	85	83	83
CH_3CCl_3	9.8	97	117	97	97
CCl_4	10.1	117	119	117	117
CH_2Br_2	10.5	95	172	93	95
1,2-$C_3H_6Cl_2$	10.5	63	65	63	63
C_2HCl_3	10.6	95	130	130	95
$CHCl_2Br$	10.6	83	85	83	83
$CHBr_2Cl$	11.8	79	127	129	127
$(CH_3)_2S_2$	11.2	79	94	94	—
C_2Cl_4	12.3	164	166	166	164
$CHBr_3$	13.5	173	171	173	173
$C_2H_4Br_2$	13.3	107	109	107	107

[a]DB-1 column, 1.0-μm phase loading, -70 °C, 1.5 min $\rightarrow$ 125 °C at 20 °C/min.
[b]Isotope peak verified on concentrated samples, *see* Reference 2.

because the computer quantitation program can measure peak height on small peaks more accurately than peak area.

Linearity was determined by duplicate analyses at three different sample volumes (concentrations). The results for six representative compounds are shown in Figure 8. Responses were linear within the uncertainty limits shown in Table V. The measurements were made over the range of concentrations normally encountered in the air and seawater samples. The signal-to-noise ratios (S/N) for these measurements are shown in Table VI for typical air and seawater concentrations. For OCS at 500 pptv the S/N ratio is better than the flame photometric detector. For $(CH_3)_2S$ the signal strength was intentionally reduced by selecting a 0.026-s MID scan time because of its very high concentration in seawater ($\sim$ 10 ppbv).

Results. Marine air and seawater samples were collected on the March–June 1983 National Oceanic and Atmospheric Administration's (NOAA) *Discoverer* cruise studying the marine sources of acid rain precursors. The general cruise track was from Alaska south via Hawaii, to the equator, east along the equator to Mexico, and return to Seattle via San Diego with sev-

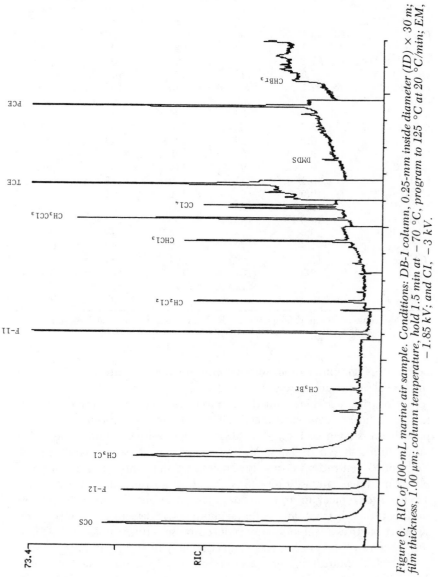

Figure 6. RIC of 100-mL marine air sample. Conditions: DB-1 column, 0.25-mm inside diameter (ID) × 30 m; film thickness, 1.00 µm; column temperature, hold 1.5 min at −70 °C, program to 125 °C at 20 °C/min; EM, −1.85 kV; and CI, −3 kV.

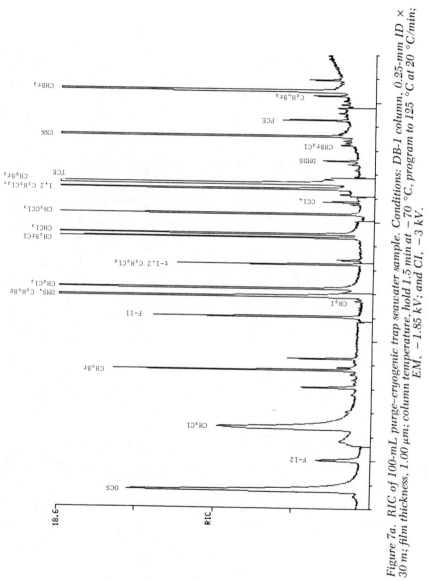

Figure 7a. RIC of 100-mL purge–cryogenic trap seawater sample. Conditions: DB-1 column, 0.25-mm ID × 30 m; film thickness, 1.00 μm; column temperature, hold 1.5 min at −70 °C, program to 125 °C at 20 °C/min; EM, −1.85 kV; and CI, −3 kV.

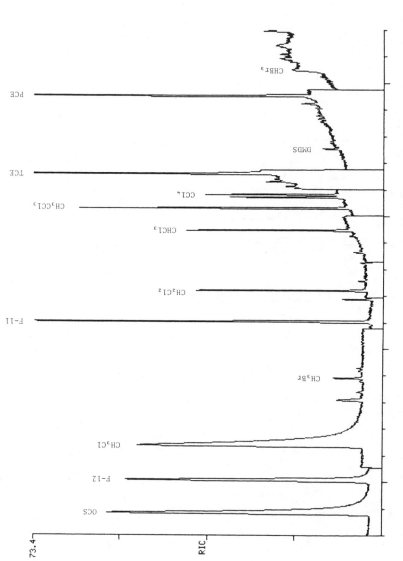

Figure 7b. RIC of 100-mL equilibrated seawater headspace sample. Conditions are the same as in Figure 7a.

Table V. Percent Relative Standard Deviation of
Compounds on a Capillary Column GC–MS–DS System

Compound	Percent Standard Deviation	
	Peak Height	Peak Area
OCS	7	10
CCl_2F_2	8	8
CH_3Cl	13	8
CH_3Br	11	28
CCl_3F	2	4
C_2H_5Br[a]	6	6
$(CH_3)_2S$[a]	6	6
CH_2Cl_2	18	16
$trans\text{-}C_2H_2Cl_2$	7	7
CH_2BrCl[a]	26	27
$CHCl_3$	10	16
CH_3CCl_3	5	10
CCl_4	6	13
CH_2Br_2[a]	8	7
C_2HCl_3	14	16
$CHCl_2Br$[a]	5	5
C_2Cl_4	10	15
$CHBr_3$	5	6

[a]Determination made on triplicate samples.

eral intensive tracks back and forth along the coast. Results for selected samples covering the latitudinal range of 30° N to 6° S are shown in Table VII. The air and seawater data are reported as parts per trillion by volume at 25 °C for comparison of the saturation values.

The concentrations of the anthropogenic compounds (CCl_2F_2, CCl_3F, CH_3CCl_3, and CCl_4) in the air samples (Table VII) show no significant differences over the northern latitudes studied. However, the data for the southern hemispheric samples show lower values. This result is consistent with the source regions, lifetimes, and transport of these gases (26). The atmospheric OCS concentrations were too variable to show any significant trends dependent on latitude. The local variability observed may be consistent with its immediate source in the surface seawater (1). No measurable bromine compounds other than CH_3Br were observed in the air samples by GC–MS. The levels of CH_3Cl, CH_2Cl_2, and $CHCl_3$ are also consistent with earlier results and reflect their biogenic origin in the surface seawater.

The seawater data show the presence of a number of bromine compounds. These compounds were not measurable in the atmospheric samples, so biological production in the oceans is the probable source. The elevated CCl_2F_2 and the higher values in the CCl_4 data indicate that these samples may be slightly contaminated; CCl_2F_2 was used extensively in the laboratory on the ship.

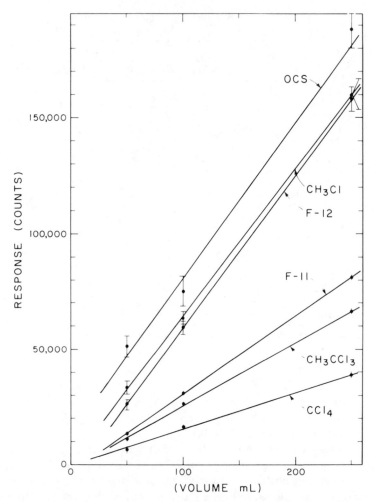

Figure 8. GC–MS response vs. volume of standard. The volume range is representative of the concentrations of compounds found in marine samples.

The air concentration, C_g, the seawater concentration, C_l, and the Henry's constant H (2) can be used to calculate the percent relative saturation S of a compound in the ocean with the equation

$$S = \left(\frac{C_l H}{C_g} - 1 \right) 100 \qquad (3)$$

A positive saturation means a compound will transfer from the ocean to the atmosphere, and a negative saturation means it will transfer from the atmosphere to the ocean.

Table VI. Signal-to-Noise Ratio (S/N) Measurements for Typical Air and Seawater Samples

Compound	Air Concentration (pptv)	S/N, Air	Seawater Headspace Concentration (pptv)	S/N, Seawater
OCS	500	250	1000	500
CCl_2F_2	380	250	50	20
CH_3Cl	600	200	500	200
CH_3Br	12	3	10	3[a]
CCl_3F	200	250	30	50
CH_3I	1	2	5	10
$(CH_3)_2S$	—	—	10,000	1000
CH_2Cl_2	30	500	30	500
$CHCl_3$	20	200	60	200[a]
CH_3CCl_3	170	600	100	600
CCl_4	155	400	100	200
CH_2Br_2	—	—	—	
C_2HCl_3	10	200	10	200
$CHCl_2Br$	—	—	—	
$CHBr_2Cl$	—	—	—	
$(CH_3)_2S_2$	—	—	100	100
C_2Cl_4	100	200	90	200
$CHBr_3$	0.7	2	1	2[a]

[a]The concentration used to calculate the S/N is the minimum concentration, and S/N ratios are often higher.

Table VIII shows the percent relative saturation of the gases measured. For OCS, CH_3Cl, and CH_3Br the ocean is supersaturated and acts as a source of these gases to the atmosphere. Positive saturations occur with C_2H_5Br, CH_2BrCl, CH_2Br_2, $CHBrCl_2$, $CHBr_2Cl$, and CH_3Br. This finding demonstrates that these compounds are produced in the ocean and released into the atmosphere. For CCl_3F the saturation percentages are small and vary from supersaturated to undersaturated; therefore, the ocean and atmosphere are in equilibrium.

Conclusions

The capillary column GC–MS–DS instrument proved to be a sensitive and universal detector for chlorine, bromine, and sulfur gases. The sensitivity for a particular compound can be varied by changing the MID scan time for the characteristic ion. This ability makes it easier to analyze samples that have compounds present at both high and low concentrations, such as $(CH_3)_2S$ and CH_3I in seawater. The mass spectrometer is also used to measure compounds that are not completely resolved on the capillary GC–EC. The GC–MS technique is reproducible and linear over the range of concentrations found in marine samples.

Table VII. Concentrations of Trace Gases in Marine Air and Seawater Headspace Samples

Latitude	Longitude	$T(°C)$	OCS	CCl_2F_2	CH_3Cl	CH_3Br	CCl_3F	C_2H_5Br	CH_2Cl_2	$CHCl_3$	CH_2BrCl	CH_3CCl_3	CCl_4	CH_2Br_2	$CHBrCl_2$	$CHBr_2Cl$	$CHBr_3$
Marine Air Samples																	
6° S	105° W		NM	347	628	9	203	—	7	15	—	131	145	—	—	—	—
0°	119° W		575	358	650	12	207	—	16	32	—	172	143	—	—	—	—
0°	136° W		572	330	657	13	209	—	18	21	—	150	151	—	—	—	—
1° N	158° W		575	339	621	15	210	—	14	16	—	144	151	—	—	—	—
9° N	158° W		600	342	665	9	207	—	38	18	—	154	148	—	—	—	—
27° N	124° W		505	344	663	18	216	—	35	25	—	179	155	—	—	—	—
Seawater Concentration in Headspace Samples																	
6° S	105° W	28	604	225	2266	69	47	—	189[a]	426	+	447	154	—	+	—	+
3° S	105° W	29	756	NM	2026	47	47	—	23[a]	NM	—	265	116	—	—	+	+
3° S	158° W	28	521	62	1760	42	34	+	23[a]	296	—	197	110	—	+	—	+
0°	115° W	29	751	280	2051	60	97	+	153[a]	260	—	265	376	—	+	—	+
0°	127° W	29	NM	194	2090	120	55	+	175[a]	260	—	302	552	—	+	+	+
0°	145° W	29	695	133	2125	45	36	+	32[a]	218	—	214	NM	—	—	+	+
2° N	158° W	29	737	225	1197	77	32	—	117[a]	359	+	298	602	—	—	—	+
3° N	105° W	28	770	116	1968	65	40	+	178[a]	432	+	212	108	—	+	—	+
13° N	105° W	29	734	1467	2291	52	30	+	75[a]	213	+	176	90	+	+	—	+
28° N	122° W	18	574	1128	1318	52	56	+	63[a]	343	+	332	160	—	—	—	—

NOTE: Surface samples were collected with a stainless steel bucket from the top meter of the ocean. Air and seawater concentrations are reported as parts per trillion by volume. For seawater this is milliliters of gas at 25 °C per 10^{12} mL of seawater. This unit is used to facilitate the air–sea exchange model. + indicates a peak was detected; —, no peak was quantitated; and NM, not measured.

[a] The headspace concentration is reported because reliable Henry's constants are not available to calculate concentration.

Table VIII. Percent Relative Saturation of Trace Gases in Seawater

Latitude	Longitude	OCS	CH_3Cl	CH_3Br	CCl_3F	C_2H_5Br	$CHCl_3$	CH_2BrCl	CH_3CCl_3	CCl_4	CH_2Br_2	$CHBrCl_2$	$CHBr_2Cl$	$CHBr_3$
6° S	105° W	+184	+65	+130	+16	—	+577	+	+221	+30	—	+	—	+
3° S	105° W	+254	+44	+18	+16	—	NM	—	+50	+4	—	+	+	+
3° S	158° W	+144	+30	-15	-17	+	+370	—	+45	0	—	+	—	+
0°	115° W	+253	+46	+50	+135	+	+102	—	+45	+224	—	+	—	+
0°	127° W	NM	+110	+200	+32	+	+196	—	+65	+375	—	+	—	+
0°	145° W	+226	+50	+13	-13	+	+148	—	+30	NM	—	+	+	+
2° N	158° W	+246	-10	+54	-24	—	+436	+	+96	+418	—	—	—	+
3° N	105° W	+261	+45	+30	-5	+	+545	+	+38	-7	—	—	—	+
13° N	105° W	+232	+58	+73	-29	+	+158	+	+7	-21	+	+	+	+
28° N	122° W	+206	-8	-20	+30	+	+160	+	+35	+8	—	+	—	—

NOTE: Percent relative saturation was calculated as $[(C_l H/C_g) - 1]\,100$. A positive result indicates supersaturation; a negative result indicates undersaturation. The symbols are as in Table VII.

The sample collection and extraction techniques were used on the NOAA *Discoverer II* 1983 oceanographic cruise. The samples were analyzed with the GC–EC–FI–FPD and GC–MS–DS instruments. Several new trace gases were identified, quantified, and routinely measured. The results show the ocean to be the source of many bromine compounds (CH_3Br, C_2H_5Br, CH_2BrCl, CH_2Br_2, $CHBrCl_2$, $CHBr_2Cl$, and $CHBr_3$), and they substantiate that the ocean is a source for OCS, CH_3Cl, and $CHCl_3$.

Literature Cited

1. Rasmussen, R. A.; Khalil, M. A. K.; Hoyt, S. D. *Atmos. Environ.* **1983**, *16*, 1591.
2. Hoyt, S. D. Ph.D. Dissertation, Oregon Graduate Center, Beaverton, Or., 1982.
3. Rasmussen, R. A.; Khalil, M. A. K.; Gunawardena, R.; Hoyt, S. D. *J. Geophys. Res.* **1982**, *87*, 3086.
4. Lovelock, J. E. *Nature* **1975**, *256*, 193.
5. Singh, H. B.; Salas, L. J.; Stiles, R. E. *J. Geophys. Res.* **1983**, *88*, 3684.
6. Lovelock, J. E.; Maggs, R. J.; Rasmussen, R. A. *Nature* **1972**, *237*, 452.
7. Bernard, W. R.; Andreae, M. O.; Watkins, W. E.; Bingemer, H.; Georgii, H.-W. *J. Geophys. Res.* **1982**, *87*, 8787.
8. Cline, J. D.; Bates, T. S. *Geophys. Res. Lett.* **1983**, *10*, 949.
9. Khalil, M. A. K.; Rasmussen, R. A.; Hoyt, S. D. *Tellus* **1983**, *35B*, 266.
10. Crutzen, P. J. *Geophys. Res. Lett.* **1976**, *3*, 73.
11. Turco, R. P.; Whitten, R. C.; Toon, O. B.; Pollock, J. B. *Nature* **1980**, *283*, 283.
12. Charlson, R. J.; Rodhe, H. *Nature* **1982**, *295*, 683.
13. Hudson, R. D.; Reed, E. NASA Report 1049, 1979.
14. Rasmussen, R. A.; Rasmussen, L. E.; Khalil, M. A. K.; Dalluge, R. W. *J. Geophys. Res.* **1980**, *85*, 7350.
15. Rasmussen, R. A.; Holdren, M. W. *Chromatogr. Newsl.* **1972**, *1*, 31.
16. Farwell, S. O.; Gluck, S. J.; Bamesberger, W. L.; Schutte, T. M.; Adams, D. F. *Anal. Chem.* **1979**, *51*, 609.
17. Keith, L. H. "Identification and Analysis of Organic Pollutants in Water"; Ann Arbor Science: Ann Arbor, Mich., 1976.
18. Dandeneau, R. D.; Zerenner, E. H. *J. High Resolut. Chromatogr.* **1979**, *2*, 351.
19. Jennings, W. *J. High Resolut. Chromatogr.* **1980**, *3*, 601.
20. Sauter, A. D.; Betowski, L. D.; Smith, T. R.; Strickler, V. A.; Beimer, R. G.; Colby, B. N.; Wilkinson, J. E. *J. High Resolut. Chromatogr.* **1981**, *4*, 366.
21. Trussell, A. R.; Moncur, J. G.; Lieu, Fang-Yi; Leong, L. Y. C. *J. High Resolut. Chromatogr.* **1981**, *4*, 156.
22. Rasmussen, R. A.; Lovelock, J. E. *J. Geophys. Res.* **1983**, *88*, 8369.
23. McAuliffe, C. *Chem. Technol.* **1971**, *1*, 46.
24. Foulger, B. E.; Simmonds, P. G. *Anal. Chem.* **1979**, *51*, 1089.
25. Rasmussen, R. A.; Harsch, D. E.; Sweany, P. H.; Krasnec, J. P.; Cronn, D. R. *J. Air Pollut. Control Assoc.* **1977**, *27*, 579.
26. Rasmussen, R. A.; Khalil, M. A. K.; Dalluge, R. W. *Science* **1981**, *211*, 285.

RECEIVED for review November 29, 1983. ACCEPTED April 23, 1984.

A Measurement Strategy for Sea-Salt Aerosols

THEODORE B. WARNER[1] and JOHN B. HOOVER

Naval Research Laboratory, Washington, DC 20375

A design concept for an automatic sampling and measuring system for aerosols is proposed, and an electrode method, using a nonflowing-junction reference electrode and capable of measuring 90 ng of chloride in a 50-μL sample, is reported. This sample size corresponds to 4 min of sampling at 20 L/ min when aerosol concentration is 2 μg/m³. Data management and communications are also discussed.

THE SEA-SALT AEROSOL is probably the largest single source of particulate matter in the atmosphere: some 10^{12}–10^{13} kg are introduced each year (*1*). Atmospheric sea salt, its origin, and its concentration in the air are reviewed elsewhere (*2*). Knowledge of the distribution of sea salt is important because this aerosol is involved in the formation of rain (*3*) and some kinds of fog (*4, 5*), and it plays a part in transfer to the atmosphere and enrichment of many species such as organic materials (*6, 7*), heavy metals (*8*), bacteria (*9*), and viruses (*10*). Sea-salt aerosol interferes with electromagnetic propagation and can damage or degrade the performance of marine turbines (*11*). These diverse effects have stimulated research to measure and predict the effects of geographical, meteorological, and oceanographic conditions on the aerosol concentration.

The study of such complex phenomena may be roughly separated into three phases. During the first, in which a body of descriptive data is amassed, the problem is outlined in a qualitative way. Phase two, where attention focuses on possible mechanisms, leads to a quantitative description of the phenomenon and to an appreciation of the relative significance of the controlling parameters. In the final phase, a model of the phenomenon is developed, and predictions may be made and tested.

Current research concerning sea-salt aerosols is in the early stages of the second phase. A substantial body of descriptive data exists, and the

[1]Current address: Warner Associates of Glen Echo, Inc., 6002 Madawaska Road, Bethesda, MD 20816.

0065-2393/85/0209-0057 $06.00/0

magnitude of the atmospheric sea-salt load (2 and references therein) is understood. A combination of several source mechanisms, notably bubble bursting and wave breaking, is probably responsible for most of the observed sea-salt aerosol. Wet and dry deposition and even the filtering effects of vegetation (12, 13) have been considered as aerosol sinks. Attention is now turning to the mechanisms of these and other processes.

Overview of the State of the Art

Current Data and Future Needs.   WIND–SEA-SALT RELATIONSHIPS. A major question of current interest is the role of surface wind in sea-salt aerosol production. In the past, the large sample sizes required by analytical limitations have made it necessary to sample, with high-volume samplers, for several hours. The long integration times masked the short-term variations in sea-salt concentration. Typically, a surface wind speed averaged over each sample period was also reported. On the basis of these types of data, empirical relations between wind velocity and sea-salt loading were suggested. Lovett (15), for example, proposed the equation $\ln S = AU + B$, where S is the sea-salt concentration (micrograms per cubic meter), and U is the wind speed (meters per second). Lepple et al. (14) analyzed similar data from the South Atlantic by using this equation. The results are shown in Figure 1.

Lovett found slopes A in the range 0.13–0.19 for six different subsets of

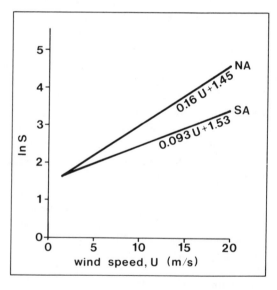

Figure 1. Wind–sea-salt correlations from the North and South Atlantic. Key: SA, South Atlantic data taken at 16 m (14); NA, North Atlantic data taken at 5–15 m (15); S, salt concentration ($\mu g/m^3$); and U, wind speed (m/s).

the data (three elevations at each of two stations in the North Atlantic). His weighted average slope is 0.16 with a relative standard deviation of 37.5%; no correlation coefficients were reported. Lepple et al. (*14*) found the slope to be 0.093, but the correlation coefficient is only 0.33. The lack of agreement between these results indicates that the proposed equation is probably not generally descriptive of the dynamic processes encountered in the atmosphere.

DATA REQUIREMENTS. The sophistication represented by the cited analyses is not adequate for investigation of mechanisms of aerosol generation and transport. The lack of agreement, poor correlation, and large standard deviation noted in the examples just discussed suggest that any attempt to investigate the mechanisms of the production of sea-salt aerosol must consider parameters other than gross aerosol concentration and average wind speed. For example, instantaneous winds must be considered because short episodes of gusting winds may inject large quantities of sea salt into the atmosphere while contributing little to the average wind. Because the wind speed can decay much faster than the particle concentration, a hysteresis effect will also occur. Accordingly, the wind speed history must be considered.

Furthermore, measurement of the gross sea-salt aerosol concentration is not sufficient. Because the aerosol residence time depends on the particle size, the magnitude of the hysteresis effect will be sensitive to the particle size. Thus, the instantaneous particle-size distribution and the history of this distribution must be known.

At the opposite extreme, the long-term wind velocity must be considered because it controls the movement of the air mass being sampled. The occurrence of recent precipitation events alters the sea-salt aerosol concentration, as does the passage of air over a large land mass.

These parameters are some of the most important needed for developing an aerosol model, but by no means all of them. Most likely, temperature of the sea surface and air, height of the mixing layer, and sea surface roughness also need to be known.

The measurements of all parameters must be made on appropriate time scales. Measurements at intervals significantly longer than the time constant of the process in question will not contribute to an understanding of the dynamics of the process. Thus, wind speed should be monitored on a scale of minutes rather than hours. Ideally, the aerosol concentration data should also be acquired in the same time interval. This ideal will require advances in air sampling techniques—the traditional methods are far too slow.

Critical Evaluation of Sampling Methods. Available sampling techniques are summarized in Table I. Diffusional methods are not considered in detail here because they are not well adapted to fast processing of large volumes of air.

Table I. Comparison of Sampling Methods

Method	Advantages	Disadvantages
Diffusion (thermal, electrostatic, gravitational)	Works very well for separating particles of different masses. Especially effective for small particles ($D < 1$ μm).	Slow, depends on stability of flow patterns. Often requires a large collection surface.
Cyclone separators	Can be designed to separate very small particles and to give good mass segregation.	Particle cutoff size depends on inlet flow rate. Instrument size is proportional to $1/D_{min}^2$ (D_{min} = minimum particle diameter).
Impactors/Impingers	Simple and rugged. Work very well for large particles.	Do not work well for small particles. May have particle bounce–blow off problems, especially for dry particles.
Virtual impactors	Eliminates bounce and blow off. Separated particles are left in suspension and may be transported accordingly. Sharp cut points and low losses are possible.	Existing designs do not operate well at $D < 2.5$ μm.
Filtration	Simple, cheap, and rugged. Cutoff may be easily changed.	Filter blanks may be a problem if chemical analysis is intended. Samples are difficult to remove from some types of filters.

Cyclone separators produce continuous size distributions, which can be advantageous, but they also produce widely dispersed samples. Most analytical techniques are not well adapted to such samples. In addition, cyclones are larger and more complex than some of the alternatives.

Impactors and impingers do not work well for the small particles that are most important in long-range transport because these particles tend to follow the streamlines around obstacles. Further, the high velocity of the jets causes particles to bounce off hard targets. The latter effect is not critical at the high humidities encountered over the oceans, but might be significant in relatively dry air. Finally, particles may be blown off the target and re-entrained in the air stream.

Virtual impactors retain the advantages of real impactors and, because they do not have a solid collecting surface, particle bounce off and blow off are eliminated. Unfortunately, they also retain the standard impactor's inability to handle small particles. Currently, dichotomous virtual impactors are limited to particles greater than 2.5 μm (*16*). Reducing the cutoff point to the desired 0.5–1.0-μm range is a difficult task (*17*). If these engineering problems could be overcome, then virtual impactors would become much more attractive. The possibility of manipulating samples while still in suspension would open up new avenues for transporting the sample to the sensor.

All of the foregoing approaches have some advantages in particular circumstances. However, filtration appears to be the most versatile approach currently available. With proper selection of the filter medium, low backgrounds for chemical analysis can be obtained easily. The ability to change the particle-size range by a simple filter change is also a significant advantage.

Proper selection of the filter medium is more of an art than a science. The filter cutoff must be chosen to capture the smallest particles of interest. Other factors that must be considered are the type of filter (bulk or surface), the required flow rate, and the size of the membrane. These parameters are not independent and the best choice will usually involve trade-offs. Finally, the material from which the filter is made must be considered. It must be selected for compatibility with the intended postfiltration processing. Glass-fiber filters, for example, often have very high blanks for common ions such as chloride and sodium.

Future Measurement Strategies

The preceding discussion indicates that the future trend in aerosol measurement must be toward faster determinations of a much larger number of variables. Many variables have not been considered at all in previous work. The suggested aerosol monitor, described in this section, is a step in that direction.

Automatic Aerosol Monitor Concept. FUNCTIONAL REQUIRE-
MENTS. An ideal monitor would automatically collect and analyze aerosol samples and record geographical, meterological, and oceanographic
data. It would be capable of operating unattended for the length of a typical cruise (2–4 weeks), and would vary sample collection to adapt to widely
differing conditions. It would store data in a standard format accepted by
interested laboratories, and display selected results in real time. Finally,
the ideal monitor would be constructed mostly from commercially available components.

AUTOMATIC OPERATION. Automation will allow data to be accumulated and time-correlated much more rapidly than any manual system
could achieve, and thus would attain the needed sampling frequencies discussed earlier. Automation will also allow operation in places where an
operator cannot easily go. For example, Blanchard and Woodcock (2)
found that for studying vertical distributions of the aerosol (2), an automatic device carried aloft in a balloon could be most useful.

The device must be able to measure aerosol concentrations from about 2
to 1200 $\mu g/m^3$. This range would cover calm wind conditions up to wind
speeds over 35 m/s. Currently, little is known below wind speeds of about 3
m/s. Few measurements of high-concentration aerosols have been made
because they generally occur during very severe weather conditions when it
is dangerous or impossible for people to service instruments. Attempts at
modeling aerosol concentrations have been seriously hampered because of
the lack of data at both extremes. The small particles, which remain in
suspension at low wind speeds, are the most important in global transport
processes because they have long residence times.

RUGGEDNESS AND UNTENDED OPERATION. Oceanographic equipment
must withstand a corrosive and hostile environment. An ideal instrument would operate with no attention at all between ports, thus making
operation possible on a wide range of ships with no demands on the ship's
personnel.

SELF-ADAPTING CAPABILITY. The range of aerosol concentrations
that can be handled in a fully automated unit can be expanded markedly if
sampling times and conditions can be adjusted in accord with the actual
aerosol content of the atmosphere or on other measured parameters. With
such a feature, automatic isokinetic adjustments become possible. Such adjustments allow a marked increase in the quantitative accuracy of the calculation of actual aerosol concentrations in the air from the measured aerosol amounts.

DATA STORAGE AND DISPLAY. Very careful preplanning of eventual
use of data and collaborative discussions with other groups can facilitate
exchange of data in raw form. This topic will be discussed in detail later.
Real-time displays could be changed rather easily to meet the specific needs
of a given cruise or experiment by using menu-driven display routines.

Rather than having generalized displays of all the measured quantities, the computer can extract the relationship being studied and display the diagnostic variables.

COMMERCIALLY AVAILABLE COMPONENTS.   Very large amounts of labor, talent, and money are required to develop new computer-based systems (*18*). Major system components must be commercially available if designs made in one laboratory are to be readily adaptable to help solve problems in others (*19*). The design concept proposed here will maximize use of off-the-shelf components. Thus, a system developed in one laboratory could be recreated with little additional effort in another. The main cost in making a working system will be software development, which can cost more than 80% of the total production cost of the average computer-based product (*18*). This cost could be spread over a much larger base by standardizing data formats and sharing the software among collaborating laboratories.

OVERALL CONCEPT.   The overall concept is to have units for control, measurement of environmental data, aerosol sampling, and data storage and display. These functional units should be modular so that modifications may easily be made.

The heart of the monitor can be a personal computer: widely available, inexpensive, and with relatively large memory and computational ability. Although relatively slow, its speed far exceeds the demands of most analyses. Its software will control automatic sample collection, sample manipulation and transport to the measuring chamber, measurement, collection and correlation of other needed environmental data, real time display, and data storage.

Ideas and strategies concerning critical elements of the process—sample collection, manipulation, and transport—are discussed later. The requirement for reliability under field conditions dictates that a reasonably simple system be adopted. The sampler represents the major piece of custom-constructed equipment. Although remote-controlled pumps, valves, and motors can be purchased, the collector is still a major design challenge.

A measurement method based on an ion-selective electrode determination of chloride is suggested. This method is reasonably simple, rugged, and can be run for long times without attention. Several variations of this technique are considered. One variation is a conventional electrode measurement except that the flowing-junction reference electrode can be replaced by a fluoride ion selective electrode as a nonflowing reference. Another method was developed and evaluated for the measurement of very small aerosol samples; the measurement of chloride ion concentration is made on 50-μL samples. When the major-constituent composition of sea salt is known, the total aerosol weight can be calculated. The small volume allows measurable concentrations to be attained with short sample collection times. This technique is described in detail later.

Choices regarding collection and correlation of other environmental data depend on the requirements of the experiment. In general, a suite of sensors that easily connect with the computer through commercial interfaces can be developed. The desired subset of this suite could then be installed for use in solving the problem at hand. Similarly, the type of result displayed in real time depends on the purposes of the experiment.

Mass data storage represents something of a problem. Floppy discs are probably inadvisable under the severe conditions at sea (vibration, large motion, dirty air, etc.). Some forms of data storage that do appear desirable are Winchester discs (if head crashes can be avoided), digital tape recorders, electronic memory systems (protected from power interruptions), or bubble memories (a bubble memory for the Apple II has become available commercially).

Sample Collection Strategies.   In this section we will consider some of the problems posed by this system and our proposed solutions. As previously discussed, filtration appears to be the best collection technique.

ADAPTING TO LARGE CONCENTRATION RANGE.   One potential difficulty is the large dynamic range of sea-salt aerosol concentration. Figure 1 shows that aerosol mass loading may span three orders of magnitude, from 2 to 1200 $\mu g/m^3$. A sample time and flow rate suitable for low concentrations may overload and clog the filters when high concentrations of aerosol are encountered. Conditions appropriate for sampling high-concentration aerosols will not provide a sufficiently large sample during low concentration episodes. The solution to this problem has been to vary the volume of air that is sampled.

Because total air volume is the product of air-flow velocity (meters per second), orifice cross-sectional area (square meters), and time (seconds) any of these parameters could be changed in theory. In practice, a continuously variable orifice is difficult to build; therefore, this value is fixed. The requirement for isokinetic flow will dictate the air-flow velocity. Therefore, only sample time can be independently controlled.

ISOKINETIC FLOW.   To understand why the requirement for isokinetic flow dictates inlet air-flow velocity, consider a parcel of air contained within a cylinder whose cross section is defined by the inlet orifice area. Under isokinetic conditions the mass flow within this cylinder is equal to the flow through the orifice so that all entrained particles are carried into the sampler (Figure 2A). For underpumping conditions, the flow through the orifice is less than the incident flow, and some of the air bypasses the orifice. The more massive entrained particles will continue on their original trajectories, enter the sampler, and result in the capture of more mass than was originally contained in the sampled volume of air (Figure 2B). In Figure 2C, air that would not normally be sampled is diverted into the orifice. This time the high-mass particles miss the orifice, and low-mass collection results.

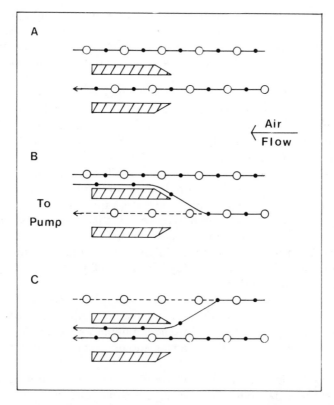

Figure 2. The effects of sampling under isokinetic (A), low-flow (B), and high-flow (C) conditions. The cross-hatched regions represent the sampling orifice; large and small symbols represent large and small aerosol particles, respectively; solid lines are air streamlines; and dashed lines are heavy particle trajectories.

Isokinetic flow can be achieved if the sampling system monitors the ambient wind velocity and adjusts the pumping speed as appropriate. Unfortunately, this system exacerbates the dynamic-range problem because the system samples faster in high winds, when the aerosol concentration tends to be higher.

SAMPLING TIME.   The only parameter that can be independently controlled is the sampling time. If the amount of aerosol collected can be estimated in real time, then the sampling time may easily be adjusted as required. One approach is to monitor the pressure drop across the filter to indicate the approximate degree of filter loading. Sampling could be continued as long as necessary, yet terminated before filter clogging occurs. Of course, in such a variable-time system the actual collection time for each sample must be recorded. This value is needed to calculate the ambient atmospheric aerosol concentration. Another, possibly simpler approach is to use the amount of aerosol found on the previously measured filter to compute the proper collection time for the present sample. This method assumes fairly small changes in aerosol concentration over short periods of time.

SAMPLE MANIPULATION. If the sample has been collected on a filter membrane of some type, then the next steps are to put the sample into a form appropriate for the detection method and to bring the sample and detector into contact. In a fully automated system these steps may present a significant design challenge.

For many methods, including the electrochemical method currently under consideration, the only required sample preparation is to dissolve the salts in an accurately known volume. One approach builds the sensor(s) into the filter chamber. Computer-controlled, fluid-delivery systems can be used to add the required amount of water to the cell. An in situ stirring device might be required, and certainly some provision for flushing the chamber must be made. Such a system is conceptually relatively simple but does have a number of disadvantages.

One problem is loss of sampling time. With this approach, the entire solubilization–measurement–rinse cycle must be completed before a new sample can be acquired. This requirement will place an upper limit on the rate at which data may be taken, and it could prevent the study of transient phenomena.

A modification of this method would use multiple chambers so that the system could be collecting one sample while analyzing the previous one. Possibly, a third chamber could be flushed and prepared for the next collection cycle at the same time.

Another problem is loss of sample after the measurement has been made. Often, it may be desirable to save the sample for replicate analyses, determination of other parameters, or archival storage. Certainly, extra samples could be arranged with the addition of a carousel of sample vials, but only at the cost of considerable additional complexity and size.

FILTER-PAPER TAPE. A technique that inherently conserves the sample uses a filter-paper roll. The sample would be deposited at a collecting station and then advanced to a measuring station. Ultimately, it would be wound onto a takeup reel and saved for further investigation.

An automated atmospheric radon analyzer that operates in exactly this fashion was developed at Naval Research Laboratory (20), and the use of a filter tape to collect aerosols for later analysis was described previously (21). These systems collect aerosols and have been operated successfully on many cruises during the past several years. An analytical method that appears to have great promise for such a system is discussed in detail in the next section.

A difficulty common to all methods is the danger of cross contamination. Any system that uses an analytical chamber is likely to be more difficult to clean than a system using only surface contact electrodes. At the very least, the use of an analytical chamber would require a larger volume of rinse water between measurements. For a system that is intended to operate unattended for long periods, this could be a significant limitation.

A Small-Volume, Ion-Selective Electrode Method for Chloride Determination

The technique described here, which permits determination of chloride on a filter paper, involves pressing the filter between two electrodes: a chloride electrode and a nonflowing reference electrode. The chloride-sensitive electrode (Corning Model 476126) is reported (manufacturer's data) to have a linear calibration down to 5×10^{-5} mol/L. Because only about 50 μL of solvent is required to moisten the filter, the sample will probably not be removed from the filter during the measurement.

Figure 3 shows that a usable, though nonlinear, calibration may be obtained down to 1×10^{-6} M chloride. Table II shows sampling requirements calculated on the basis of reasonable estimates of pumping speed and aerosol concentration. These requirements assume a 10-mL sample size. Useful sea-salt concentration data with 8-min sampling periods can be obtained even at wind speeds well below 1 m/s. Apparently, then, the requirement for collecting data at a rate sufficient for investigation of short-term variations can be met.

Ion-Selective Electrode as a Nonflowing Reference. For long-term, unattended operation at sea, a conventional reference electrode with flowing junction is undesirable. Junction clogging can cause drift and unstable readings, and internal filling solutions must be replenished. Therefore, the fluoride ion selective electrode was selected as a nonflowing reference because of its long-term stability and reliability. Its voltage is controlled by using 0.1 M NaF as the ionic-strength adjustor.

Fifty-Microliter Samples. For small sample volumes, Durst (23) showed that with a gelled fluoride electrode in an inverted position and a conventional calomel reference electrode, the fluoride electrode could be made the sample "container," and measurements can be made in samples as small as 50 μL. This concept was applied here with the chloride electrode in the inverted position because all connections in the electrode are solid state; no liquids are inside the body. The fluoride electrode (Orion Model 94-09), used here as the reference, was directly opposed to the chloride electrode with an air gap of about 1 mm (Figure 4). Both flat membranes were lightly coated with silicone oil to make them hydrophobic, and the test solution was pipetted into the gap.

Between samples the gap was blotted dry because washing with distilled water caused abrupt changes in ionic strength and fluoride concentration at the reference electrode, and the changes slowed attainment of

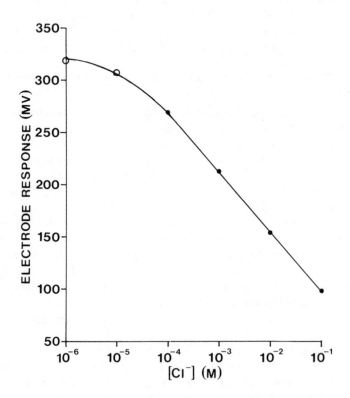

Figure 3. Typical response curve for chloride ion-selective electrode.

stable readings. These abrupt changes could be avoided by washing with 0.1 M NaF solutions. Such washing, although feasible for manual operation, would add undesirable complexity to an automated system, so it is avoided by a double-measuring technique designed to minimize errors caused by cross contamination from the preceding sample, which could be several orders of magnitude different in chloride concentration.

Standard solutions containing 50–5000 μmol of NaCl per liter of 0.1 M NaF were used to obtain a calibration curve. A randomly selected standard was then introduced to simulate a field unknown. With actual field unknowns, the standard nearest in potential to the unknown would then be measured to correct for drift in the calibration curve, and the unknown concentration would be computed. In laboratory tests the same standard had to be remeasured to correct for drift in the calibration curve. Potentials were measured after 7-min contact times. The data (Table III) show that,

Table II. Sea-Salt Monitor Performance Requirements

Wind Velocity (m/s)	Salt Concentration[a] (μg/m³)	Cl⁻ Concentration[b] (μg/m³)	Sample Volume[c] (m³)	Sample Time[d] (min)
0.0	4.3	2.3	0.15	7.7
1.0	5.0	2.8	0.13	6.5
4.4	8.6	4.7	0.08	3.8
6.7	12	6.8	0.05	2.6
9.3	19	10	0.03	1.7
15.5	51	28	0.01	0.6
35.0	1153	634	0.0006	0.03

[a] Ambient air salt concentrations based on Lovett's salt loading–wind speed relation: $\ln S = 0.16U + 1.45$, where S = salt load ($\mu g/m^3$), and U = wind speed (m/s).

[b] Chlorine, in the form Cl⁻ ion, constitutes 55% by weight of the total dissolved material in the ocean (22).

[c] Assumes a detection limit of 0.36 μg of Cl⁻ (10 mL of solution with a detection limit of 10^{-6} mol/L chloride).

[d] Based on a pumping speed of 20 L/min (0.02 m³/min). This result does not take into account corrections for isokinetic sampling.

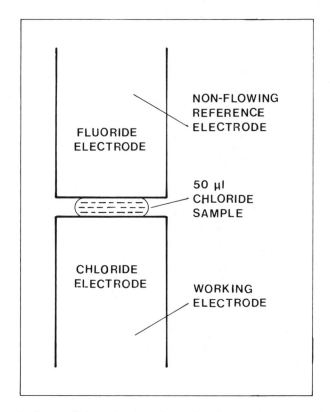

Figure 4. Opposed ion-selective electrodes for measurement of small amounts of chloride.

with 50-μL volumes and the nonflowing reference, uncertainties are 2–6% (relative standard deviations) and accuracy is well within the uncertainty range.

If sampling times must be shortened and/or if very low ambient concentrations must be measured, this method can reduce the amount of chloride needed from 0.36 μg (Table II) to 0.09 μg while restricting measurements to the linear calibration range (5 × 10^{-5} M). The cost is introduction of considerable manual effort, because samples would be collected on filters and analyzed by hand in a separate operation.

Dilute Samples and Manual Method. As already discussed (Figure 3), a usable but nonlinear response that is obtained can allow measurements down to 1 × 10^{-6} M chloride. If 50-μL volumes are coupled with measurements down to 10^{-6} M, then the amount of chloride needed can be reduced further, from 0.09 to 0.002 μg. Imprecision will increase below 5 × 10^{-5} M.

For manual measurement of chloride in aerosols the following proce-

Table III. Response of Chloride Electrode with Ion-Selective Reference Electrode (Fluoride)

Chloride Concentration (M)	Chloride Mass[a] (μg)	Average Concentration Found[b] (M)	Relative Standard Deviation (%)
5×10^{-3}	8.9	4.99×10^{-3}	1.8
1×10^{-3}	1.8	9.84×10^{-4}	1.5
5×10^{-4}	0.89	4.99×10^{-4}	3.8
1×10^{-4}	0.18	9.5×10^{-5}	5.5
5×10^{-5}	0.089	5.1×10^{-5}	5.6

[a] Sample volume = 50 μL.
[b] Number of samples = 5.

dure appears feasible: (1) Collect the aerosol on a standard filter membrane having a diameter chosen to match that of the electrodes; (2) dry the sample; (3) place it between the electrodes; (4) remoisten the sample with 50 μL of 0.1 M NaF; and (5) press the electrodes together and measure the potential.

A method for measuring samples as small as 10 μL by pressing a combination fluoride electrode on confined spot test filter paper (S&S Yagoda 211-Y) has been described (24). An adaptation of this technique may permit measurements with as little as 10 μL of remoistening solution in our application.

Filter-Tape Method.   A filter-tape system may be envisaged for automating even the 50-μL method. An automatic tape system for collection of aerosols on filter paper was discussed earlier. The concept could be used here to transport samples to a drying chamber and then to an electrochemical cell where the sample is remoistened with a known volume of ionic-strength adjuster and measured by pressing the opposed electrodes together as shown in Figure 5. The paper must be pretreated by addition of hydrophobic barriers to confine the sample to the desired area of the paper. All of these steps could be automated under control of the microprocessor.

Data Management

Large Data Sets.   We have discussed data acquisition problems that must be addressed, but we have not yet considered the question of managing the resultant data. The use of automated aerosol monitors and ancillary instruments will result in massive quantities of data. A typical 4-week experiment in which 10 parameters are monitored at 10-min intervals would require about 0.25 Mbytes of storage. The problems of data storage, organization, and retrieval are likely to be comparable to or greater than the difficulties of data analysis.

To use such large data sets, computerized data management systems will be needed. These systems must communicate with data acquisition systems and transcribe the received data onto permanent bulk storage media.

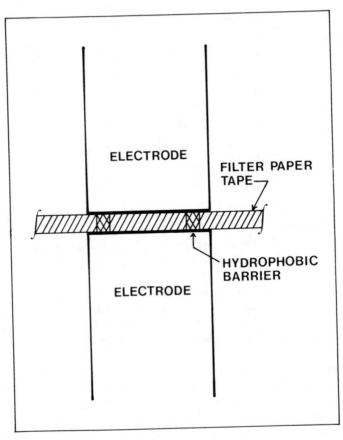

*Figure 5. Proposed system for measurement of aerosol collected on continu-
ous filter-paper tape.*

Data storage must be accomplished so that applications software can recover and use any specific datum at a later time; therefore, the format and
coding scheme for data storage must be clearly defined.

 Data Format and Coding. First, the format for the files must be defined. The format specifies the file size and the arrangement of data within
the file. Different files will, in general, contain different amounts and types
of information, so the format must be flexible yet unambiguous. A standard procedure for this kind of situation is to preface the data with a block
of system information, a file header. The header must conform to a rigidly
defined format if it is to be intelligible. The header will contain values for
variables that specify the type, amount, and arrangement of the actual
data. Often, the data are presented as an array, in which case the header
would give the dimensions of the array. Such a system provides high versa-

tility at the expense of somewhat increased memory. This approach has been used for the development of data base standards (25, 26). In these cases (25, 26), the data-formatting programs are parts of software packages that include applications programs. Applications programs provide data search, plotting, and other functions.

The second requirement is a coding scheme that allows digital data to represent alphanumeric characters. Several such standards already exist, notably the American Standard Code for Information Interchange (ASCII) and the Extended Binary Coded Decimal Interchange Code. ASCII is almost universally used in microcomputer systems and is probably the better choice for development of new systems.

Data Standards. As a practical matter, no single research group is likely to have the resources required to conduct all of the experiments needed for a thorough understanding of aerosol behavior. Because cooperative efforts will be required, mechanisms should be established by which different laboratories could readily exchange data. Therefore, data file standards should be established. These standards would ensure that data formats and coding schemes would be mutually intelligible. In some large projects, such as the Sea–Air Exchange Program (SEAREX), data repositories have already been established for use by all project members. These efforts could serve as models for the development of other data exchange agreements.

The problem then becomes one of making these disparate data sets easily accessible. Accessibility would be facilitated if one or more transmission standards were chosen to specify the protocols for exchanging data. These might include standards for electronic data transmission (baud rates, word lengths, parity, etc.) for exchangeable storage media (disk or tape sizes, bit densities, etc.) or for some combination. In any case, the medium through which data exchange would occur must be well defined.

The data exchange capability just outlined would be entirely voluntary. Something similar is likely to evolve around a set of de facto standards established by those groups that first begin using computerized data management systems. Sharing this software early would be to their advantage so as to minimize duplication of expensive development efforts.

Acknowledgments

We thank F. K. Lepple and D. J. Bressan for sharing unpublished data with us and for numerous helpful discussions.

Literature Cited

1. Duce, R. A. *Pure Appl. Geophys.* **1978**, *116*, 244–73.
2. Blanchard, D. C.; Woodcock, A. H. *Ann. N. Y. Acad. Sci.* **1980**, *338*, 330–47.
3. Woodcock, A. H.; Blanchard, D. C. *Tellus*, **1955**, 7, 437–48.
4. Woodcock, A. H. *J. Atmos. Sci.* **1978**, 35, 657–64.

5. Woodcock, A. H.; Blanchard, D. C.; Jiusto, J. E. *J. Atmos. Sci.* **1981**, *38*, 129–40.
6. Barger, W. R.; Garrett, W. D. *J. Geophys. Res.* **1970**, *75*, 4561–66.
7. Blanchard, D. C. In "Applied Chemistry at Protein Interfaces"; Baier, R. E., Ed.; ADVANCES IN CHEMISTRY SERIES No. 145, American Chemical Society: Washington, D.C., 1975; pp. 360–87.
8. Piotrowicz, S. R.; Duce, R. A.; Fasching, J. L.; Weisel, C. P. *Mar. Chem.* **1979**, *7*, 307–24.
9. Blanchard, D. C.; Syzdek, L. D.; Weber, M. E. *Limnol. Oceanogr.* **1981**, *26*, 961–4.
10. Baylor, E. R.; Baylor, M. B.; Blanchard, D. C.; Syzdek, L. D.; Appel, C. *Science* **1977**, *198*, 575–80.
11. Ruskin, R. E.; Jeck, R. K.; Lepple, F. K.; Von Wald, W. A. "Salt Aerosol Survey at Gas Turbine Inlet Aboard USS SPRUANCE"; NRL Memorandum Report 3804, Naval Research Laboratory: Washington, D. C., 1978; pp. 1–113.
12. Byers, H. R.; Sievers, J. R.; Tufts, B. J. In "Artificial Stimulation of Rain; Proceedings of the First Conference on the Physics of Cloud and Precipitation Particles"; Cloud Physics Conference; Weickman, H.; Smith, W., Eds.; Pergamon: London, 1957; pp. 47–72.
13. Toba, Y. *Tellus*, **1965**, *17*, 131–45.
14. Lepple, F. K.; Bressan, D. J.; Hoover, J. B.; Larson, R. E. "Sea Salt Aerosols, Atmospheric Radon and Meterological Observations in the Western South Atlantic Ocean (February 1981)"; NRL Memorandum Report 5153, Naval Research Laboratory: Washington, D. C., 1983; pp. 1–63.
15. Lovett, R. F. *Tellus* **1978**, *30*, 358–64.
16. Loo, B. W., Adachi, R. S.; Cork, C. P.; Goulding, F. S.; Jaklevic, J. M.; Landis, D. A.; Searles, W. L. "A Second Generation Dichotomous Sampler for Large-Scale Monitoring of Airborne Particulate Matter"; LBL Report 8725, Lawrence Berkeley Laboratory, University of California: Berkeley, Ca., 1979; pp. 1–18.
17. Loo, B. W., Personal communication.
18. Dessy, R. *J. Chem. Ed.* **1982**, *59*, 320–27.
19. Osteryoung, J. *Science* **1982**, *218*, 261–65.
20. Larson, R. E.; Bressan, D. J. *Rev. Sci. Inst.* **1978**, *47*, 965–69.
21. Bressan, D. J. "Maritime Atmosphere Salt Load Monitoring System"; NRL Memorandum Report 4581, Naval Research Laboratory: Washington, D. C., 1981; pp. 1–41.
22. Sverdrup, H. U.; Johnson, M. W.; Fleming, R. H. "The Oceans, Their Physics, Chemistry, and General Biology"; Prentice-Hall: Englewood Cliffs, N. J., 1942; p. 175.
23. Durst, R. A.; Taylor, J. K. *Anal. Chem.* **1967**, *39*, 1483–85.
24. "Instruction Manual, Chloride Electrode Model 94-17B, Combination Chloride Electrode Model 96-17B," Orion Research Form 94-171M/1820, 1981; p. 26.
25. Trusty, G. L.; Haught, K. M. "An Aerosol Data Base Format"; NRL Memorandum Report 4605, Naval Research Laboratory: Washington, D. C., 1981; pp. 1–15.
26. Hoover, J. B. "The METEOR Software Package for Analysis of Meteorological Data"; NRL Memorandum Report 4674, Naval Research Laboratory: Washington, D. C., 1982; pp. 1–83.

RECEIVED for review July 5, 1983. ACCEPTED February 22, 1984.

Dependence of Sea-Salt Aerosol Concentration on Various Environmental Parameters

D. J. BRESSAN and F. K. LEPPLE

Naval Research Laboratory, Washington, DC 20375

An automated aerosol sampler coupled to a programmable environmental monitoring system was used to collect nearly continuous data on the salt aerosol loading 16 m high in the atmospheric boundary layer in the western South Atlantic Ocean during February 1981. The relationship between sea-salt aerosol concentration and local wind speed was emphasized. The overall correlation of their time series was poor (r = 0.3). However, when the data are separated by meteorological and oceanographic conditions, the short-term correlations are improved significantly. Extensive interpretation of our data indicates that condensation processes, hysteresis effects, sudden air-mass changes, atmospheric boundary layer structure, and sea water temperature must be considered when modeling marine atmospheric salt load from wind speed.

RELIABLE SEA-SALT AEROSOL MODELS are required for predicting atmospheric optical propagation, for understanding precipitation and fog formation, and for protecting materials on ships and aircraft operating in the marine environment. An initial step in developing global models of sea-salt aerosol populations is identification and evaluation of some relationship between measurable properties of sea-salt aerosols and various meterological parameters. Wind stress on the ocean surface is a generator of bubbles that later burst to produce both jet and film droplets. Several investigators (1–5) have studied the relationship between local wind speed and sea-salt aerosol concentration. However, nearly all data were collected at wind speeds less than 12 m/s, and the wind and sea-salt aerosol concentration data were often taken intermittently and averaged over relatively long time periods (4–24 h). Lovett's extensive study (5) in the North Atlantic (more than 1800

0065-2393/85/0209-0075/$07.00/0

samples over 11 months) showed that the mathematical relationship between a given wind speed and the measured atmospheric sea-salt aerosol concentration can vary by as much as a factor of 10.

To date, most investigations have attempted to relate the generation of sea-salt aerosol to local wind speed alone. However, sea-salt aerosol concentrations should also be affected by previous wind conditions and airmass histories, thus determinations of both advective and locally produced aerosol concentrations are required. Relevant modeling parameters might include the thermal structure of the sea and air, relative humidity, atmospheric turbulence and mixing height, wind duration and fetch, and the directional spectrum and steepness of the waves. The best gains to be made in salt aerosol research are through the use of an automatic, unattended, short-duration aerosol sampler. This device would yield a nearly continuous and representative time series of relatively closely spaced samples. Removing operator dependency also ensures sampling at the higher wind speeds, which are usually accompanied by rough seas that make shipboard work difficult. Our system also provides for the automatic recording of the presumed relevant parameters for which we could provide sensors.

Since 1979 we have developed two similar, but not identical, systems to accomplish the objectives just discussed. Tests on the North Atlantic in winter have proven the reliability of the sampler and data recording units, which have continuously operated for up to 5 weeks without need for a technical operator. Only the ship's speed and heading entries require manual updating.

During a February 1981 oceanographic cruise in the western South Atlantic Ocean, we used our system aboard the USNS *Hayes* to sample marine aerosols automatically for 16 days at 1.5-h intervals along the cruise track shown in Figure 1. The sea-salt aerosol concentrations were determined for correlation with wind speed and other relevant meteorological parameters. This chapter identifies the most important variables and the extent to which they affect modeling of salt loads in the marine atmosphere.

Experimental

Our sampling device, the automatic radon counter and aerosol sampler (ARCAS), was used to collect the continuous set of aerosol samples. This instrument also counted the radon daughter-product decay and printed a nearly real-time record for each sample (*see* Reference 6 for photographs and details of operation). The ARCAS I system consists of a deck-mounted sampling unit and an indoor electronic unit containing sampler controls, calendar clock with display, thumbwheels for manual data entry, and a thermal printer.

The indoor unit controls both sampling and data recording. It accepts analog or digital data from other sensors, records the data on magnetic tape, and provides a printed record on paper tape. Figure 2 illustrates the basic aerosol sampler deck unit. For shipboard sampling, the short inlet tube shown in Figure 2 was replaced with a 2-m long tube. This unit was mounted forward on the flying bridge at the midline of the USNS *Hayes*. The inlet height was approximately 16 m above sea level. Wind-

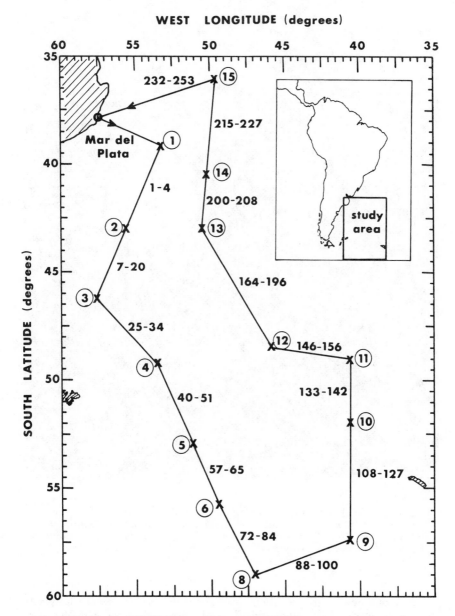

Figure 1. Track of USNS Hayes leaving Mar del Plata, Argentina on February 1, (Julian Day 32) and returning February 19, 1981. Circled numbers refer to oceanographic stations; numbers along cruise track designate aerosol samples collected underway.

Figure 2. ARCAS deck unit showing filter transport mechanism (A), aspirating motor (B), motor controllers (C), radiation-counting photomultiplier tubes (D), pulse amplifier transmitter units (E), high- and low-voltage power supplies (F), calibration counter (G), filter supply roll (H), and takeup roll (I).

tunnel tests showed nearly horizontal air flow at this location, which has unobstructed trajectories ±120° from straight ahead. For this cruise, the ARCAS was set to sample air for the first 20 min out of every 90 min, and then increment the filter paper and count the decay of radioactive radon daughter products for 9–10-min intervals. The 5-cm diameter, discrete samples are taken and stored on glass-fiber filter paper (Hollingsworth and Vose HE 1022). After each sample is moved from between the pair of radiation-counting tubes (7), it is covered with a layer of Mylar film, which prevents sample cross-contamination as the aerosol samples are rolled onto the filter-paper takeup spool. These rolls are returned to the laboratory where the sea-salt content of each sample is determined. The system's air-flow rate was 1.33 m^3/min, yielding a sampled volume of approximately 27 m^3 of air for each aerosol collection.

Radon (^{222}Rn) measurements made by the ARCAS provide a simple, reliable, real-time indicator of the relative maritime or continental nature of the air over coastal or oceanic areas (8). With a half-life of 3.8 days, ^{222}Rn originates from the decay of ^{226}Ra, a member of the ^{238}U decay chain. At least 98% of ^{222}Rn originates from land masses (9). The radon flux at the surface depends on the radium content of the soils and rocks, the permeability of the source materials, atmospheric pressure, soil moisture, and vegetative cover (10). Relatively abrupt changes in the radon concentration over the ocean usually indicate changes in air masses and the passage of frontal systems.

For aerosol sampling, the inlet shown in Figure 2 was designed to keep out rain and large fog or spray drops, yet allow smaller droplets and sea-salt particles to enter from any direction. Estimations of the most likely air-flow patterns with computations (11) of viscous drag forces indicate that for wind speeds up to 31 m/s (60 knots), salt particles up to at least 40-μm diameter should follow the air-flow pattern down the inlet tube; particles of 100-μm diameter impact on the opposite inner wall of the vertical inlet tube above this wind speed. Biasing against extremely large droplet sizes will minimize collection of any spray generated by ship motion.

The upper particle-size cutoff of the ARCAS inlet was tested on other cruises, and the overall sampling efficiency was compared to membrane filters (Nuclepore) operated under essentially isokinetic conditions. To determine the size of salt particles admitted into the ARCAS inlet and carried to the filter stage, we operated a single-stage impactor (lower size cutoff of 2 μm), which had an identical inlet configuration and the same flow rate as the ARCAS. Microscopic examination of the impaction surfaces revealed a few salt particles with diameters in excess of 40 μm collected at wind speeds above 10 m/s. These results support the theoretical calculations. Comparisons of the ambient salt aerosol concentrations determined by the ARCAS and the filter system showed an average agreement of ± 15% for nine North Atlantic aerosol samples collected during wind speeds ranging from 2 to 20 m/s (Lepple and Bressan, unpublished data).

In the laboratory, aerosol samples were individually cut from the filter-tape roll and extracted on a 47-mm filter holder (Millipore) with 200 mL of distilled deionized water maintained at 50–60 °C. The leachates were analyzed for sodium content by flame atomic absorption with a Perkin-Elmer Model 373 spectrophotometer. The soluble sodium was assumed to be derived only from sea salt. Estimates of total sea salt in each sample were obtained by multiplying the sodium value by 3.25, which is the salinity-to-sodium ratio in bulk seawater (12). The sodium content of the 5-cm diameter, glass-fiber filter blanks averaged 18 μg ± 12% (20 samples). Based on an average sample volume of 27 m^3, the background "salt" level produced by the sodium blank in each filter would be equivalent to 2.2 μg/m^3.

A specially designed high-volume air sampler (HV) (13) was also operated on the *Hayes* to collect marine aerosols over a longer time interval (7–18 h each). This HV sampler is constructed of poly(vinyl chloride) and has one 2.5 × 25-cm inlet slit that can be manually adjusted to obtain approximately isokinetic sampling conditions. The blower motor was separated from the filter holder unit by a 1-m length of flexible hose to minimize possible aerosol contamination. The HV was mounted on the forward edge of the flying bridge bulkhead about 15 m above sea level and within 3 m of the ARCAS inlet. The flow rate through the 20 × 25-cm glass-fiber filters was approximately 1.7 m^3/min. For these HV filters, the sea-salt component was measured by extracting the soluble sodium in one-half of the filter with the same technique as for the ARCAS samples.

In conjunction with the older ARCAS I system, an environmental monitoring system (EMS) (designed at the Naval Research Laboratory) was used to obtain meteorological information at 15-min intervals and output the data to printed paper and magnetic tapes. For this cruise, the EMS consisted of a Fluke Model 2240B data logger, a Columbia Data Products Model 300C data cartridge recorder, and several sensors and signal conditioners. Measurements included relative wind speed and direction (Teledyne Geotech Model 201), relative humidity (General Eastern Model 450), barometric pressure (Texas Electronics Model 2012), and ambient temperature (General Eastern and Fluke platinum resistance probes). The temperature and relative humidity sensors were mounted in an aspirated shield (Climet Model 016-2) located on the rear mast of the *Hayes* approximately 21 m above sea level. Estimated accuracies of the meteorological sensors are as follows: wind speed, ± 0.5 m/s or 2% of true air speed, whichever is greater; threshold velocity, 1 m/s; wind direction, ± 3°; and air temperature, ± 0.5 °C for General Eastern platinum resistance thermometer (PRT), ± 0.2 °C for Fluke PRT, and ± 0.5 °C for sea surface temperature.

Ship speed and heading were entered into the EMS with thumbwheel switches. Data are then available for computer calculations of true wind speed and direction. Power supply voltages for the system sensors were also routinely monitored to help screen for potentially bad or missing data.

Radon decay counts and EMS data were processed on a DEC-10 computer by using the METEOR FORTRAN programs (*14*).

Unfortunately, many of the wind measurements being fed into the Fluke data logger from the Geotech wind system were compromised because of electrical interference from the ship's power distribution system. Thus, instead of always having automated wind data recorded four times per hour as planned, we also had to utilize data manually recorded in the ship's log. At 2-h intervals, relative wind was read from the ship's anemometers (at 26-m height) and then converted to true wind values by the ship's officers and logged. From our past experience, the accuracy of their data is approximately ± 1.5 m/s and ± 5°. Fortunately, strip-chart recordings of the meteorological data were also taken via an isolated circuit. These records were used to check and linearly interpolate the values of true wind speed and direction between the ship's log entries. The wind speed corresponding to each value of salt aerosol concentration was obtained by averaging the interpolated values during the 20-min sampling interval. This situation illustrates that continuously operating backup systems are not a redundancy; they are a necessity. More complete documentation on all of the sensors and the time-series graphs for relative humidity, barometric pressure, water vapor concentration, and radon daughter-product concentration for this cruise can be found elsewhere (*15*).

Results and Discussion

Salt Aerosol Observations. The average salt aerosol concentration for the entire cruise was 21.7 ± 30.9 μg/m^3 (arithmetic mean and standard deviation, respectively, for 253 samples from the ARCAS). Salt aerosol concentrations were also measured by Prospero (*16*) from ships in the South Atlantic Ocean at an elevation of 15 m. His overall statistics for 35 samples taken in the tropical and central South Atlantic (5° S to 35° S) are 9.1 ± 5.3 μg/m^3. When he includes five additional samples collected near the Cape of Good Hope, these values become 11.3 ± 9.2 μg/m^3. A higher salt aerosol concentration was expected in our more southerly sampling area (35° S to 60° S) because of the higher average wind speeds there.

Before proceeding to an examination of the salt aerosol and wind-speed time series, the aerosol particle capture characteristics of the HV sampler and ARCAS I should be compared empirically. Table I lists the instances where those comparisons and contrasts can be made between these samplers. Over the continuous HV sampling period, the ARCAS is sampling 22% of the total time period (20 min out of every 90 min). For HV Samples 4 and 5 when the wind would be expected to produce salt particles less than 60 μm in diameter at 98% relative humidity (*17*), the mass ratios are in acceptable agreement. For HV sample intervals 7 and 8, obtained during conditions of ship spray and light drizzle, comparison implies that much of the salt is contained in droplets larger than those accepted by the ARCAS inlet. Although this salt is airborne, it seems questionable to regard it as an aerosol because of the limited lifetime of the large particles involved.

COMPARISONS AND DATA SCATTER. Overall, the sea-salt aerosol concentrations (Figures 3–5) show considerable variation with time. Furthermore, salt load response to increasing or decreasing local wind speed (also shown in Figures 3–5) is not consistent, nor is the wave-height response shown in Figure 6. The fine structure observed in this data is not evident in the extensive data set plotted by Lovett (*18*) because he sampled approximately four times per day, and we sampled 16 times per day. Both data sets show instances when salt aerosol levels either do not change with increasing wind speed or even show rapid increases with decreasing wind speed. These instances exemplify some of the vagaries that can lead to low point-by-point correlations between wind speed and salt aerosol concentration. These vagaries are apparently due to salt-load hysteresis effects occurring during the aerosol generation and dissipation phases, as well as to salt aerosol advection. In this chapter, we refer to hysteresis as the lag in response of salt aerosol concentration to changes in wind speed. This lag often results in lower than expected (i.e., steady state) salt `aerosol concentrations during

Table I. Comparison Between HV and ARCAS I Under Differing Environmental Conditions

Sampler		Total Airborne Salt Concentration (μg/SCM)		Ratio	
HV	ARCAS	HV	ARCAS	HV/ARCAS	Wind, Weather
4	25–37	45	62	0.7	7–10 m/s, clear
5	73–84	32	23	1.4	5–8 m/s, cloudy
7	132–142	297	16	19	10–13 m/s, bow spray
8	148–156	115	5	23	0–7 m/s, drizzle

NOTE: SCM≡standard cubic meter.

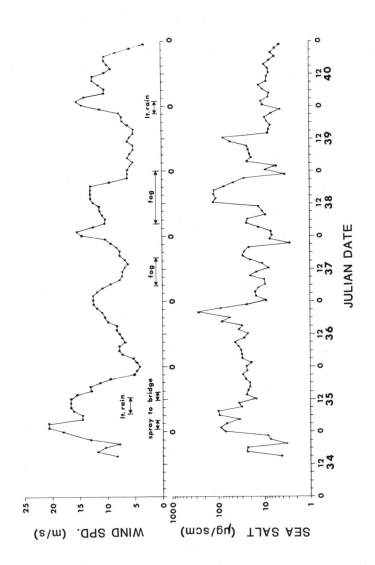

Figure 3.   Time-series plot of sea-salt aerosol concentration and local wind speed from Julian Day 34 to Julian Day 40, 1981.

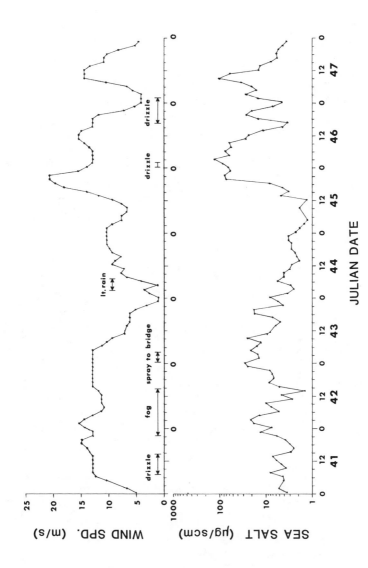

Figure 4. Time-series plot of sea-salt aerosol concentration and local wind speed from Julian Day 41 to Julian Day 47, 1981.

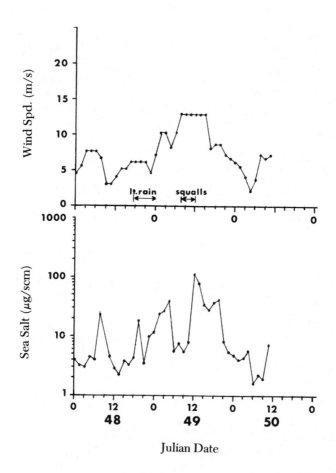

Figure 5. Time-series plot of sea-salt aerosol concentration and local wind speed from Julian Day 48 to Julian Day 50, 1981.

increasing wind speed conditions and higher salt aerosol concentrations during decreasing wind speed conditions.

RESIDENCE TIME.   If the salt peak on Julian Day 45 at 0900 is due to continually sampling the same air mass, the residence time is ~ 19 h. [Residence time is defined as the time for the salt concentration to drop to "low values" following a wind speed drop (5).] Lovett's (5) observations of the time variation of sea-salt concentration and wind speed indicated residence

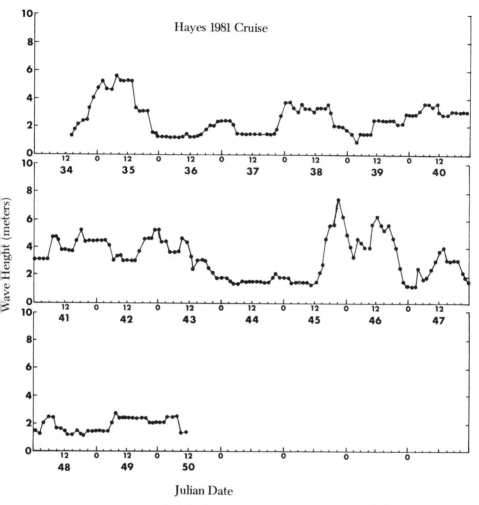

Figure 6. Surface wave height time series from Julian Day 32 to Julian Day 50, 1981.

times between 12 and 24 h. However, for much of our data, the residence time of the salt aerosol appears to be about 6 h, as indicated by the decrease in concentration with decreased wind speed. An apparent decrease in the residence time might be caused by passing into an air mass with a lower salt aerosol concentration. For whatever reason then, variations in the salt-load decay rate do occur (i.e., residence time is a variable).

RADON AS A PREDICTOR.   Because radon is almost exclusively land derived and salt aerosol is essentially all sea derived, relatively high radon concentrations should be associated with relatively low salt aerosol values in a given area. Therefore, a negative correlation should exist between the short-term trends in radon concentration and salt aerosol concentration. Examination of the trends in both time series shows negative correlations only 60% of the time (39 of 64 comparisons). This percentage hardly provides a reliable predictive capability for atmospheric salt loadings in this region. Variations in the air-mass trajectories over South America could certainly complicate radon-based distinctions between marine and continental air. Continuously monitoring the salt aerosol particle-size distributions would have yielded complimentary information, possibly by indicating the age of the aerosol, to differentiate between local and advected salt components. However, this scheme is not "predictive," but it does help interpret the current local situation.

RAIN, FOG, AND SHIP SPRAY.   During this cruise we encountered periods of fog, precipitation, and ship-generated spray. These occurrences are shown along the time axis of the wind-speed time series (Figures 3–5). Scavenging or washout of salt aerosol by precipitation is evident (0900, Day 35; 2200, Day 39; 0900, Day 41; and 0900, Day 49), particularly during the rain squalls on Day 49. ARCAS data show an apparent scavenging or washout effect, more pronounced for drizzle than for fog conditions, due to the particle-size capture characteristics of the ARCAS inlet discussed previously. Reductions in salt aerosol concentration from expectedly higher values were also reported in the presence of precipitation (5).

Salt spray from the bow reached the wheelhouse below the ARCAS on three occasions. Except for the first case on Day 35, salt aerosol peaks did not coincide with the reports of spray. Again, the ARCAS discriminates against aerosol particles or droplets larger than fog (i.e., ship-generated spray).

Data Interpretation.   The wind speed was correlated to the natural log of the salt load with the equation, $\ln \theta = au + b$, where θ is the sea-salt aerosol concentration in micrograms per SCM, a is the slope of the linear-regression line, b is the intercept of the linear-regression line, and u is the wind speed in meters per second (1, 5).

Table II lists slope and intercept values for the linear-regression equations and Gaussian statistics, both for the full data set and for subsets categorized by various sampling, meteorological, or oceanographic conditions. The overall statistics for the cruise (Case I in Table II) indicate that the geometric mean salt aerosol concentration was 11.5 ± 3.0 μg/SCM, and the arithmetic mean wind speed was 9.8 ± 3.9 m/s. Generally, data are included in the table to show that condensation processes, hysteresis effects, and advection impose difficulties when trying to match the time series of local wind speed and salt aerosol concentrations.

Condensation processes may play a decisive roll by incorporating salt aerosol into larger droplets. Under these conditions the upper size limit of particles being sampled will determine the observed salt load. Thus, we

Table II. Sea-Salt Aerosol-Wind Speed Relationships

Case	Slope (a)	Intercept (b)	Correlation Coefficient (r)	Mean Values		Standard Deviation		Data Points	Conditions
				ln θ (μg/SCM)	u (m/s)	$\sigma_{\ln\theta}$	σ_u		
I	0.093	1.53	0.33	2.44	9.8	1.11	3.9	249	All pairs θ > 1 μg/SCM
II	0.098	1.53	0.36	2.53	10.1	1.11	4.0	189	Sampling underway
III	0.049	1.71	0.16	2.15	8.9	1.05	3.4	60	Sampling on station
IV	0.096	1.57	0.35	2.54	10.1	1.11	4.0	187	Wind in forward sector
V	0.097	1.82	0.25	2.82	10.3	1.08	2.8	62	Wind in aft sector
VI	0.038	2.50	0.10	2.91	10.7	0.90	2.4	40	Wind in aft sector, underway only
VII	0.100	1.44	0.38	2.43	10.0	1.14	4.3	150	Wind in forward sector, underway only
VIII	0.173	0.75	0.56	2.13	7.9	1.09	3.5	52	Interval 1 ($T \geq 17\,°C$)
IX	0.148	0.87	0.44	2.53	11.3	1.44	4.3	76	Interval 2 ($11\,°C \leq T < 17\,°C$)
X	0.105	1.73	0.39	2.60	8.3	0.95	3.5	44	Interval 3 ($5\,°C \leq T < 11\,°C$)
XI	−0.001	2.32	0.01	2.31	10.3	0.94	3.3	82	Interval 4 ($T < 5\,°C$)
XII	0.127	1.08	0.28	2.64	12.4	1.18	2.6	153	u > 8 m/s only
XIII	0.096	1.44	0.31	2.38	9.8	1.16	3.7	231	Rain removed
XIV	0.108	1.33	0.37	2.33	9.4	1.15	4.0	206	Fog removed
XV	0.110	1.30	0.36	2.35	9.5	1.17	3.9	189	Rain and fog removed
XVI	0.135	0.72	0.53	2.26	11.4	1.24	4.9	44	Rising winds
XVII	0.135	1.11	0.60	2.51	10.4	0.97	4.3	43	Falling winds
XVIII	0.097	1.44	0.35	2.40	9.8	1.01	3.7	63	6-h averaging
XIX	0.112	1.29	0.37	2.41	9.9	0.86	2.9	15	24-h averaging

tried statistical runs with rain and/or fog removed from the data set. These runs showed no significant improvement in statistics (Table II, Cases XIII–XV). Therefore, the statistical manifestation of condensation–precipitation effects may often be overshadowed by other variables.

Figure 7 is a scatter diagram of the ARCAS salt data versus wind speed. It shows the data points used in the analysis and the resulting linear-regression equation. Four point sets whose salt values are below 1 $\mu g/m^3$ have been omitted because they are within 3σ of the blank for salt determination. The point scatter in Figure 7 is typical for such plots, as shown previously (18, 19). In fact, optical measurements of the total volume density of marine aerosol particles (20) show nearly a three-order-of-magnitude range at a given wind speed (5 m/s). Also, any delay in the sea-salt aerosol concentration's response to increases or decreases in wind speed would further broaden the envelope of values beyond possible scatter caused by drizzle and advection. A delay effect is seen in Figure 4 for Day 45 at 1400–2300 h.

DATA RELIABILITY.   In Cases II–VII of Table II, the data have been subdivided to examine possible aerosol sampling bias due to forward ship motion and/or relative wind direction. In Cases II and III, the wind data do have a bias toward lower speeds while on station compared to underway. This bias is to be expected because stations were not taken in rough weather conditions. Overall, the similarity in the regression analysis and Gaussian parameters for Cases I, II, IV, V, and VII reaffirms the omnidirectional sampling capability of the ARCAS, as well as minimal ship effect under the conditions encountered during this cruise.

WIND EFFECTS AND HYSTERESIS.   Advected salt is presumably the hardest to predict from local observations. Yet, this mechanism seems to be the best to explain the parts of our time series that show salt peaks when the wind is low or steady. Apparently, either the ship is approaching a zone of generation, or a previously more active air parcel has slowed and the observations are of residual salt in the process of settling out.

Higher correlations were obtained when trying to separate effects of wind generation from removal by settling (Table II, Cases XVI and XVII). Salt loads at decreasing wind speeds were separated from the data set as were salt loads at increasing wind speeds. Separation was done because the mechanisms controlling salt loading should be different at these times. The salt concentrations can increase quite rapidly but seem to decay at about 10–16%/h measured over 12–24 h. (Because of these apparent log-linear changes, we continued to correlate the natural log of salt with wind speed.) Combined subsets of increasing wind speeds (totaling 44 data points) had a correlation coefficient of 0.53, with a slope of 0.14 and an intercept of 0.71, although 43 points with decreasing wind speeds gave a correlation coefficient of 0.60, a slope of 0.10, and an intercept of 1.46. These results are markedly better correlated than the overall data. The importance and vari-

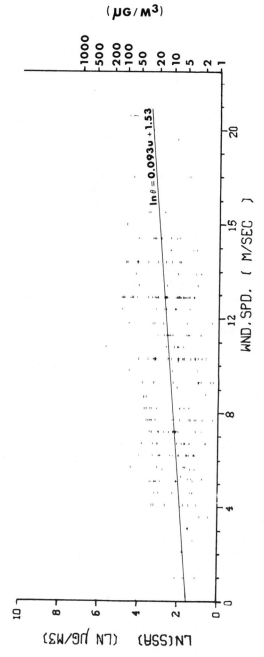

Figure 7. *Natural log of sea-salt aerosol concentration vs. wind-speed data plot with linear-regression equation for entire cruise.*

ability of hysteresis effects is shown by the relatively higher salt loads at lower wind speeds (larger *b* value) during times of decreasing winds.

To test the hypothesis that sampling the salt aerosol for time intervals comparable to previous investigators might improve correlations, we averaged the data over 6 and 24 h. These statistics are listed in Table II, Cases XVIII and XIX. Only a slight improvement of questionable significance is observed in the correlation coefficient for the 6-h average data (combining four adjacent data points) as compared to the full data set (Case I). Similarly, the 24-h average data set shows only a slight improvement over the 6-h average data set. The previous results indicate that separating the meteorological processes improves the predictability of the salt aerosol concentration to a better degree than by long-term time averaging.

SEA SURFACE TEMPERATURE EFFECTS.   The cruise track was divided into four intervals according to sea surface temperature. Each interval spanned approximately 6 °C with decreasing sea surface temperatures (SST) from Stations 1–8. Linear-regression parameters for the four intervals are presented in Cases VIII–XI; they show that the relationship between local wind speed and sea-salt aerosol concentration varies drastically. Linear-regression slopes range from near zero to approximately 0.17; the respective correlation coefficients vary from near zero to 0.6. The lowest correlation coefficients and slope values relating salt aerosol to wind speed are associated with the coldest SST values.

Mechanisms by which the SST might affect the measured sea-salt aerosol concentration are difficult to infer because of large-scale meterological differences among the four sampling intervals. However, in addition to directly affecting such bulk water properties as kinematic viscosity, temperature differences between the seawater and the overlying air affect the mesoscale meteorology through stability, which in turn influences wave height and vertical mass flux. Thus, SST appears to be an important parameter.

Our data set definitely shows high variability for any simple relationship between local wind speed and sea-salt aerosol concentration. This occurrence may only reflect unique meteorological conditions during one summer season because frequency plots of our wind-speed data show values outside the ranges of the expected seasonal averages prevailing in this region of the South Atlantic (*21*). The data set was also filtered (Table II, Case XII) to exclude samples collected at wind speeds less than 8 m/s. This trial exercise results in an increase in the slope when compared to Case I, but the correlation coefficient is unexpectedly lower. However, this manipulation does not represent true winter conditions because the temperature difference between the sea and the air would be different in the winter than it was for our February cruise.

STABILITY EFFECTS.   Although wind speed is the driving force for salt generation, other factors complicate a direct link between anemometer readings and salt aerosol concentrations. These factors include condensation effects, advection, and hysteresis effects. Also, the constancy of salt-load equilibrium level at any one measured wind speed seems to be questionable because of variations in the stability structure of the atmosphere.

Moreover, the temperature difference between air at 21 m and the sea is apparently not always a sufficient predictor of atmospheric instability, as during the time series between Julian Day 44 at 0000 and Day 45 at 1200 (Figure 4). The problem is one of attaining stable conditions when the air is 5° cooler than the seawater. To solve this problem, we propose consideration of the mechanism described by Ling et al. (22). The air must have low relative humidity with sufficient surface stress to eject sea-spray droplets into the air. The key to the mechanism is that the evaporating airborne droplets are extracting latent heat (of evaporation) from the atmosphere instead of the sea. Most likely, this mechanism will selectively cool the low turbulent eddies containing evaporating droplets. These eddies will thereby become relatively stable with respect to the average surrounding and overlying air, and vertical salt mixing will be inhibited. (Our computations show that, under ideal conditions, evaporating 1 g of water could cool 1 m^3 of air by 2.4 °C). Although expected to be a temporary phenomenon, the induced stability would also inhibit the vertical transport of momentum (note the lack of wave-height response for this interval in Figure 6). Operating below 1 or 2 m, this effect occurs far below the usual anemometer heights.

Because stability is affected by the SST, it is noteworthy that the lowest dependence of salt loads on wind speed occurred during Interval 4 (when the SST values were lowest). In the mean, most likely stable atmospheric conditions exist with lower sea surface temperatures.

Comparing our results with those of other investigators whose data has been fit by linear regression, we obtained slopes, intercepts, and other pertinent measurement information shown in Table III. In the high-temperature segments of our data (Intervals 1 and 2) where the SST values are comparable to those of Woodcock (1) and Lovett (5), the agreements in slope and intercept values are much better. As previously noted, these higher correlations may be attributable to a direct effect of higher water temperatures, a greater degree of atmospheric instability, or both.

Woodcock (1) obtained his well-correlated salt versus wind-speed data by estimating wind force (which is done by observing wave and white-cap conditions) from his airplane. Because anemometer measurements were not used, variability in the downward mixing of wind momentum (from anemometer height) did not affect his estimates because the overall result of the wind field plus the downward mixing processes was being evaluated. Apparently, variations in atmospheric stability affect both upward salt transport and wave height. Observations that include wave height (as does wind force) should yield better estimates of salt loads than would anemometer-measured wind speeds alone. Wind force versus wind-speed

Table III. Comparison of Linear-Regression Parameters for the Equation $\ln \theta = au + b$
from Various Sea-Salt Aerosol Studies

Reference	Slope (a)	Intercept (b)	Conditions and Location
1	0.16	0.94	600–800 m (cloudbase); 20–25 °C SST; Hawaii
5	0.16	1.45	5, 10, 15 m; 10–15 °C SST; North Atlantic, yearly average
2	0.12	—	sea surface, production rate calculated from others
This work all data	0.09	1.53	16 m; 0–22 °C SST; South Atlantic, summer
Interval 1	0.17	0.75	16 m; 17–22 °C SST; South Atlantic, summer
Interval 2	0.15	0.87	16 m; 11–17 °C SST; South Atlantic, summer

plots give a somewhat awkwardly shaped curve showing logarithmic tendencies over part of its range. Therefore, it is significant that Blanchard and Woodcock (17) showed the 1953 salt-load data to be log linear with respect to wind speed. This result reinforces our conclusion that the natural log of the salt load should be correlated with wind speed, or $\theta = be^{au}$.

HISTOGRAMS OF SALT AEROSOL AND WIND DISTRIBUTIONS.   As an alternative to point-by-point correlation, we examined relationships between the salt-load frequency distribution and those of other relevant parameters. With a sufficiently long time series, we could presume representative sampling of wind speeds, salt loads, wave heights, and meteorological conditions for an area. The probability distributions and variances of salt loadings, wind speeds, fetches, and atmospheric stabilities are expected to be geographical variables in any statistical model resulting from this approach.

During the analyses, histograms of our data were constructed to visualize the shapes of the distributions of salt load, wind speed, and wave height. The histograms of Figure 8 show that the wind-speed distribution is approximately bimodal, with an additional tail on the right. The $\ln \theta$ is also approximately bimodal, with a larger tail to the right. This extra tailing on $\ln \theta$ would be difficult to predict from the wind-speed distribution. The distribution of ln (salt load) mimics the wind-speed distribution fairly well, in spite of poor point-by-point correlation of their time series.

The distributions can be compared by ordering the data values and computing the correlations between equally ranked pairs from the ordered data. Various functions relating the distributions of salt aerosol concentration to wind speed were tested to determine the best fit of the data. Our baseline correlation coefficient (0.79) was for a direct linear relationship between the ordered pairs. Relating salt concentration to the square of

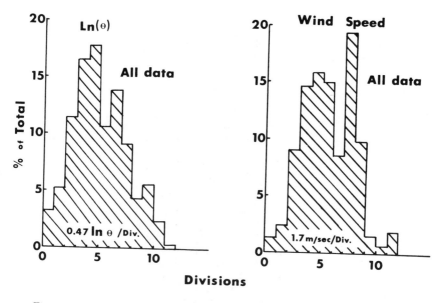

Figure 8. Frequency histograms of salt aerosol and wind speed for all time series data.

Histograms show the percentage of total sample per division. For the salt aerosol histogram, the first division indicates concentration values up to ln (0.47) (i.e., 0–1.6 μg/m³), the second division encompasses ln (0.46) to ln (0.94) (i.e., 1.6–2.6 μg/m³), etc. The wind-speed distribution is divided into linear intervals of 1.7 m/s each.

wind speed (wind stress) and to wind speed to the 3.3 power (whitecapping; *see* Reference 23) yielded correlation coefficients of 0.89 and 0.94, respectively. However, at higher powers of wind speed, the correlation begins to decrease. The best correlation ($r = 0.99$) was found for ln θ versus wind speed (Case XX in Table IV). The ln θ versus wind speed appeared to be the best fitting function for the unordered data set as well. Also, comparison of the ln θ distribution with the wave-height distribution yielded a correlation coefficient of 0.97 (Case XXI in Table IV). The wave-height distribution was tested here because sea surface features represent a step in the energy-transfer process between wind forcing and the subsequent concentration of sea-salt aerosol.

Computer runs on our ordered data of ln θ versus wind speed and wave height (shown in Table IV) may be compared to coefficients of 0.33 and 0.27 for the nonordered ln θ versus nonordered wind speed or wave-height data, respectively. These results lead us to the following conclusions. First, the near-perfect correlations of the ordered data indicate that ln (salt load)

Table IV. Sea-Salt Aerosol–Wind Speed–Wave-Height Relationships (Ordered Data)

Case	Slope (M)	Intercept (B)	Correlation Coefficient (r)	Mean Values (m)		Standard Deviation		Number of Points	Conditions
				y	x	σ_y	σ_x		
XX	0.291	−0.42	0.99	2.41	9.8	1.10	3.8	250	ln θ vs. wind speed
XXI	0.895	−0.08	0.98	2.39	2.76	1.15	1.25	253	ln θ vs. wave height
XXII	0.31	−0.27	0.97	2.76	9.77	1.25	3.91	254	Wave height vs. wind speed

NOTE: Relationships were determined from the equation, $y = Mx + B$ where $y = \ln \theta$ (μg/SCM) for Cases XX and XXI or wave height (m) in Case XXII, and $x = u$ (m/s) for Cases XX and XXII or wave height in Case XXI.

distributions for this area can be reliably estimated from the wind speed or wave-height *distributions*. Second, the relationship between salt loading and wind speed (or wave height) is better fit by a log function than by a linear or power function. Therefore, no matter what dynamic model of salt loading versus wind speed is chosen, it should produce salt-load values such that the distribution of ln θ is linearly related to the distribution of u values that were put in as the driving force for the model. For instance, if the input wind speeds form a normal distribution, then the output salt loadings should form a ln (normal distribution).

So far no measurements show how the coefficients for these functional relationships would vary from one part of the ocean to another. However, those coefficients should be sought because general prediction by area (and possibly by season) would alleviate problems of continuous in situ measurements necessary for point-by-point, time-wise prediction. Once an area is modeled, the problems due to advection, variations in stability, and precipitation removal would be taken care of statistically. Thereby, for specific oceanic areas and seasons, a specific cumulative salt loading should be predictable from the distribution of wind speed for that area.

A further analytical step can be taken by making linear plots of the cumulative percent distributions of each ordered data set versus percentage of time spent sampling. Both axes are thereby normalized. Figure 9 shows that the ln θ plot best corresponds to the wind and wave curves as previously discussed. Other interesting features can also be seen. For example, 90% of the time was spent collecting 56% of the total salt aerosol, while only 10% of the sampling time accounted for the remaining 44% of the salt aerosol. A large proportion of the energy-transfer events in turbulent processes occur in a relatively small proportion of the time. Apparently, our salt aerosol concentration data fits that pattern to some degree but not to the extreme observed by Gordon (*24*).

The upper distributions in Figure 9 show a simple curvature that could be nearly removed by taking logs of the variables. This results in log u and log [ln θ] or $\theta = e^{BuM}$, where $B = 0.16$ and $M = 1.17$ are the intercept and slope of the line fitted to the ordered data above 5 m/s.

Conclusions

A simple physical relationship does not exist between synoptic measurements of wind speed and sea-salt aerosol concentrations in the marine atmosphere because of advection, hysteresis, condensation processes, and the varying stability of the marine boundary layer. In the region of the South Atlantic Ocean discussed in this chapter, the low correlation between the time series for sea-salt aerosol concentration and local wind speed is attributed to the high variability of the effects just mentioned. Removing the temporal constraint by ordering both data sets results in an extremely high ($r = 0.99$) correlation coefficient. This result provides promise for the

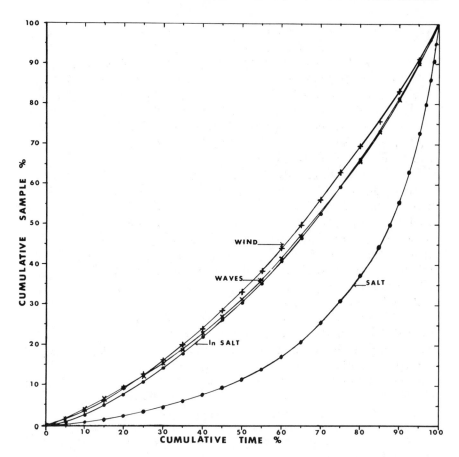

Figure 9. Cumulative curves, with normalized axes, showing the percentage of total acquired units vs. total time spent sampling and/or total number of samples taken.

general estimation of salt aerosol distributions from wind-speed distributions for specific oceanic areas and seasons because, given an input distribution for wind speed (u), the output distribution of salt aerosol concentration (θ) will be proportional to e^u. For our study area, the equation form $\theta = e^{Bu^M}$ (where $B = 0.16$ and $M = 1.17$ for wind-speed data above 5 m/s) seems to best encompass the statistical and deterministic parameters for the aerosol concentration distributions.

   Systematic differences appear to exist in sea-salt aerosol concentrations related to locations having significantly different sea surface temperatures. The predominant effect of sea surface temperature may pertain to the probability of formation of stable or unstable conditions in the marine boundary layer.

Wave features may be useful for enhancing the accuracy of real-time estimates of the salt aerosol concentration because the composite wave spectrum contains a history of the previous wind and reflects the degree of coupling of the sea surface with the local winds.

Measurements using two samplers that had different upper range particle-size cutoffs indicated that during rain and fog the airborne salt shifts from aerosol-sized particles to droplets of approximately 50-μm diameter (upper range of fog droplets) or greater. Consequently, the determination of the total concentration of atmospheric salt would require special sampling techniques to deal with both wet and dry particles of all sizes.

Recommendations

More research in diverse regions is necessary to obtain quantitative information on the variability in sea-salt aerosol concentration and related parameters before reliable predictive models can be formulated. For the near future, the prediction of salt aerosol concentration should be modeled on a statistical basis with available data. This exercise may lead to reliable generalizations for easily categorized oceanic regimes.

A real-time prediction capability for salt aerosol concentration is expected to take longer to develop. Eventually, such a capability could be based on regional statistical models whose parameters are modified from synoptic satellite observations.

Any future field studies in pursuit of deterministic models should include similar meteorological measurements at two or three elevations to determine vertical fluxes, atmospheric stability, and mixing parameters near the sea surface. Other factors, such as water temperature, directional wave-height spectra, and whitecap coverage, should be monitored almost continuously and automatically throughout the experiment. Unfortunately, reliable measurements of some of these parameters are extremely difficult to make from a moving ship. For shipboard sampling, the height of the sampler inlet relative to the wave tops should also be monitored. Additionally, much beneficial ancillary information could be provided by continuous measurement of particle sizes from at least two elevations.

Acknowledgments

We acknowledge the efforts of our colleagues, Bud Larson and John Hoover, in analyzing the larger data set from which we obtained the material for this chapter. We are also indebted to Andre Robinson, Katherine Schwarz, and Joseph Liu for laboratory assistance in processing and analyzing the aerosol samples and in meteorological data reduction. Earlier versions of the manuscript have benefitted from comments by Kingsley Williams, Duncan Blanchard, and Joseph Prospero. We thank Clifford Gordon, Clifford Trump, Gaspar Valenzuela, Richard Mied, William

Plant, Benn Okawa, and John Dugan for instructive discussions and for the loan of valuable reference materials.

We acknowledge the long-term encouragement and support provided by Conrad H. Cheek, the former head of the Chemical Oceanography Branch at Naval Research Laboratory. He served ably as Senior Scientist during several key research cruises. Typing of the manuscript was performed by Peg Szymczak, Norma Shaffer, and Janet Smith. Their diligence is greatly appreciated.

Literature Cited

1. Woodcock, A. H. *J. Meteorol.* **1953**, *10*, 362–71.
2. Toba, Y. *Tellus* **1961**, *17*, 131–45.
3. Tsunogai, S.; Saito, O.; Yamada, K; Nakaya, S. *J. Geophys. Res.* **1972**, *77*, 5283–92.
4. Savoie, D. L.; Prospero, J. M. *J. Geophys. Res.* **1977**, *82*, 5954–64.
5. Lovett, R. F. *Tellus* **1978**, *30*, 358–64.
6. Bressan, D. J. Naval Research Laboratory Memorandum Report 4581, 1981, 41 pp.
7. Larson, R. E.; Bressan, D. J. *Rev. Sci. Inst.* **1978**, *47*, 965–69.
8. Wilkniss, P. E.; Larson, R. E.; Bressan, D. J.; Steranka, J. *J. Appl. Meteorol.* **1974**, *13*, 512–15.
9. Wilkening, M. H.; Clements, W. E. *J. Geophys. Res.* **1975**, *80*, 3828–30.
10. Larson, R. E.; Bressan, D. J. *Proc. Second Conf. Coastal Meteorol.* **1980**, 84–100.
11. Hochrainer, D., In "Analysis of Airborne Particles by Physical Methods"; Melissa, H., Ed.; CRC Press: W. Palm Beach, Fla., 1978; chap. 2.
12. Riley, J. P.; Chester, R. "Introduction to Marine Chemistry"; Academic Press: New York, 1971; 465 pp.
13. Lepple, F. K., Ph.D. Thesis, Univ. of Delaware, Newark, Del., 1975.
14. Hoover, J. B. Naval Research Laboratory Memorandum Report 4674, 1982, 83 pp.
15. Lepple, F. K.; Bressan, D. J.; Hoover, J. B.; Larson, R. E. Naval Research Laboratory Memorandum Report 5153, 1983, 63 pp.
16. Prospero, J. M. *J. Geophys. Res.* **1979**, *84*, 725–31.
17. Blanchard, D. C.; Woodcock, A. H. *Ann. N. Y. Acad. Sci.* **1981**, *338*, 330–47.
18. Lovett, R. F. M. S. Thesis, Heriot-Watt Univ., Heriot-Watt, United Kingdom, 1975.
19. Blanchard, D. C.; Syzdek, L. *J. Phys. Oceanogr.* **1972**, *2*, 255–62.
20. Trusty, G. L.; Cosden, T. H. Naval Research Laboratory Report 8497, 1981, 46 pp.
21. "Atlas of Pilot Charts," Defense Mapping Agency, Publ. No. 105, 1981.
22. Ling, S. C.; Saad, A.; Kao, T. W., In "Turbulent Fluxes through the Sea Surface: Wave Dynamics and Prediction"; Favre, A.; Hasselmann, K., Eds.; Plenum: New York, 1978; pp. 185–97.
23. Monahan, E. C. *J. Phys. Oceanogr.* **1971**, *1*, 139–44.
24. Gordon, C. M. *Nature* **1974**, *248*, 392–94.

RECEIVED for review August 15, 1983. ACCEPTED March 5, 1984.

6

Continuous Sediment Sampling System for Trace Metal Surficial Sediment Studies

J. E. NOAKES, R. A. CULP, and J. D. SPAULDING

Center for Applied Isotope Studies, University of Georgia, Athens, GA 30605

The continuous sediment sampling system (CS³), designed to provide both continuous seafloor sediment samples to a vessel underway and rapid elemental analyses while still at sea, has undergone successful sea trial tests. The system consists of a towed sled for sample collection, a fully programmable on-board sample processor, and X-ray fluorescence elemental analysis instrumentation. Ground-truth studies show good agreement between analyses of seafloor sediment samples collected by box coring with those of samples collected with the CS³ system.

THE LEVELS OF INDUSTRIAL CHEMICAL WASTES BEING DUMPED into our nation's coastal waters are constantly being monitored by the National Oceanic and Atmospheric Administration (NOAA). To carry out the frequent marine monitoring programs necessary to stay abreast of this problem, it was determined in 1977 that more expedient shipboard sampling and analyses systems would have to be developed. One area of prime interest for sampling and scrutiny was the surficial seafloor sediments. To enhance rapid data collection, the ability to analytically measure samples while at sea was essential. This capability would allow areas of interest to be delineated and studied in detail before the survey vessel returned to port.

Developing a seafloor sediment sampling capability for NOAA was assigned to researchers at the University of Georgia. Their previous marine sediment work had utilized pathfinder elements in marine sediments for identifying nonindigenous materials (1–7). The basic concept of this approach is that the fine-sediment silt and clay of the seafloor represent the ultimate sink or permanent reservoir for a large spectrum of organic and inorganic materials injected into the marine environment (1–7). This reservoir

0065-2393/85/0209-0099 $06.00/0

can result from atmospheric, alluvial, or subseafloor emanations, of either natural or man-made origin. Therefore, the sampling system needed to address its capabilities to preferential selection of the fine-sediment fraction.

To obtain as much data as possible, the system needed a continuous sampling capability. Also, to fulfill the requirement of rapid data analyses while at sea, nondestructive elemental analysis was selected for the method of shipboard analysis. A sampling and analysis system called the continuous seafloor sediment sampler (CS^3) was developed from this work.

CS^3 Design and Development

The CS^3 system is made up of three major components: a seafloor sediment sampler, shipboard sample processor, and nondestructive elemental analysis instrumentation. It was described in detail previously (8). Design requirements for the seafloor sediment sampler are that it be in constant contact with the seafloor while being towed at speeds up to 6 knots, agitate only the upper surficial seafloor sediment and create a plume, contain a pumping means to sample the sediment slurry plume, and be capable of transporting the sediment slurry to a surface ship. These conditions were achieved by designing a towable sled that contained, within its structure, a submersible pump that was hose-connected to the surface ship.

Several configurations of the seafloor sediment sampler were fabricated and tested. The optimum design consisted of a cylindrical, stainless steel container of 25.4-cm outside diameter (OD) by 1.14 m long, with an inner mesh envelope housing a 10-stage motor-driven 0.56-kW, 220-V(AC) submersible pump. The sled housing and pump weight in air is 68 kg (see Figure 1). A tow line of 12.7-mm diameter, multistrand steel cable was used to pull the sled on the seafloor. Also attached to the sled was a 25.4-mm inside diameter (ID) rubber transfer hose and an electrical cable supplying power to the submersible pump. To reduce the chance of entanglement of the lines attached to the sled, all three lines were lashed together with nylon quick connectors at time of deployment of the sled over the stern of the vessel. The seafloor sampler was extensively tested up to 30.5-m depths at speeds approaching 4 knots.

To ensure that only the finer fraction of the sediment slurry was processed, a shipboard centrifugal cone separator was connected to the slurry transfer hose to remove the coarse-sediment fraction. The cone separator was a 101.6-mm diameter, urethane-coated centrifugal cone (Demco 275). Under a normal operational pressure of 221 kPa the cone separator is capable of delivering 57.0 L/min of sediment slurry, whose sediment particle size ranged from 2 to 32 μm as measured on the particle data Model 111 analyzer.

The shipboard sample processor is a self-contained, fully automated, sample preparation system that fits into a shipping container measuring 1.02 m long by 1.02 m high by 0.5 m wide, and weighs about 91 kg. It is de-

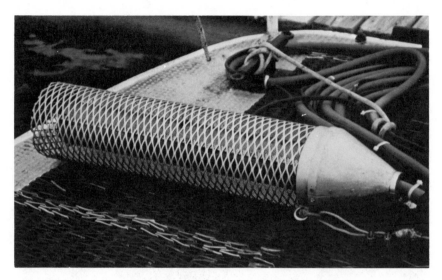

Figure 1. Seafloor sediment sampler.

signed to continuously accept the sediment slurry without obstruction of the flow so as not to interrupt the sampling sequence with respect to time and location. The processor is fully programmable for regulating sample volume and frequency of collection to enable collection of samples reflecting optimum seafloor sediment conditions (*see* Figure 2).

Sample preparation involves ejecting a selected sediment slurry aliquot onto a continuous roll of filter paper, dewatering by filtration, drying by heated, forced air, coating with a plastic spray, redrying by heated, forced air, and then storing.

When the sediment slurry is ejected onto the filter paper, date, time , and sample number are recorded on the edge of the filter paper for sample identification. The filter paper, which unwinds as needed for sample collection, is in a 92.0 m long by 76 mm wide roll and acts as both a filter and support media. Samples are prepared in circular 35-mm diameter sediment wafers that vary in thickness; the maximum thickness is less than 1 mm. Wafers are collected on the continuous filter paper sheet, 10 to a sheet, before being automatically cut for storage. Of the filter papers evaluated, a National Filter Media mat 300–851–000 filter paper was selected because of its low metal content and the high filter rate achievable with particle size ranges as small as 2–3 μm. Borden's Permaclear acrylic plastic 5-500 was the plastic spray source used because of its nondetectable metal content and its availability in aerosol containers.

The shipboard elemental analysis system consists of X-ray fluorescence (XRF) equipment, a nuclear thickness gauge, and a Hewlett Packard 9825-S computer. The XRF method of analysis was selected as most suit-

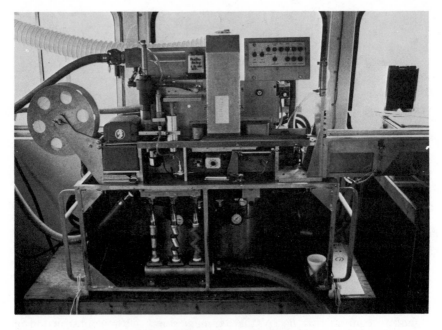

Figure 2. Shipboard sample processor.

able to shipboard use for the following reasons: (1) It is capable of rapid, nondestructive analysis; (2) it has high sensitivity to a large number of elements; (3) little or no sample preparation is necessary; and (4) analysis is based on surface area, thus the requirement of weighing samples at sea is eliminated.

Two different XRF analytical systems were used in the initial tests. The first equipment tested was a Kevex Series 3000 energy dispersive system that employed an Am-241 nuclear source and dysprosium target for generating X-rays. A solid-state, liquid nitrogen cooled, surface barrier radiation detector measured the excited X-rays from the elements making up the sediment sample (*see* Figure 3). Problems in maintaining an ample supply of liquid nitrogen at sea and transporting nuclear materials necessitated replacement of this XRF system with a preferred X-ray tube-excited unit that does not require a nuclear source or a cooled detection system. The Portaspec Model 2501 XRF spectrograph was selected as meeting these requirements and also because of its lightweight portability, ease of operation, and sensitivity for analysis of a broad spectrum of elements (Figure 4). This equipment consists of a probe head housing either a tungsten or molybdenum target X-ray tube, a collimator, a lithium fluoride crystal goniometer, and a proportional tube radiation detector. This system is po-

Figure 3. Source-excited XRF analytical system.

Figure 4. Tube-excited XRF analytical system.

tentially capable of detecting and analyzing over 60 elements at various concentration levels.

Because most, but not all, emitted X-rays come from the surface of the wafer, a wafer thickness measurement is made prior to each XRF analysis to enable a wafer thickness correction factor to be applied. The equipment for measuring wafer thickness is made up of an Fe-55 nuclear source and a Geiger tube radiation detector (Figure 5). Thickness measurements of the known standards and shipboard sediment wafers are carried out by positioning the wafer between the nuclear source and the radiation detector and recording the attenuation of the X-ray radiation as it passes through the sediment sample. A comparison of the wafer data to that of the calibrated samples enables rapid, nondestructive determination of the thickness of the sediment samples (Figure 6).

The Hewlett Packard 9825-S computer with internal tape drive and printer was used for shipboard data storage and elemental analyses calculations. Data handling involves input of sample identification number, wafer thickness values, and XRF measurements. The computer is programmed to use this data for elemental analyses calculation and hard-copy printout.

Prior to carrying out ground-truth studies, a more quantitative understanding of several aspects of the CS[3] sediment slurry delivery system was needed. Of major importance was the need to know, as precisely as possible, as many aspects as possible of the sediment slurry from the time it enters the sea-floor-sled pumping means to its extraction at the shipboard sample processor.

Direct measurement of the residence time of the 92-m hose under normal pumping rates gave a surprisingly constant reading of 47 s with three separate measurements. A sediment entrainment volume from the total hose system was calculated to be 46 L with a slurry flow rate of 59 L/min under a cone operating pressure of 221 kPa. Knowledge of the residence time of the sediment within the hose delivery system enabled the precise location of the site of sediment sampling to be calculated by using the ship's location and the length and angle of the trailing hose connected to the undersea sled.

In addition, determining the operational parameters of the centrifugal cone separator was important. The function of the centrifugal cone is to separate the finer sediment fraction of interest from the heavier or larger particles (typically sand size) and deliver the sorted fine fraction to the processor. As part of this study, particle-size analyses were carried out on sediment slurries collected from the centrifugal cone separator operating at various pressures. A minimum pressure of 221 kPa is required to facilitate good separation of the fine particles from the heavy sand particles. Higher pressures achieved by restricting the output of the centrifugal cone excessively reduced the slurry flow rate to the processor but provided little advantage in

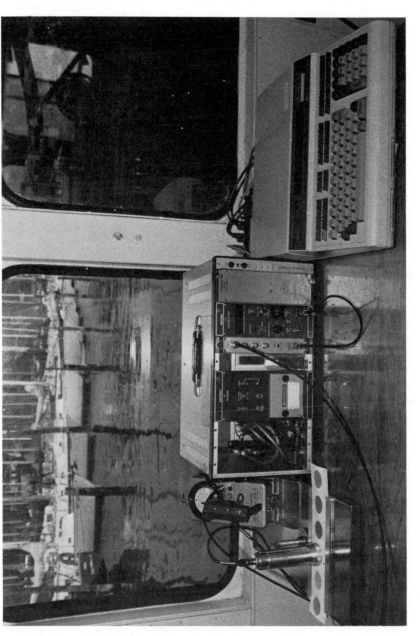

Figure 5. Sediment wafer-thickness gauge.

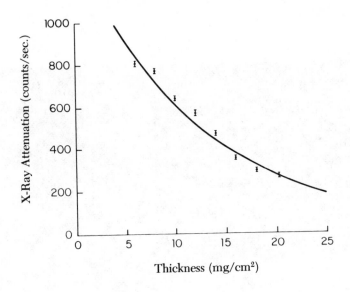

Figure 6. Sediment thickness calibration curve.

particle-size sorting. To achieve a more diverse particle-size separation a variable pressure pump must be incorporated into the sampling system in front of the centrifugal cone. The variable pressure pump increases the operational pressure to the centrifugal cone separator without restricting the flow rate.

After the testing and evaluation of the operational parameters of the sediment retrieval system, the ground-truth study of the CS^3 system was initiated to determine whether the CS^3 samples reflected the true sediment composition and, therefore, its ability to correctly and rapidly survey the seafloor sediment.

Baltimore Harbor Ground-Truth Studies

The final phase in the development of the CS^3 system was to test and evaluate the capability of the CS^3 system to analytically portray the true elemental content of the seafloor sediments. To carry out this task, a joint Center for Applied Isotope Studies (CAIS)–NOAA ground-truth study was initiated. A site for the study was selected in the Baltimore Harbor (Maryland) along the Patapsco River, south of North Point and adjacent to the Brewerton Channel (*see* Figure 7). This site is a discontinued spoil area with documented high levels of heavy metals in the sediments. Water depths of this site are 15–25 m, well within the operational range of the CS^3 system, and it is located close to available NOAA facilities.

The ground-truth sediment sampling program was carried out by

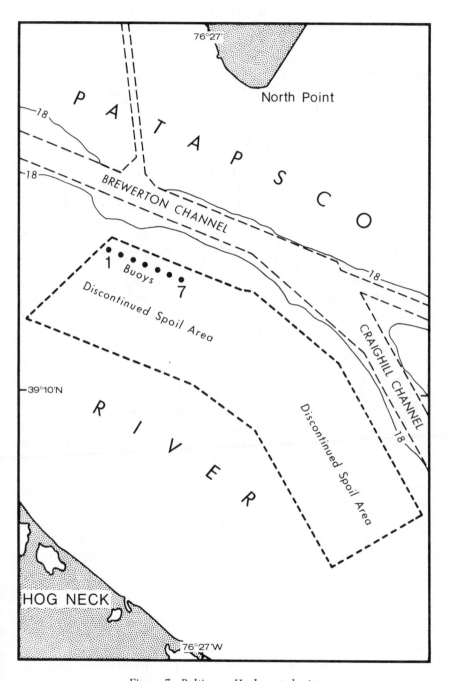

Figure 7. Baltimore Harbor study site.

establishing seven stations in the spoil area. The stations were laid out linearly, approximately 75 m apart; each was marked with a buoy. At each of the stations a box core was taken of the bottom sediments. Box coring was selected over grab or core sampling as the preferred method of obtaining undisturbed surficial and subsurface seafloor sediments. From each of the box cores a cylindrical core was taken. This core was then sectioned from top to bottom into 30 mm thick slices (Figure 8). Each 30 mm thick slice was evaluated for its elemental content by two independent analytical methods. Additional cores from Stations 1 and 7 were sectioned into linear quadrants for compositional profile studies.

Sediment sampling of the seven stations using the CS^3 equipment was carried out by running transects with the survey vessel parallel to, and as close as possible to, the marker buoys. The CS^3 underwater seafloor sediment sampler was pulled at a speed of three knots and, when abreast of each buoy, the sediment collected was recorded as being from that station. The sediment wafers prepared aboard ship from the collected slurries were immediately analyzed by XRF for three elements (Mn, Fe, and Ti) and were stored for further land-based analyses of other elements. A comparison of the elemental content of the sediments collected from the seven stations by box coring and with the use of the CS^3 equipment constituted the basis for ground-truth evaluation of the CS^3 system.

Analytical Results

The 30-mm sediment slices of the segmented cylindrical cores obtained from box coring at the seven stations were dried, pulverized, and thoroughly mixed to yield a uniform sample for analysis. Sediment from each of these slices was analyzed by two independent methods. The first method used a Perkin-Elmer model 5000 atomic absorption spectrophotometer (AA) for the elements Fe, Mn, Ti, Pb, Zn, Cu, Cr, Ni, Co, Hg, and Cd (9). The second method utilized a Philips PW 1410 X-ray fluorescence spectrometer for the analysis of elements Fe, Mn, Ti, Ca, K, P, Si, Al, Mg, Na, Pb, Zn, Cu, Cr, V, and Ba (10). The AA analysis was chosen because of the known accuracy and sensitivity to a wide spectrum of elements. The XRF analysis was chosen for its accuracy and similar nondestructive mode of analysis equivalent to the shipboard XRF analysis. Good agreement between the AA and the XRF values was felt to be imperative because the Philips XRF equipment was to be used in the land-based multielement analysis of the CS^3-collected sediment samples.

The AA results of selected sediment segments of the cylindrical cores are listed in Table I. The elements were chosen for their potential environmental impact, and because many heavy elements were known to be of higher concentrations in the spoil area than normally found in marine or estuarine sediments. The data in Table I have a $\pm 5\%$ relative standard deviation and detection limits on the order of 10 ppm for most of the analyses.

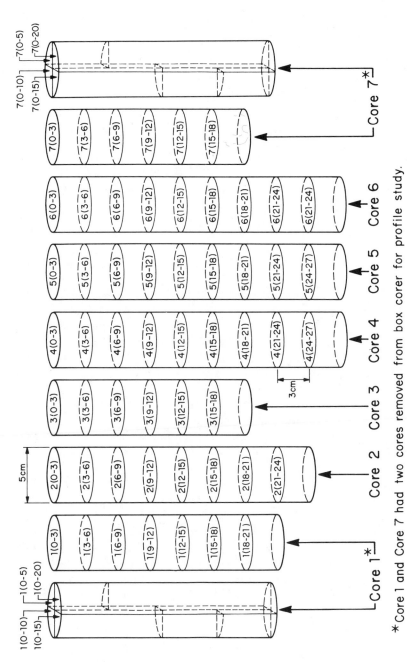

Figure 8. Baltimore Harbor surficial sediment cores.

*Core 1 and Core 7 had two cores removed from box corer for profile study.

Table I. Skyline Labs' Atomic Absorption Analysis of Baltimore Harbor Sediment Cores

Sample	Fe	Mn	Ti	Pb	Zn	Cu	Cr	Ni	Co	Hg	Cd
1 (0–3)	6.1	0.255	0.51	100	450	65	90	45	40	0.15	0.6
2 (0–3)	6.2	0.240	0.53	95	470	70	145	45	35	0.17	0.6
3 (0–3)	6.1	0.240	0.48	90	420	65	135	40	35	0.12	0.4
4 (0–3)	6.5	0.455	0.51	95	460	65	125	60	45	0.16	0.4
5 (0–3)	6.5	0.345	0.54	95	470	75	135	55	45	0.17	0.4
6 (0–3)	6.2	0.420	0.51	100	475	70	100	55	45	0.18	0.4
7 (0–3)	6.1	0.215	0.52	100	455	70	100	50	40	0.17	0.4
1 (0–5)	6.2	0.255	0.49	95	440	65	90	50	35	0.21	0.4
1 (0–10)	6.2	0.290	0.53	95	445	65	80	45	40	0.21	0.4
1 (0–15)	5.9	0.275	0.51	80	390	60	70	45	35	0.15	0.4
1 (0–20)	6.0	0.285	0.48	80	375	60	110	40	35	0.13	0.2
4 (24–27)	7.7	0.125	0.59	105	520	80	195	50	30	0.21	0.4
7 (0–10)	6.2	0.360	0.53	90	455	70	120	55	40	0.21	0.4
7 (0–15)	6.4	0.380	0.49	95	460	70	120	55	45	0.20	0.4
7 (0–20)	6.3	0.380	0.53	95	455	70	105	50	45	0.20	0.4

NOTE: Results for Fe, Mn, and Ti are given in percents; all others are in parts per million.

The AA analyses were used as a guide for selecting the elements to be analyzed with the Philips XRF system (Table II). Major elements not usually of environmental concern were analyzed as well as trace elements because of their effect on the X-ray absorption of the heavier elements. Elements below the XRF analysis detection limits, such as Hg and Cd, were not analyzed. Samples prefixed with the letter F were mechanically separated so as to bias the fine-particle size. This procedure indicated to what degree the shipboard centrifugal cone separator, used to select the finer sediment particles, influenced the elemental composition.

Table III shows the XRF analysis of the fine-sediment fraction of the sediment samples listed in Table II made into 100-mg wafers. These samples were biased to the fine fraction, as were the F samples in Table II, and were analyzed on filter paper, identical to what would be prepared by the CS[3] system. This information was used primarily to relate the concentration of the elements in the standard pellets of Table II of known thickness and composition to the CS[3] wafers shown in Table IV.

Table IV shows the Philips XRF elemental analysis of the CS[3] sediment wafers collected adjacent to the seven buoys in the Baltimore Harbor spoil area. These samples were compensated for thickness variations exactly like the 100-mg wafers listed in Table III. The thickness values of some of the heavier elements (Fe, Pb, Zn, and Ba) needed to be multiplied by a constant factor to compensate for their proportional decrease in X-ray absorption when the higher energy Philips XRF system was used. For Pb the corrected values fell below acceptable detection limits and were not recorded.

Table II. Philips X-Ray Fluorescence Elemental Analysis of Baltimore Harbor Sediment Cores

Sample	Fe	Mn	Ti	Ca	K	P	Si	Al	Mg	Na	Pb	Zn	Cu	Cr	V	Ba
1 (0–3)	6.2	0.25	0.58	0.36	1.59	0.15	24.4	8.13	1.10	1.85	132	530	121	155	158	399
2 (0–3)	6.5	0.26	0.61	0.32	1.67	0.12	25.3	8.46	0.99	1.14	128	567	124	161	168	423
3 (0–3)	6.0	0.23	0.55	0.44	1.58	0.14	24.0	8.02	1.15	1.92	118	513	112	155	157	362
4 (0–3)	6.5	0.44	0.59	0.39	1.76	0.11	24.4	9.00	1.05	1.15	113	606	129	147	165	452
5 (0–3)	6.8	0.34	0.62	0.37	1.70	0.14	24.9	8.89	1.03	1.14	135	618	132	185	185	441
6 (0–3)	6.5	0.43	0.59	0.40	1.71	0.13	24.9	8.88	1.04	1.12	122	540	122	149	166	433
7 (0–3)	6.4	0.22	0.58	0.36	1.64	0.16	24.3	8.60	1.08	1.90	123	543	102	164	167	421
2 (21–24)	6.7	0.37	0.61	0.38	1.71	0.12	25.0	7.83	1.02	1.06	120	630	114	165	177	439
4 (24–27)	8.1	0.12	0.68	0.27	1.61	0.16	24.7	7.33	0.97	1.11	149	643	119	233	234	406
6 (24–27)	8.0	0.12	0.71	0.27	1.64	0.15	24.1	7.39	0.96	1.05	176	675	134	236	264	428
1 (0–5)	6.3	0.25	0.58	0.56	1.63	0.13	25.0	8.32	1.04	1.37	114	523	106	164	175	410
1 (0–10)	6.4	0.30	0.59	0.49	1.66	0.13	25.3	8.58	1.02	1.19	103	514	101	160	171	422
1 (0–15)	6.3	0.29	0.59	0.40	1.71	0.12	25.5	8.00	1.05	1.20	95	491	101	138	174	434
1 (0–20)	6.3	0.28	0.58	0.49	1.73	0.11	25.4	8.18	1.09	1.20	79	448	98	146	176	394
7 (0–5)	6.6	0.33	0.59	0.37	1.70	0.14	24.4	8.15	1.06	1.25	105	565	103	152	179	398
7 (0–10)	6.6	0.36	0.60	0.35	1.73	0.13	24.9	8.89	1.02	1.19	109	550	115	150	187	421
7 (0–15)	6.7	0.36	0.61	0.35	1.72	0.13	24.7	8.83	1.01	1.17	106	559	91	165	172	415
7 (0–20)	6.6	0.36	0.61	0.34	1.72	0.14	24.7	8.68	1.03	1.19	135	565	106	158	170	410
F1 (0–15)[a]	6.5	0.29	0.61	0.59	1.74	0.11	25.1	8.08	0.97	0.65	95	488	94	145	170	408
F2 (21–24)[a]	6.8	0.38	0.61	0.50	1.72	0.11	25.3	8.84	1.01	0.65	114	670	127	179	185	478
F7 (0–15)[a]	6.8	0.37	0.62	0.51	1.71	0.13	25.0	8.51	0.95	0.62	135	573	124	159	176	448

NOTE: All samples are 4-g pellets. Results for Pb, Zn, Cu, Cr, V, and Ba are given in parts per million; all others are in percents.
[a]F denotes fine fraction of sediment only.

Table III. Philips X-Ray Fluorescence Elemental Analysis of Baltimore Harbor Sediment Cores

Sample	Fe	Mn	Ti	Ca	K	P	Si	Al	Mg	Na	Pb	Zn	Cu	Cr	V	Ba
W1 (0–3)	6.3	0.22	0.56	0.32	1.61	0.28	22.4	6.47	0.81	0.49	106	526	90	130	150	388
W2 (0–3)	5.8	0.19	0.49	0.26	1.42	0.10	20.6	6.73	0.81	0.42	144	515	61	125	143	406
W3 (0–3)	7.3	0.24	0.63	0.35	1.79	0.17	26.2	8.16	1.03	0.54	97	578	94	143	174	416
W4 (0–3)	6.1	0.35	0.53	0.31	1.65	0.10	23.2	7.86	0.90	0.49	65	477	71	118	157	417
W5 (0–3)	6.3	0.27	0.58	0.29	1.68	0.13	24.0	7.98	0.93	0.32	99	515	95	148	164	410
W6 (0–3)	5.9	0.34	0.55	0.31	1.68	0.11	24.2	8.05	0.93	0.51	a	507	77	126	151	415
W7 (0–3)	6.4	0.18	0.58	0.30	1.65	0.16	23.8	7.59	0.93	0.48	80	552	75	153	161	399
W2 (21–24)	6.1	0.29	0.52	0.34	1.50	0.09	22.4	7.11	0.88	0.47	47	578	56	123	143	431
W4 (24–27)	7.1	0.10	0.59	0.26	1.44	0.14	21.3	6.60	0.81	0.45	78	481	74	172	208	407
W6 (24–27)	7.5	0.11	0.73	0.26	1.68	0.15	23.3	7.17	0.93	0.48	211	1285	143	236	276	418
W1 (0–10)	6.2	0.24	0.56	0.29	1.64	0.12	23.8	7.08	0.92	0.49	146	512	68	139	152	401
W1 (0–15)	5.8	0.23	0.53	0.29	1.64	0.11	23.1	7.02	0.92	0.49	49	446	28	119	132	363
W1 (0–20)	5.7	0.23	0.53	0.31	1.71	0.10	24.2	7.32	0.97	0.51	60	410	21	118	134	363
W7 (0–10)	6.2	0.29	0.56	0.30	1.68	0.13	23.2	7.24	0.92	0.48	91	531	54	125	147	395
W7 (0–15)	6.3	0.30	0.55	0.29	1.62	0.12	22.4	7.03	0.88	0.47	69	560	68	124	151	374
W7 (0–20)	6.4	0.29	0.56	0.29	1.64	0.12	22.7	7.17	0.91	0.48	a	524	86	130	154	379

NOTE: All samples are 100-mg wafers. Results for Pb, Zn, Cu, Cr, V, and Ba are given in parts per million; all others are in percents.
a Below detection limit when thickness compensated.

Table IV. Philips X-Ray Fluorescence Elemental Analysis of CS³ Samples from Baltimore Harbor

Sample	Fe	Mn	Ti	Ca	K	P	Si	Al	Mg	Na	Pb	Zn	Cu	Cr	V	Ba
September 15, 1982 A.M.																
C125	6.3	0.22	0.62	0.34	1.98	0.20	27.5	10.21	1.17	0.92	—[a]	469	66	171	165	419
C126	6.6	0.19	0.59	0.32	1.93	0.19	27.3	10.17	1.15	0.92	—[a]	584	71	193	190	495
C127	6.0	0.16	0.53	0.31	1.83	0.19	26.9	9.88	1.13	0.90	—[a]	512	65	155	171	444
C128	6.6	0.24	0.65	0.33	2.03	0.22	27.2	9.32	1.19	0.94	—[a]	552	66	157	177	432
C129	6.5	0.22	0.64	0.37	1.97	0.22	27.2	9.72	1.18	0.93	—[a]	565	90	168	184	435
C130	6.2	0.22	0.64	0.37	1.95	0.21	27.1	9.71	1.17	0.91	—[a]	545	82	148	165	404
C131	6.6	0.19	0.62	0.37	1.91	0.22	26.6	9.51	1.15	0.90	—[a]	563	83	150	184	446
September 15, 1982 P.M.																
C286	6.2	0.24	0.65	0.35	1.97	0.20	27.8	10.16	1.19	0.98	63	519	54	133	153	416
C287	6.0	0.23	0.64	0.39	1.95	0.20	27.6	9.98	1.16	0.97	116	477	72	139	160	409
C288	6.0	0.18	0.56	0.37	1.90	0.18	27.2	9.92	1.15	0.94	—[a]	474	74	145	169	475
C289	6.3	0.25	0.65	0.39	1.98	0.22	26.9	9.11	1.16	0.96	120	486	64	127	133	435
C290	6.7	0.26	0.65	0.43	1.95	0.22	27.1	9.91	1.18	1.00	128	435	64	113	115	457
C291	6.0	0.24	0.65	0.38	1.96	0.23	27.1	9.89	1.17	0.99	126	546	67	132	149	403
C292	6.0	0.23	0.64	0.34	1.96	0.22	27.2	9.88	1.16	0.99	84	519	74	138	154	397
September 16, 1982 A.M.																
C305	5.6	0.16	0.53	0.49	1.80	0.19	26.2	9.33	1.16	1.06	—[a]	454	74	156	160	422
C306	5.7	0.16	0.53	0.47	1.80	0.18	26.2	9.50	1.17	1.05	—[a]	472	73	150	156	442
C307	5.4	0.15	0.50	0.52	1.74	0.19	25.7	9.30	1.16	1.11	—[a]	433	78	145	151	419
C308	5.9	0.15	0.50	0.43	1.70	0.21	24.8	8.51	1.13	1.00	—[a]	474	65	153	172	438
C309	6.1	0.17	0.57	0.40	1.90	0.22	26.5	9.14	1.18	1.01	—[a]	471	41	162	171	469
C310	5.5	0.16	0.53	0.34	1.82	0.22	26.3	9.09	1.16	0.96	—[a]	432	71	151	166	433
C311	5.8	0.16	0.56	0.39	1.88	0.21	26.4	8.99	1.18	0.98	—[a]	434	48	161	157	445

NOTE: Results for Pb, Zn, Cu, Cr, V, and Ba are given in parts per million; all others are in percents.
[a]Below detection limit when thickness compensated.

Discussion

This study entailed the comparison of the sediment obtained by both box coring and CS[3] sampling by the three different analytical systems described.

The AA analyses shown in Table I were of samples from the uppermost sections of all seven cores, from the lowest section within Core 4, and from each quadrant of Cores 1 and 7. Table II summarizes the elemental analysis performed with the high-energy XRF system. The samples were identical to those for AA analysis except for their preparation. Comparison of Tables II and III indicates a very good agreement between the two analytical systems. Variations in composition from one station to another were exhibited by both systems for some of the major elements. In addition to the close agreement between the more common elements Fe, Mn, and Ti, very good agreement was found between the trace elements. These results confirmed the applicability and accuracy of both analytical systems for this type of sample measurement.

As soon as the sediment composition was established, a method to compare these results to the CS[3] samples was devised. First, a calibration curve was constructed by making numerous samples of identical composition and varying thickness. The thickness value for each sample was computed by using the attenuation of X-rays from an Fe-55 nuclear source as they passed through the sample. This procedure eliminated the need for weighing samples prior to analysis. The calibration curve was then used to correct compositional value when the thickness was less than infinitely thick with regard to emitted X-rays. The curve was also used to find the minimum thickness required to stop all of the primary X-rays for a given element during XRF analysis (i.e., the minimum infinite thickness). To substantiate this procedure, 100-mg samples were made from the same sediment sections as those listed in Table II. The XRF analysis of these 100-mg samples are listed in Table III. Because the composition of each sample is equivalent, the accuracy of the thickness correction calibration curve is represented in the comparison of Tables II and III. Except for a few specific elements, the XRF analysis of the 4-g pellets and the 100-mg wafers showed good agreement. For the heavier elements (Fe, Pb, Zn, and Ba) an additional correction was required. Because of the agreement between Tables II and III the same calibration curve for thickness correction was applied to the CS[3] samples as shown in Table IV.

The data in Table IV indicate very good analytical agreement with earlier presented data. This agreement is very important because the sediment samples used were separated into the finer fraction by passage through the centrifugal cone prior to analyses. Because the data agree well, they strongly support the use of the finer sediment sample to accurately represent the total sediment sample. The data in Table IV also present some evidence to help evaluate the maximum depth of sediment sampling of the CS³ sled. This value can be approximated by extrapolation of the Fe and Mn values from Tables I–IV. For example, the Fe and Mn analyses by AA and XRF for samples from the lowest cylindrical core sections, that is, 4(24–27), indicate a distinct increase and decrease, respectively, of these two elements at this depth. The values in Table IV do not indicate this change and are most likely caused by the sled's higher location within the surficial sediment layer.

Table IV shows good agreement between the CS³ heavy metals values and the ground-truth data shown in Tables I and II. With the exception of Pb, all the elements were determined with good accuracy and indicated a ground-truth agreement between the surficial sediments and the CS³ wafers produced from the same material. The XRF analysis made use of six different standards that were compared to NBS standard reference material 1646. This standard is a marine sediment of nearly identical mineral composition to that of the Baltimore Harbor samples.

Table V shows the comparative elemental analyses of the surficial sediments obtained from the seven stations by box coring and the CS³ system. Three elements (Fe, Mn, and Ti) were selected for comparison because of their presence in high levels in the marine sediments and the good sensitivity for detection by the Perkin-Elmer AA, the Philips XRF, and the ship-

Table V. Ground-Truth Comparative Elemental Analysis of Baltimore Harbor Surficial Sediment

Sample No.	Skyline AA	Philips XRF		Portaspec XRF		
		Pellet	Wafer	CS_1^3	CS_2^3	CS_3^3
1 (0–3) Fe	6.1	6.2	6.5	6.8	6.8	5.9
1 (0–3) Mn	0.25	0.25	0.29	0.29	0.28	0.27
1 (0–3) Ti	0.51	0.58	0.55	0.60	0.61	0.55
2 (0–3) Fe	6.2	6.5	6.3	6.9	6.6	6.0
2 (0–3) Mn	0.24	0.26	0.27	0.30	0.28	0.27
2 (0–3) Ti	0.53	0.61	0.51	0.62	0.58	0.56
3 (0–3) Fe	6.1	6.0	7.1	6.4	6.3	5.6
3 (0–3) Mn	0.24	0.23	0.31	0.28	0.29	0.24
3 (0–3) Ti	0.48	0.55	0.58	0.60	0.57	0.52
4 (0–3) Fe	6.5	6.5	6.1	7.0	6.7	6.1
4 (0–3) Mn	0.46	0.44	0.43	0.29	0.27	0.26
4 (0–3) Ti	0.51	0.59	0.53	0.61	0.56	0.55
5 (0–3) Fe	6.5	6.8	6.6	6.9	6.7	6.3
5 (0–3) Mn	0.35	0.34	0.35	0.29	0.27	0.27
5 (0–3) Ti	0.54	0.62	0.56	0.59	0.55	0.58
6 (0–3) Fe	6.2	6.5	6.1	6.8	6.8	6.0
6 (0–3) Mn	0.42	0.43	0.41	0.29	0.28	0.26
6 (0–3) Ti	0.51	0.59	0.54	0.59	0.57	0.54
7 (0–3) Fe	6.1	6.4	6.5	6.9	6.7	6.2
7 (0–3) Mn	0.22	0.22	0.26	0.28	0.27	0.25
7 (0–3) Ti	0.52	0.58	0.57	0.61	0.58	0.57

NOTE: All results are given in percents.

board Portaspec XRF systems. Of the three analytical systems used in this study, the Portaspec XRF equipment was the least sensitive to the wide spectrum of elements measured, largely because of the restrictive 30-kV operational voltage of the equipment. Because the Portaspec XRF's primary role was to be used aboard ship in a qualitative role to rapidly identify sediments of high metal content while at sea, this lack of sensitivity was not felt to be a deterrent to its use, especially because the CS^3 wafers could be readily stored after the shipboard XRF analyses and later reanalyzed by the more sensitive land-based Philips XRF instrumentation.

The results of this study support three basic conclusions: (1) The CS^3 system is a valid means for collecting surficial seafloor samples for pollution studies; (2) the basic assumption of this study—that the fine-sediment fraction of marine sediments are the primary host of heavy metal contaminants injected into the marine waters—is correct; and (3) the shipboard XRF system formerly designed as a qualitative means for assisting in selecting more precise sampling areas while at sea can now be considered as a valid analytical tool capable of producing quantitative data.

Literature Cited

1. Bertine, K. K.; Mendeck, M. F. *Environ. Sci. Technol.* **1978**, *12*, 201–7.
2. Gambrell, R. P.; Khalid, R. A.; Patrick, W. H., Jr. *Environ. Sci. Technol.* **1980**, *14*, 431–6.
3. Helmke, P. A.; Koons, R. D.; Schomberg, P. J.; Iskander, I. K. *Environ. Sci. Technol.* **1977**, *11*, 984–9.
4. Lu, J. C. S.; Chem. K. Y. *Environ. Sci. Technol.* **1977**, *11*, 174–82.
5. Rabitti, S.; Boldrin, A.; Vitturi, L. M., unpublished data.
6. Schroeder, B.; Thompson, G.; Sulanowska, M. *Am. Lab.* **1983**, March, 66–72.
7. Sinex, S. A.; Heiz, G. R. *Environ. Sci. Technol.* **1982**, *16*, 820–5.
8. Noakes, J. E., Harding, J. L. *Proc. 14th Annual Oceanogr. Tech. Conf.* **1982**, 4294, 731–5.
9. Ward, F. N.; Nakagawa, H. M.; Harms, T. F.; Van Sickle, G. H., *Geol. Surv. Bull. (U.S.)* 1289, 45p.
10. Bertin, E. P. "Introduction to X-Ray Spectrometric Analysis"; Plenum: New York, 1978; pp. 327–89.

RECEIVED for review July 5, 1983. ACCEPTED March 1, 1984.

Development of Shipboard Copper Analyses by Atomic Absorption Spectroscopy

RICHARD W. ZUEHLKE and DANA R. KESTER
Graduate School of Oceanography, University of Rhode Island, Kingston, RI 02881

This chapter describes a technique for automatically precon-centrating dissolved copper in seawater for subsequent anal-ysis by graphite furnace atomic absorption spectroscopy. Copper and other trace metals are chelated by 8–hydroxy-quinoline either immobilized on silica in a chromatographic column or in solution with the oxine complexes isolated on a C_{18} reverse-phase liquid chromatography column. Elution with an appropriate solvent provides concentration factors in the range of 25–30 times the concentration in the initial sam-ple. The system is adaptable for removal and preconcentra-tion of organically complexed copper. The methods are eval-uated in terms of their analytical chemistry and ability to produce valid profiles through an oceanic water column. The technique is promising if pH and column-loading rates are optimized, and if attention is paid to the role of naturally oc-curring organic ligands that can affect the retention of cop-per by the columns.

REAL-TIME CHEMICAL MEASUREMENTS IN THE OCEAN are receiving increas-ing attention as a means of advancing our understanding of chemical pro-cesses in the sea. The traditional approach for many types of chemical mea-surements has been to collect a suite of samples at various depths and locations in the ocean, to return the samples to a laboratory ashore, and to chemically analyze them. This process produces results months or years af-ter sampling. Measurements of biologically active nutrients (phosphorus, nitrogen, and silicon) and gases (dissolved oxygen and the carbon dioxide system parameters) have been exceptions to this general approach; ship-board techniques have been used extensively for these variables for many

0065-2393/85/0209-0117 $06.50/0

years. Several advantages are realized by shipboard analytical measurements: biological and chemical effects during storage can be avoided; when the data are available in nearly real time, the sampling can be modified in view of the results obtained; and experimental studies at sea become possible. This chapter reports our efforts with techniques to measure transition metals at low concentration in ocean water on board a research vessel.

Several criteria that were set forth at the beginning of our work determined our choices in the analytical approach to this problem. We ultimately wish to perform measurements with a precision and accuracy of about 5% at concentrations on the order of 10^{-9} mol/kg. Although we have been focusing on a single metal to optimize the techniques in this stage of the work, we plan to extend the capability to a broad range of metals. When the methods are perfected, the interpretation of the analytical results should be relatively straightforward and comparable to results obtained by using shore-based measurements. The techniques should be sufficiently rapid to allow nearly real-time data acquisition and high spatial resolution of metal distributions in the ocean. In practice, this requirement sets a limit of 30–45 min of total processing time, including any reagent addition, volume measurement, filtration, preconcentration, and dissolution or other phase change.

Copper was selected as the first metal for which to attempt to optimize the shipboard analyses because considerable information is available about the marine chemistry of copper, and because this new analytical capability would greatly enhance our ability to study copper in the ocean. The concentration of copper in the ocean varies from 0.5 to 5 nmol/kg in response to biological and geochemical processes (Table I). The chemical speciation of copper has received considerable attention because the biological effects of copper depend on its chemical form (1–3). The principal forms of copper include inorganic complexes such as $CuCO_3^0$, $CuHCO_3^+$, $CuOH^+$, and organically bound copper (4, 5).

All of the available analytical methods in seawater at concentrations on the order of 1 nmol/kg require a preconcentration step prior to analyses. The preconcentration step can serve two purposes: It brings the amount of metal to be detected into the analytical range of the measuring technique, and it can place the metal into a medium more favorable for analysis than seawater. Two methods are being widely used for copper measurements in seawater: anodic stripping voltammetry (ASV) and heated graphite atomization atomic absorption spectroscopy (HGA AAS). With ASV, the preconcentration step consists of reducing copper [generally Cu(II) in seawater] to an amalgam in a mercury thin-film or drop electrode. Detection is performed electrochemically by oxidizing the copper back into the solution in response to a varying applied voltage at the electrode. With HGA AAS various preconcentration procedures can be used. Some of the most common are coprecipitation of the metal by a scavenging solid phase (6), solvent

Table I. Seawater-Dissolved Copper Distributions

Regime	Major Processes	Typical Concentrations (nmol/kg) and Locations	Reference
Surface waters	Aeolian and riverine input	0.5–1.5 (N. Pacific)	32
	Biological uptake	4.5–5.5 (Arctic)	35
		1.5–4.5 (Indian)	36
		0.5–2.0 (E. Atlantic)	37
		1.0–1.9 (E. Atlantic)	38
		1.2 (N. Atlantic, Sargasso Sea)	39
		1.0 (N. Atlantic, Sargasso Sea)	25
		1.3–1.6 (N. Atlantic, Slope Water)	25
Mid-depths	Particulate scavenging	up to 8.7 (Middle Atlantic Bight)	25
	Regeneration from sinking	1.5–5 (N. Pacific)	32
	biogenic debris (in deep waters)	2–5 (Arctic)	35
		1.5–5 (Indian)	36
		1.1–5.0 (E. Atlantic)	38
		1.5–2 (N. Atlantic, Sargasso Sea)	39
		1.0–2.1 (N. Atlantic, Sargasso Sea)	25
		0.7–1.5 (N. Atlantic, Slope Water)	25
Bottom waters	Remobilization from sediment	up to 5.3 (N. Pacific)	32
	pore waters	5.2 (Indian)	36
		4–10 (E. Atlantic)	38
		2.3 (N. Atlantic, Sargasso Sea)	25
		1.4 (N. Atlantic, Slope Water)	25

extraction after chelation (7, 8), ion exchange followed by acid elution (9), and retention on an immobilized complexing phase (10, 11).

The recovery of a metal from nearly all preconcentration techniques can depend on the chemical speciation of the metal. With ASV this dependence has led to the distinction of a metal as being electrochemically active or inactive. The electrochemical method can also yield information that distinguishes among different active forms of a metal (12). For copper (II) in chloride media such as seawater, the ASV technique is kinetically limited by Cu(I) chloro-complexes; therefore, information is not obtained about the equilibrium species in solution (13). Surface–active organic substances may accumulate at the electrode surface and modify the chemical behavior of the metal being analyzed. The preconcentration methods used in conjunction with HGA AAS are assumed not to depend on the chemical species present. Comparisons of Chelex extraction and solvent extraction of metals from seawater indicate that Chelex resin [iminodiacetate–styrene ion exchange resin (Bio-Rad Laboratories)] probably does not retain some forms of metals (14, 15). Demonstrating that a preconcentration method can recover all chemical forms of a metal is an important exercise. The ability to recover essentially all of an added spike of metal from a sample is an insufficient test because some forms of the metal may not be in equilibrium with the chemical form of the spike (16, 17). The speciation of a metal becomes an important factor in evaluating an analytical procedure, especially in natural waters analyzed in real time. Speciation has been less important in previous extraction analyses when all samples have been preserved in strong acid for periods of days to months. Even subsequent adjustment of pH with ammonia or acetate simplifies a metal's speciation because the acid digestion destroys or alters such strong ligands as dissolved organic substances and a large fraction of the carbonate species.

Considerable progress has been made concerning the speciation of copper in seawater. We developed an equilibrium model for the speciation of copper in seawater. This model is based on an internally consistent set of inorganic stability constants at the ionic strength of ocean water and an empirically derived characterization of copper–organic complexes (5). At pH 8 the distribution of dissolved copper in a typical seawater sample is 56.0% $CuCO_3^0$, 1.4% $CuHCO_3^+$, 38.9% organic copper, 1.3% $CuOH^+$, and 1.4% Cu^{2+}. This distribution of species will change, of course, depending on the concentrations of organic ligands and on pH.

In designing a preconcentration scheme to operate in real time, consideration must be given to kinetic factors involved in the preconcentration step. In general, any preconcentration step will alter the trace metal's chemical equilibrium and will involve a time-dependent reaction or mass-transfer step. Of those species likely to be affected during preconcentration, the most thermodynamically and kinetically stable are typically the organically bound forms. Organically bound copper has been carefully studied and provides the best example.

Exchange of radiolabeled copper with binding sites on dissolved organic matter in seawater shows that a group of sites representing some 70 % of the binding capacity exhibits very rapid exchange (within seconds); the remaining sites exchange only slowly, with time constants on the order of hours (4). This observation suggests that naturally occurring organic copper may tend to resist complete conversion to another form in a preconcentration step, even if thermodynamically favored.

Previously (18), dissolved copper was removed from seawater in a batch extraction onto Chelex-100 ion-exchange resin in several kinetically distinct steps: The first represents virtually instantaneous uptake of kinetically labile forms such as inorganically and some organically bound copper; the second is a much more slowly removed form (time constant on the order of hours, representing about 35 % of total copper, and presumably involving an organic form of copper); and the third is an inert (probably organic and/or colloidal) form that is not removed even after several days, and accounts for approximately 15 % of total copper. Organically bound trace metals (especially copper) should, therefore, be regarded as species with a strong potential for introducing kinetic limitations into any real-time preconcentration process.

For at-sea analyses of copper, we chose HGA AAS as the primary analytical method. Although ASV had been used at sea (13), we regarded that method as an auxiliary technique to provide species-dependent information rather than unambiguous total- or dissolved-metal concentrations. In previous work (19) using shore-based analyses, comparisons were done of a cobalt–ammonium pyrrolidine dithiocarbamate (Co–APDC) coprecipitation method (6) and an APDC–methyl isobutyl ketone (MIBK) solvent extraction method (7). Both techniques gave comparable results on test samples. The coprecipitation method was chosen because all manipulations could be carried out in a single container from the initial sampling at sea to the injection into the furnace ashore (20). For at-sea analyses with high spatial resolution producing nearly real-time results, we wanted to use a preconcentration method that was amenable to automation. In principle, nearly any manual method could be automated, but, in practice, most experience with automated methods has been based on some type of flow system. The most convenient way to concentrate a constituent from a large volume of sample into a small volume of analyte is to use an extraction onto an immobilized phase. The immobilized phase must be available in very fine mesh to promote maximum contact between it and the seawater, to permit use of small columns, and to operate effectively with a short contact time. In addition, the immobilized phase must have the highest possible affinity for the trace metals of interest.

General Approach to Automated Preconcentration

Although the automated, trace-metal preconcentration system has undergone a number of design changes, the major characteristics are summarized in Figure 1. Most fluids in the all-Teflon system are moved with nitrogen pressure, and paths for both solutions and gas are selected by solenoid valves that are controlled by a 16-circuit programmable timer. In almost all cases, the preconcentration is totally free of operator intervention, and the sample is never exposed to the atmosphere during the process.

In the design in Figure 1, pressure is applied to a 1-L, acid-cleaned Teflon bottle that contains seawater drawn directly from a 12- or 24-bottle sampling rosette system; the seawater sample is then forced into a measuring reservoir. When the reservoir overflows, nitrogen pressure is applied to it, and the seawater is directed through the chromatographic column. A 0.4-μm Nuclepore filter, when wetted, passes and filters the seawater; when it is dry, it will not pass nitrogen. Thus, any nitrogen and/or air initially above the filter is released by opening a solenoid valve momentarily. The open valve allows the filter assembly to fill with and pass seawater through the column and a solenoid valve to a waste container. When the measuring reservoir is emptied, the filter stops nitrogen before it reaches and dries the column.

The column is next washed with Milli-Q water to remove sea salt; the wash is also directed to the waste container. The elution step follows; the eluting solvent is delivered from a 50-mL syringe operated by a displacement pump (Sage Instrument Co.). In this manner, a precisely measured volume of solvent is directed through the column and through a special, low-dead-volume eluant path into a sealed, acid-cleaned linear polyethylene vial.

Following elution, the eluant-waste solenoid is switched to the waste position, and a second, comparable volume of eluant is passed to clean the

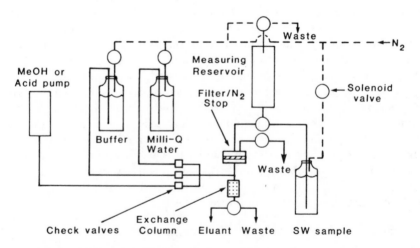

Figure 1. Schematic diagram of trace-metal preconcentration system.

column of any remaining traces of metal sample. Following this cleaning, the column is conditioned for the next seawater sample by passing a quantity of buffer solution or Milli-Q water, as the situation requires. Eluates are analyzed on-board ship with HGA AAS.

Detailed parameters for the various analyses are given in the next section, but several common concepts are discussed here. First, flow rates (especially important in the column-loading step) are all controlled in this design by nitrogen pressure and any restrictions in the fluid path. With the filter in place, no serious restriction is imposed unless the filter loads up (as may occur in high-productivity surface waters). The filter can then become a critical restriction during passage of a sample; the load rate will occasionally be slowed below that set by nitrogen pressure alone; therefore, the filter must be changed frequently. This shortcoming becomes particularly apparent when the kinetics of the chromatographic process are a factor in the recovery of metal from the sample. An improvement is to shift from individual filter membranes in a fixed holder to newly available disposable filter cartridges.

Second, the elution step introduces a dilemma: Removal of all the trace metal on the column theoretically requires an infinite volume of eluting solvent, yet the concept of preconcentration demands that eluent volume be kept to a minimum. In practice, a highly reproducible fraction of a trace metal is removed from a chelating column by the first several milliliters (*21*); thus, elution is carried out with this volume, the elution efficiency is factored into the final results, and a second portion of eluting solvent is passed to complete elution of the remaining "tail" into the waste container.

All work was carried out in a specially designed portable clean laboratory that also housed a Perkin-Elmer 5000 atomic absorption spectrophotometer with accompanying HGA-500 graphite furnace assembly and AS-40 autosampler (*22*).

Solution-Phase Copper Chelation by 8-Hydroxyquinoline

The chromatographic procedure intially investigated was reverse-phase liquid chromatography (RPLC) using C_{18}-bonded silica gel. This technique has been used previously for the preconcentration of trace metals from seawater prior to analysis by atomic absorption spectrophotometry (*23, 24*) and for the isolation of organically bound copper from seawater (*4*). These methods were modified and adapted for automation.

Total trace metals were determined after complexation in seawater by 8-hydroxyquinoline (8-HQ), isolation of the oxine complexes on the RPLC column, and subsequent elution by a small volume of methanol. As shown in Figure 2, the formation of 8-HQ complexes is strongly pH dependent. For the metals of interest, formation and isolation of the complexes at the typical pH of seawater (i.e., about 8) should, in principle, yield complete recovery. In practice, complete recovery is observed only in seawater sam-

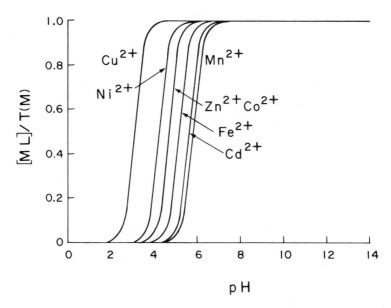

*Figure 2. Effect of pH on ratio of metal-oxine (ML) to total metal [T(M)] at
20 °C and 0.1-M ionic strength.*

ples that have been stored at pH 1.6 or lower for prolonged periods (*23, 24*)
and made basic by addition of ammonia; with fresh seawater samples,
however, the yield of copper rises as the pH is lowered.

The, the procedure adopted for determination of total copper consisted of spiking the seawater sample with an acidified 8-HQ solution, allowing the sample to digest for 15–60 min, and attaching the bottle to the
preconcentrator. The RPLC column used was a commercially available
SEP-PAK C_{18} cartridge (Waters Associates, Milford, Mass.) initially
cleaned with high-performance liquid chromatographic (HPLC)-grade
methanol, Milli-Q water, 0.3 N HCl (trace-metal cleaned), and water; because no filter was used in these experiments, column loading was stopped
manually just before the nitrogen front reached the column. Elution was
with chromatographic-grade methanol.

The methanol extract produced contained small, but varying amounts
of free and bound 8-HQ. Because this method introduced a variable background when the extract was vaporized in the graphite furnace of the
atomic absorption spectrophotometer, all extracts were spiked with an additional quantity of 8-HQ to ensure a constant matrix; the same matrix was
used in preparing the AAS standards. All pertinent parameters used in this
analysis are summarized in Table II.

Yields and recovery measurements were made by spiking acidifed, 8-
HQ-containing seawater with an additional 1 part per billion (ppb) (16

Table II. Operating Parameters for Solution-Phase Chelation and
RPLC Preconcentration of Copper

Parameter	8-HQ Method	Cu-Organic
Seawater volume	159 mL	287 mL
8-HQ concentration	3.4×10^{-4} M	—
pH	5.3	ambient
Column load rate	25–30 mL/min	25–30 mL/min
Milli-Q wash volume	10 mL	10 mL
Eluting solvent	methanol	1:1, methanol–water
Elution volume	5.0 mL	5.0 mL
Solvent wash volume	5 mL	5 mL
Milli-Q rinse volume	10 mL	10 mL

nmol/kg) of Cu^{2+}. The spiked sample was then processed through the pre-concentrator, and the extract was analyzed by HGA AAS. Comparison of the spiked and unspiked results showed that low (36–55%) recoveries were realized over a range of water-column conditions. The recoveries were only weakly flow dependent, and their averages in a given water column varied by only 2–3%. Blank determinations were carried out by passing seawater stripped of trace metals by Co–APDC coprecipitation through the precon-centrator. In the blanks, Cu recovery was not detectable; that is, the values were less than the detection limit of 0.6 nmol/kg for the 8-HQ method and less than 0.1 nmol/kg for the RPLC preconcentration method. Precision of the method is estimated as half the range of duplicate extracts taken from the same seawater sample. The precision was ±9% for the 8-HQ method and ±7% for the RPLC preconcentration method.

Isolation of organically bound copper was accomplished by using the method of Mills et al. (4). The same preconcentrator configuration employed for total metals was used, and the C_{18}-SEP-PAK cartridges were cleaned in the manner just described. Untreated seawater was passed through the column, a Milli-Q water wash followed, and elution was by a 50:50 water–methanol mixture. The column was then prepared for another run by cleaning with water–methanol and Milli-Q water.

Column-loading rates, elution volumes, and other operating parameters were similar to those used for total-metal determination, except that the volume of seawater used was 287 mL. Because a percent recovery cannot be determined for organically bound copper, only what is measured can be reported. The blank was about 0.1 nmol/kg.

Field Test of the Solution-Phase Chelation Method

Because this was a development program, field tests of the analytical system were included at several stages. Low-level atomic absorption measurements of metals had not previously been performed on board ship. Major factors that distinguish the shipboard operating environment from that found in shore-based laboratories include the prevalence of contamination from metallic structural materials, the gravitational accelerations of a rolling ship, the mechanical vibration from the ship's engine and equipment, and electrical power surges due to winches, cranes, and other large loads on the ship's generator. Thus, an analytical system that works flawlessly in a shore-based laboratory could be worthless at sea if the shipboard operating environment is not taken into account. With our system, the portable clean laboratory was designed to provide a suitable environment for the analyses.

A second reason to pursue shipboard tests along with the method development was that fresh oceanic samples may reveal chemical factors that can influence the preconcentration scheme to a different extent than can be assessed from the use of either stored oceanic samples or nearshore samples.

The first field test of the analytical system occurred on R.V. *Endeavor* cruise EN-073 in the slope-water region of the northwest Atlantic Ocean (Figure 3). The samples from this location were processed in two ways to obtain information on the chemical forms of the copper present. One preconcentrator was used with samples to which 8-HQ had been added for solution-phase chelation of the copper. A second preconcentrator was used to extract the copper on the SEP-PAK cartridge without any chelating agent added; this fraction of the copper was designated "organically bound" (4). Each sample was processed in duplicate. The observed recovery of copper at this station with the 8-HQ methods was $37.2 \pm 1.2\%$; the measured concentrations were corrected for this factor. The results (Figure 4) are shown for the total dissolved copper, the organically bound copper, and the percentage of the dissolved copper that is organically bound. The highest concentrations of copper occurred in the upper low-salinity waters. Between 100 and 400 m the copper concentration showed a slight gradient from 0.8 to 1.6 nmol/kg. The organically bound copper did not vary in the same manner as the total dissolved copper. The percentage of organically bound copper provided the clearest indication of the relationship between the chemical forms of copper and the conditions of the water column.

In the upper 70 m the organically bound copper was typically 0.7 ± 0.15 nmol/kg, whereas the total dissolved copper varied from 1 to 5 nmol/kg. Consequently, the percentage of organically bound copper increased from about 20% at 10–20 m to 80% at 70 m. The low-salinity water at 20–30 m with temperatures of 14–16 °C (Figure 3) carried relatively highly dissolved copper that was not organically bound. The maximum in the percentage of organically bound copper coincided with the seasonal oxygen minimum just beneath the chlorophyll maximum. These results indicate the ability of biological activity in the upper portion of the ocean to influence the chemical form of copper. From 100 to 400 m the copper was 30–40% organically bound. These profiles are very similar to those obtained previously (25) in the slope-water region of the North Atlantic.

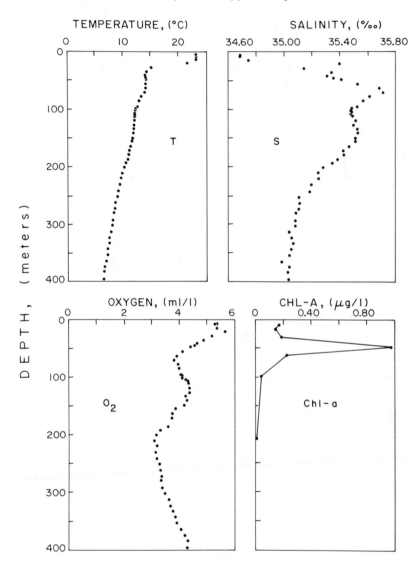

Figure 3. *Water-column characteristics in northwestern Atlantic Ocean (38° N, 71° W; August 1981).*

The results obtained in this field test demonstrated the feasibility of shipboard atomic absorption measurements for low-level metal measurements. The reproducibility of the duplicate measurements of total and organically bound copper were encouraging at the less than 1-nmol/L level. The vertical profiles show structure that is related to oceanographic factors

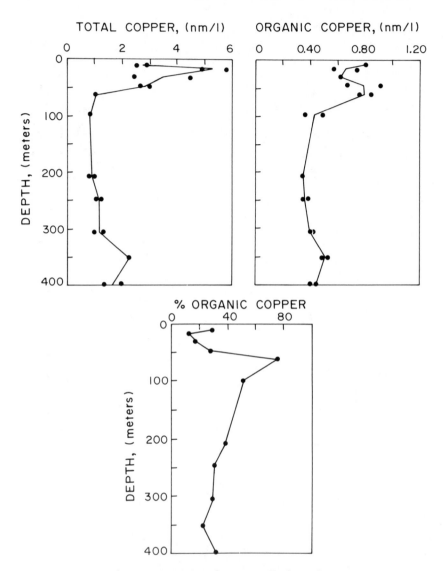

Figure 4. Distribution of total and organically bound copper at station shown in Figure 3.

with especially striking relationships indicated by the percentage of the total copper that was organically bound.

At a location in the Sargasso Sea (Station 5 of EN-073, 33° N, 72° W) a series of samples through the water column was processed to compare the solution-phase 8-HQ chelation shipboard analyses with the cobalt–APDC coprecipitation technique ashore after 2.5-months storage of the samples

(Figure 5). If the sample that gave 0.8 nmol/kg by the 8-HQ method is disregarded, the remaining values can be characterized by a slope of 1:1 between the two methods; the 8-HQ values were greater than the Co–APDC values by about 0.5 nmol/kg. This result could be explained by an uncorrected blank value in the 8-HQ method. Analytical blanks were run, and they yielded values less than the detection limit of 0.6 nmol/kg. Although the blank sample was presumably free of copper after coprecipitation, it was very high in total cobalt; such a concentration could conceivably compete for 8-HQ sites to the complete exclusion of any copper in the blank. Thus, an undetected blank probably exists in the 8-HQ method.

Immobilized-Phase Copper Chelation by 8-Hydroxyquinoline

Desirable improvements in the preconcentration method include use of a less volatile solvent than methanol and the avoidance of the need for separate addition of a chelating agent and its subsequent carryover into the eluate. The choice of a refined method includes the use of a chelating agent immobilized on a substance and used as an ion exchanger. Several immobilized chelating agent systems have been suggested and used for trace-metal preconcentrations (26–28), but the use of 8-HQ immobilized on silica gel (21) appeared to be most advantageous given the similarity to the work just described.

   Immobilization of 8-HQ on silica [or quinoline on silica gel (QSG)] was accomplished by using the method of Hill (29) as modified by Jezorek

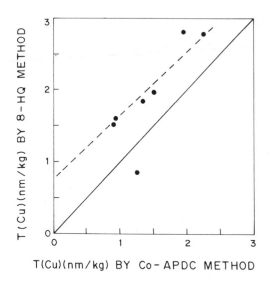

Figure 5. Comparison of 8-HQ and Co–APDC coprecipitation determinations of total copper.

and coworkers (30, 31). When finally prepared, 0.45–0.48-g quantities of the QSG were packed into SEP-PAK cartridges from which the original chromatographic substrate had been removed. Exchange capacity for copper appeared to be about 115 μmol/g, far in excess of that required to remove all trace metals from 1 L or less of seawater; under the conditions of preconcentration, negligible quantities of Ca^{2+} and Mg^{2+} are removed from seawater to serve as competitors for trace metals (21). Proton-exchange capacity was determined to be 24 meq/g.

When used with QSG, the preconcentrator configuration was that shown in Figure 1. The eluting solvent was a 2 N HCl–0.1 N HNO_3 mixture. The high proton exchange capacity of QSG necessitated the incorporation of a buffer wash. If seawater at natural pH is to be processed, the proton exchange must be eliminated; hence, a 0.68 M $NaHCO_3$ solution could be passed over the column immediately after the column's postelution cleaning with acid. For acidified samples, the buffer wash was omitted. With the exception of these modifications, the operational method was the same as that for the 8-HQ method. Samples were also taken and acidified for postcruise analysis by Co–APDC coprecipitation. The operating parameters are summarized as follows: seawater volume, 149 mL; pH, 3; column load rate, 10 mL/min; Milli-Q wash volume, 20 mL; eluting solvent, 2 N HCl–0.1 N HNO_3; elution volume, 6.1 mL; solvent wash volume, 6 mL; and Milli-Q rinse volume, 40 mL.

Blank determinations were made by rerunning previously processed "waste" samples through the preconcentrator and analyzing the eluant. The blank determination offers additional insight into the possibility that an inert or slow-to-exchange form of organically bound copper could promote low recovery. Because the blank sample was acidified seawater that had already been passed through the column, whether labile copper species were completely extracted remains questionable. If labile copper still exists, the second pass through the column should extract a significant fraction of it; the very low blank (below the detection limit of 0.28 nmol/kg), however, suggests that virtually all labile copper species were removed in the first pass, and that the copper not removed is in the form of a very inert, organic complex.

Yield and recovery experiments for copper were carried out as described in the 8-HQ method. Spike recovery was observed to be 50–75% complete when carried out at pH 3. Precision was again measured as half the range of duplicate results, or $\pm11\%$, for the same seawater sample.

Field Test of the Immobilized-Phase Chelation Method

The second field test was conducted on the R.V. *Knorr* (KN-095) in conjunction with studies of a warm core ring that formed from the Gulf Stream. Three hydrographically distinct regimes can be identified from the temperature and salinity profiles (Figure 6): a surface layer some 40–50 m

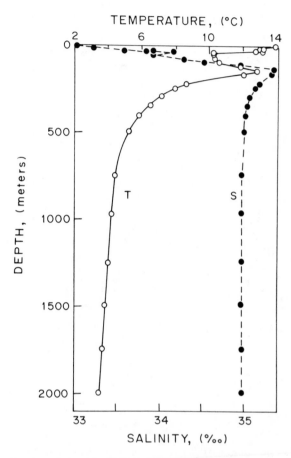

Figure 6. Temperature and salinity profile at edge of warm core Ring 82-B
(37° N, 73° W; June 23, 1982).

thick that represents entrainment of warm, low-salinity shelf water; an intermediate layer from about 50 to 150 m deep that is probably of Gulf
Stream origin; and the remaining waters suggesting a Sargasso Sea component above North Atlantic central water. The particle and chorophyll maxima were located at about 30 m.

Samples were collected throughout the upper 2000 m of the water
column, and duplicates were analyzed from most depths. Samples were
also collected, acidified, and stored for onshore analysis using the cobalt–
APDC coprecipitation method. The profile based on the shipboard analyses (Figure 7) includes a correction for the 62% recovery indicated by the
shipboard spike recoveries. The results showing a gradient in copper
through the thermocline from 2.2 nm/kg at the surface to less than 1 nm/kg
at mid-depths are consistent with other oceanic data (Table I).

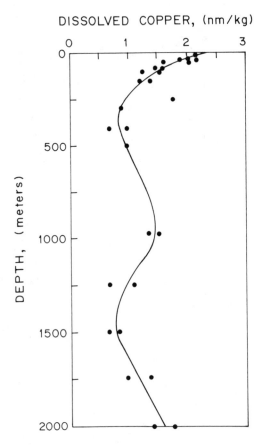

Figure 7. Distribution of dissolved copper at station depicted in Figure 6.

All samples in this series were acidified to pH 3 several hours prior to their passage through the preconcentrator. Under these conditions, free organic ligands would be expected to be fully protonated (5) and less capable of binding any added inorganic copper; free copper is expected to be readily complexed by the QSG column and, hence, entirely removed during preconcentration, if not limited by the exchange time. However, acidification to only pH 3 leaves a significant fraction (some 30%) of copper still bound to organic ligands (4). For organically bound copper to be extracted by QSG, a copper exchange between the natural organic ligand and the immobilized chelating agent must take place during contact in the column (<1 s). Because of the nonlabile character of some of the naturally occurring organic copper, some copper in seawater probably is not removed.

These factors may be used in interpreting Figure 8, which shows the

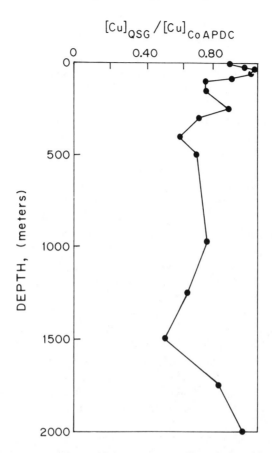

Figure 8. Ratio of copper profile in Figure 7 to that obtained by Co–APDC coprecipitation determination of total copper.

ratio of the QSG and Co–APDC profiles. In deep waters, where preset flow rates could be maintained, only 50–70% apparent recovery is noted. In the surface waters above 100 m, which show distinct shelf-water characteristics and where the active biological processes occur, an entirely different regime appears; recoveries approach 100%. Here flow rates were often limited by filters loading up in the high-productivity, highly particulate waters. Four undigested North Pacific surface samples were processed (*32*) at sea by using a Chelex ion-exchange technique (*9*). As with our surface results, good agreement was obtained (*32*) between the at-sea and postcruise digested sample results. No other data have been reported from shipboard-processed, undigested deep-water samples with which to compare our results.

Comparison of Methods

Our data and observations suggest that a complex set of variables operate to determine the effectiveness of the automated 8-HQ and QSG methods. The relative merits of the solution and immobilized phase chelation of copper by 8-HQ can be evaluated as a basis for an automated procedure preconcentration.

Chelation and Metal Recovery. The competition of protons with trace metals for chelation sites is well established. Figure 2 was constructed from stability constants determined in simple, well-defined solution media; the effect of seawater medium on the stability constants and the influence of other cations, both major (e.g., Ca^{2+} and Mg^{2+}) and minor (e.g., trace metals), was neglected. The effect of pH on the retention of metals by the preconcentration cartridges probably differs somewhat from that shown in Figure 2 because of medium effects and because of 8-HQ distortion upon immobilization.

Thermodynamically, the quantitative conversion of any naturally occurring forms of copper to the oxine form at pH 5 is favored. When labile forms of copper were converted and passed through a similar column at load rates in excess of ours (23), copper was completely recovered. Our incomplete recovery suggests that some nonlabile forms of copper do not form oxines on a time scale of several hours, and that some fraction of this residual copper is not removed by the C_{18} column at pH 5. This conclusion is consistent with the observations of Huizenga (18) which were based on retention of copper by Chelex resin. The recovery was sufficiently constant, however, that reasonable and recognizable copper profiles could be determined.

The QSG method is much less straightforward. Because of the limited time for copper exchange between ligands in the QSG column, it is desirable to work at as low a pH as possible to minimize copper binding by naturally occurring organic ligands. Manual experiments with similar columns (33) showed that copper recovery rises gradually as pH is lowered from 8 to 3, but below pH 3 falls precipitously (*see* Figure 2). That the QSG column functions well in the absence of organically complexed metal is shown by the fact that inorganic copper spikes added to pH 8 seawater could be completely recovered if the extraction was carried out before the spike could equilibrate with organic ligands. Further, as multiple spikes are added, the recovery of copper increases at higher total-copper levels. This observation is similar to the results of complexing capacity experiments (34) and suggests that free organic ligands quickly convert added copper to a nonlabile form that does not exchange readily with immobilized or dissolved 8-HQ.

The fact that the QSG method is apparently more sensitive to loading rates than the 8-HQ method suggests that, after several hours of digestion with dissolved 8-HQ, the weakly nonlabile forms of copper (18) may be converted fully to the oxine, and only the very inert form may persist. The

QSG method misses all of the inert form, but, as contact time in the cartridge is increased, an increasing fraction of the weakly nonlabile form is extracted.

Blanks. Blanks are an important consideration in evaluating the effectiveness of an analytical trace-metal method. The 8-HQ method yielded an apparent undetectable blank and a detection limit of 0.6 nmol/kg. Comparison of the 8-HQ and Co–APDC profiles suggests that the blank may be as high as 0.6 nmol/kg, and may result from the extremely high cobalt concentration of the blank sample and the background signal introduced by the high 8-HQ levels in the eluate. Although an alternate blank sample choice could eliminate any potential cobalt interference, the presence of 8-HQ in the matrix remains an unsolved problem. In contrast, the QSG method provides an almost ideal acid matrix, as shown by the low detection limit (0.28 nmol/kg) and an even lower blank.

Summary

The observations just discussed suggest that the organically bound fraction of copper plays an important role in kinetically controlling its preconcentration for near real-time measurement. They also demonstrate the need for evaluating the analytical chemistry of trace metals with freshly obtained samples, and why many of the above features would be missed in onshore laboratory experiments. Fresh samples of locally available Narragansett Bay water, for instance, contain so much particulate matter that, to run them through the preconcentrator, they would have to be prefiltered, thereby effecting changes in speciation.

Sampling in a well-defined hydrographic regime, with its characteristic biological and geochemical processes, permits use of the trace-metal profile itself as a diagnostic tool for developing new analytical methods. A certain amount of analytical method development can be done in the laboratory, but oceanographic consistency of the data provides a powerful test for assessing the reliability of the method. A field component is useful in the development process, particularly if the method is to be used for generating near real-time data.

Current results show that shipboard measurements of metals by atomic absorption spectroscopy are feasible at the nanomole per kilogram level. Additional work is needed to refine and improve an automated preconcentration technique. The effects of metal speciation on preconcentration yields may be more significant when dealing with recently collected oceanic samples than after such samples have "digested" following acidification and storage for weeks or months. Further progress can be made in the development of the automated preconcentration method by examining the behavior of several metals, in addition to copper, whose speciation in ocean waters is distinctive, and by considering some types of digestion to liberate organically bound metal prior to preconcentration.

Acknowledgments

The authors thank Alfred K. Hanson, Jr. and Francis J. Pickles for their cooperation and help in developing the automated trace-metal preconcentration procedure. The support of the U.S. Office of Naval Research through Contract N00014–81–C–0062 is gratefully acknowledged.

Literature Cited

1. Sunda W.; Guillard, R. R. L. *J. Mar. Res.* **1976**, *34*, 511–29.
2. Anderson, D. M.; Morel, F. M. M. *Limnol. Oceanogr.* **1978**, *23*, 283–95.
3. Gavis, J.; Guillard, R. R. L.; Woodward, B. L. *J. Mar. Res.* **1981**, *39*, 315–33.
4. Mills, G. L.; Hanson, A. K.; Quinn, J. G.; Lammela, W. R.; Chasteen, N. D. *Mar. Chem.* **1982**, *11*, 355–77.
5. Zuehlke, R. W.; Kester, D. R. In "Trace Metals in Sea Water"; Wong, C. S.; Boyle, E.; Bruland, K. W.; Burton, J. D.; Goldberg, E. D., Eds.; Plenum: New York, 1983; pp. 773–88.
6. Boyle, E. A.; Edmond, J. M. *Anal. Chim. Acta* **1977**, *91*, 189–97.
7. Kinrade, J. D.; Van Loon, J. C. *Anal. Chem.* **1974**, *46*, 1894–98.
8. Danielsson, L.; Magnusson, B.; Westerlund, S. *Anal. Chim. Acta* **1978**, *98*, 47–57.
9. Kingston, H. M.; Barnes, I. L.; Brady, T. J.; Rains, T. C.; Champ, M. A. *Anal. Chem.* **1978**, *50*, 2064–70.
10. Riley, J. P.; Taylor, D. *Anal. Chim. Acta* **1968**, *40*, 479–85.
11. Sturgeon, R. E.; Berman, S. S.; Desaulniers, J. A. H.; Mykytiuk, A. P.; McLaren, J. W.; Russell, D. S. *Anal. Chem.* **1980**, *52*, 1585–88.
12. Nurnberg, H. W.; Valenta, P. In "The Nature of Seawater"; Goldberg, E. D., Ed.; Dahlem Konferenzen: Berlin, 1975; pp. 87–136.
13. Huizenga, D. L.; Kester, D. R. *J. Electrochem.* **1984**, *164*, 229–36.
14. Bruland, K. W.; Franks R. P.; Knauer, G. A.; Martin, J. H. *Anal. Chim. Acta* **1979**, *105*, 233–45.
15. Sturgeon, R. E.; Berman, S. S.; Desaulniers, A.; Russell, D. S. *Talanta* **1980**, *27*, 85–94
16. Bernhard, M.; Goldberg, E. D.; Piro, A. In "The Nature of Seawater"; Goldberg, E. D., Ed.; Dahlem Konferenzen: Berlin, 1975; pp. 43–68.
17. Corsini, A.; Wan, C. C.; Chiang, S. *Talanta* **1982**, *29*, 857–60.
18. Huizenga, D. L., Ph.D. Thesis, Univ. of Rhode Island, Kingston, R.I., 1982.
19. Kester, D. R.; Hittinger, R. C.; Mukherji, P. In "Ocean Dumping of Industrial Wastes"; Ketchum, B. H.; Kester, D. R.; Park, P. K., Eds.; Plenum: New York, 1981; pp. 215–32.
20. Huizenga, D. L., "The Cobalt–APDC Coprecipitation Technique for the Preconcentration of Trace Metal Samples", Technical Report 81-3, Univ. of Rhode Island, Kingston, R.I. 1981; 93 pp.
21. Sturgeon, R. E.; Berman, S. S.; Willie, S. N.; Desaulniers, J. A. H. *Anal. Chem.* **1981**, *53*, 2237–40.
22. Zuehlke, R. W.; Pickles, F. J.; Kester, D. R., unpublished data.
23. Watanabe, H.; Goto, K.; Taguchi, S.; McLaren, J. W.; Berman, S. S.; Russell, D. S. *Anal Chem.* **1981**, *53*, 738–39.
24. Sturgeon, R. E.; Berman, S. S.; Willie, S. N. *Talanta* **1982**, *29*, 167–71.
25. Hanson, A. K., Ph.D. Thesis, Univ. of Rhode Island, Kingston, R.I., 1981.
26. Burggraf, L. W.; Kendall, D. S.; Leyden, D. E.; Pern, F. J. *Anal. Chim. Acta* **1981**, *129*, 19–27.
27. Sugawara, K. F.; Weetall, H. H.; Schucker, G. D. *Anal. Chem.* **1974**, *46*, 489–92.
28. Jezorek, J. R.; Freiser, H. *Anal. Chem.* **1979**, *51*, 366–73.
29. Hill, J. M. *J. Chromatogr.* **1973**, *76*, 455–58.
30. Fulcher, C.; Crowell, M. A.; Bayliss, R.; Holland, K. B.; Jezorek, J. R. *Anal. Chim. Acta* **1981**, *129*, 29–47.

31. Jezorek, J. R.; Fulcher, C.; Crowell, M. A.; Bayliss, R.; Greenwood, B.; Lyon, J. *Anal. Chim. Acta* 1981, *131*, 223–31.
32. Bruland, K. W. *Earth Planet. Sci. Lett.* 1980, *47*, 176–98.
33. Hanson, A. K., personal communication.
34. Plasvic, M.; Krznaric, D.; Branica, M. *Mar. Chem.* 1982, *11*, 17–31.
35. Moore, R. M. *Geochim. Cosmochim. Acta* 1981, *45*, 2475–82.
36. Danielsson, L. G. *Mar. Chem.* 1980, *8*, 199–215.
37. Boyle, E. A.; Huested, S. S.; Jones, S. P. *J. Geophys. Res.* 1981, *86*, 8048–66.
38. Moore, R. M. *Earth Planet. Sci. Lett.* 1978, *41*, 461–68.
39. Bruland, K. W.; Franks, R. P. In "Trace Metals in Sea Water"; Wong, C. S.; Boyle, E.; Bruland, K. W.; Burton, J. D.; Goldberg, E. D., Eds.; Plenum: New York, 1983; pp. 395–414.

RECEIVED for review September 6, 1983. ACCEPTED March 12, 1984.

On-Line Shipboard Determination of Trace Metals in Seawater with a Computer-Controlled Voltammetric Instrument

CESAR CLAVELL and ALBERTO ZIRINO

Marine Environment Branch, Naval Ocean Systems Center, San Diego, CA 92152

Improvements on a computer-controlled instrument for performing trace-metal analysis by anodic stripping voltammetry are presented and discussed. The ease of operation of the instrument has been improved by the use of carbon-disc electrodes and spool-type Teflon valves. The device has been used to measure Zn, Cd, Pb, and Cu in estuarine waters; recently an attempt was made to measure Cu in surface oceanic waters. Although the sensitivity and accuracy of the instrument appear insufficient for the measurement of Cu in oceanic surface waters, the approach appears promising for future work.

ELECTROANALYTICAL INSTRUMENTS THAT WOULD DIRECTLY MEASURE heavy metals in seawater seemed to us over a decade ago (1) to be the only approach that could adequately provide the spatial and temporal resolution necessary to study the dynamic properties of metals at or near the ocean's surface. Thus, for a number of years, we have worked on flow-through analytical techniques based on anodic stripping voltammetry (ASV). This chapter updates earlier methods (2–4) and discusses several improvements in the design of the automated anodic stripping voltammetry (AASV) instrument (5). These improvements were implemented to simplify and standardize previously difficult tasks. Since its original construction in 1977, the instrument was tested on two open ocean voyages and on several estuarine surveys on both coasts of the United States. A sample of the results obtained during two of these surveys has been included to illustrate the potential use of this instrument. The data are not interpreted in terms of environmental or oceanographic significance.

0065-2393/85/0209-0139$06.00/0

The application of the ASV technique to the analysis of seawater has been reviewed (6–9). Its advantages over other trace-metal methods are several: (1) sea salts do not interfere; (2) a high degree of sensitivity and selectivity can be achieved through electrolytic preconcentration, thus the analysis can be carried out directly in seawater, and few or no reagents need to be added; (3) the measurement is easily automated; (4) relatively inexpensive, rugged, and portable instrumentation can be built for field use; (5) more than one metal may be measured at one time with a sensitivity approaching parts per trillion.

The principles of both anodic and cathodic stripping voltammetry are well known, and detailed discussions can be found elsewhere (10, 11). However, a brief description of the ASV technique as applied to seawater will be presented here.

ASV Technique Applied to Seawater

A basic voltametric instrument consists of a cell, a potentiostat, and a recorder. Chemical reactions are caused by applying a potential to an electrode in a cell. The current, which flows under the experimental conditions, is monitored with the recorder. The cell generally consists of three electrodes placed in seawater: (1) the working electrode, usually a hanging drop of Hg or a Hg-plated graphite or glassy-carbon surface; (2) the reference electrode, made of either calomel or Ag wire anodized to form AgCl; and (3) a counterelectrode made of pure Pt wire.

In a typical analysis, oxygen is initially removed from the seawater sample by bubbling with an inert gas, then a preselected potential is applied to the working electrode with respect to the reference. At this potential metals in the seawater are induced to amalgamate with the Hg in or on the electrode. During this step, the transfer of metals to the electrode is increased by either stirring the solution or rotating the electrode. Following accumulation, generally called pre-electrolysis, the potential is linearly decreased to allow for the spontaneous oxidation of metals out of the amalgam. (The potential is controlled by allowing the current to flow through the counterelectrode.) As the metals oxidize (strip out of the amalgam), the current being monitored peaks at or near the oxidation–reduction potential of the metal in the medium being analyzed. The height of the peak is proportional to the concentration.

In seawater, Zn, Cd, Pb, and Cu amalgamate quasi-reversibly with Hg. Thus, these metals are routinely analyzed. Nonroutine methods for the analysis of many more transition metals in seawater by either cathodic or anodic stripping are being developed daily. Also, because of the low concentration of trace metals in oceanwater ($\ll 1$ μg/L), the stripping process is routinely carried out in the differential pulse (DP) mode, a process that enhances the metal-related component of the current over the background. Calibration of the cell (and of the analysis) is carried out by adding known quantities of the metal in the ionic form to the seawater sample and repeating the analysis. Complications and problems peculiar to the analysis of seawater have been discussed elsewhere (6–9).

The automated ASV instrument described in this chapter is of a flow-through design. The advantages of flow systems for on-line monitoring have been discussed elsewhere (12). They include a reduction of contamination and adsorption errors (a result of minimum sample handling and of the self-cleaning properties of flow systems), effective and reproducible mass transport, and inherent precision. These advantages stem from the fact that flow systems inherently tend to combine sampling and analysis procedures. Also, flow systems are generally closed environments that make it possible to analyze for minute quantities of trace metals in notoriously "dirty" metal environments such as ships.

Experimental

Our computer-controlled instrument is housed in two carrying cases (Figure 1). The "dry" box (Figure 1, right) houses the computer, the potentiostat, electronic solid-state displays of the various instrumental parameters measured, valve override switches, and maybe a small dot-matrix printer and a small recorder. The "wet" box (Figure 1, left) houses the electrolysis cell, valves, a cell pump, a standard addition pump, and containers for sample, standard, and $Hg(NO_3)_2$ solutions. The design of the "wet" box (Figure 2) essentially follows an earlier design (4). Improvements include the substitution of isotropic carbon discs in a flow-through holder for the original tubular graphite electrodes and Teflon spool valves

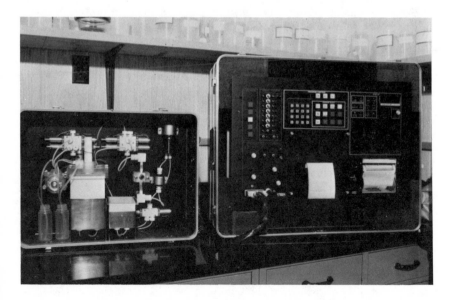

Figure 1. Computer-controlled flow-through ASV system.

for the Teflon diaphragm valves. Minor improvements include light-emitting diode (LED) displays for cell potential, lapsed time, and cell pump speed.

Electrolysis Cell.   The cell (Figure 2) consists of an electrode assembly; two rectangular Teflon reservoirs, 1- and 0.5-L, for the sample and the 2×10^{-5} M $Hg(NO_3)_2$ solution, respectively; a cell pump (peristaltic, variable speed, Cole Parmer 7545-15); and three in-house-built, solenoid-operated spool valves that allow for the transfer of solutions via Teflon tubing. The cell consists of essentially two solution loops: one recirculates the Hg solution through the electrodes, and the other recirculates the sample through the electrodes. Circulation of the Hg solution and sample through the electrodes is sequential and mutually exclusive. Two more peristaltic pumps are also external to the "wet" box. They connect the seawater reservoir to input and exhaust lines.

In a typical analysis cycle, the seawater reservoir is filled, rinsed, and refilled by alternately activating the input and exhaust pumps. Pure CO_2 is continually passed through the sample to remove O_2 and to lower the pH to 4.9. Near the end of this fill–rinse–refill–O_2 removal step, the electrodes are charged to their preselected potential, and the Hg solution is directed through them for about 6 s. Subsequently, Valves 1 and 2 (Figure 2) are switched to the "B" side, and the seawater sample is pumped through the electrode assembly and recirculated to the sample reservoir via Valve 3. During this "pre-electrolysis" step metals are amalgamated with the Hg film previously plated on the working electrode. Typically, we use a 4-min pre-electrolysis. At the end of this period, the cell pump is stopped, and the sample and electrodes are allowed to equilibrate for 15 s followed by electrolytic stripping. During this period, which takes about 3 min, an addition of standard may be made to the sample in the reservoir and the analysis repeated, or the sample may be flushed if a standard addition has already been made. Details of the analysis were described previously (3, 4).

Electrode Assembly.   This device consists of a specially machined Teflon electrode holder, two disc electrodes (only one is energized), and a clamp machined from acrylic plastic (Figure 3). The electrode discs are of low-temperature isotropic carbon alloyed with SiC (Carbo-metics, Austin, TX). They were originally developed for use in artificial heart valves (14), and are approximately 1.6 cm in diameter and 1.25 mm thick, and have the surface properties of glassy carbon. Treatment of the discs requires only polishing to a high lustre with diamond grinding compounds of 14,000 and 50,000 mesh.

The discs are seated in the electrode holder approximately 2 mm apart. Electrical contact to the back of the discs is made by inserting a Pt wire through a rubber stopper with the aid of a 16-gauge hypodermic needle. When the needle is removed, about 1 cm of wire is left exposed from each end of the stopper. One of the exposed wire tails is then folded in a loop against the face of the rubber stopper (Figure 3). The stopper is then pressed against the disc by the clamp, and connections to the potentiostat are made at the back of the stopper via alligator clips. The acrylic clamp presses against the rubber stoppers, which in turn provide a liquid-tight seal as well as pressure for the electrical connection. Only one electrode at a time was connected to the potentiostat although both electrodes can be connected to double the active surface area. However, whether connecting both electrodes would actually result in an increase of signal over background is unclear. A more logical use of the electrodes is in a differential application (13).

The reference electrode was previously described in detail (4, 5). The effectiveness of its configuration has been proven by over 6 years of stable operation during which it has been cycled through many periods of complete dryness followed by rehydration.

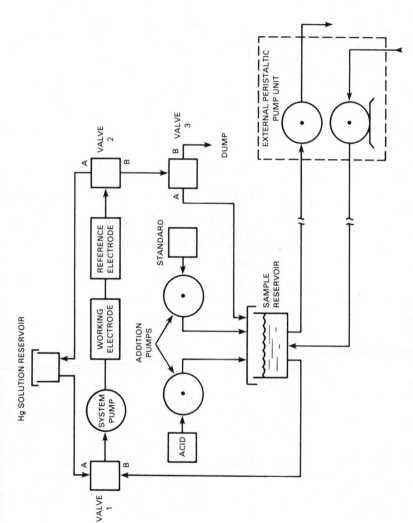

Figure 2. Diagrammatic representation of the "wet" box.

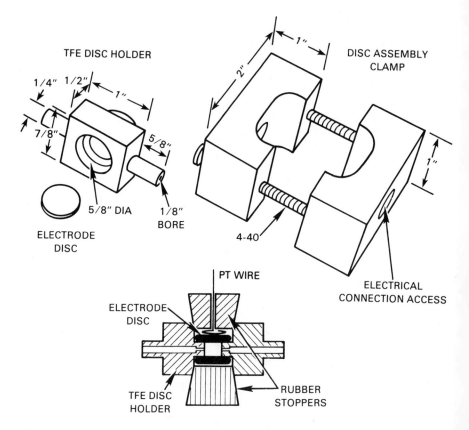

Figure 3. Diagram of the electrode assembly.

Valves.   The commercially available diaphragm valves described elsewhere (3, 4) have been replaced with in-house-built spool valves. The advantage of a sliding spool design is its resistance to damage by hard particulates in the sample, such as sand and silt, which can be present in large amounts during estuarine profiling operations. The spool valves, shown in Figure 4, consist of a rod (spool) that travels in a carbon-filled Teflon block. The block itself has three ports, two on one side and one at the center of an adjoining side. The rod has been machined down at the center (hence, "spool") to allow for the passage of fluids. By positioning the spool at either end of the block, seawater is channeled from either the left or the right port to the center port.

The valve body is machined from a carbon-filled Teflon block (75% Teflon, 25% carbon) (Ebolon), and the spools are made of polycarbonate with Viton O rings for seals. Teflon tubing is attached to the ports via Teflon O-ring compression fittings. Two 24-V DC push-type solenoids position the spool.

Hardware*.   The electronics portion of the instrument (Figure 1) consists of a microcomputer configured on three circuit boards: a central processing unit

---

*Details on both the hardware and the software are available from the authors on request.

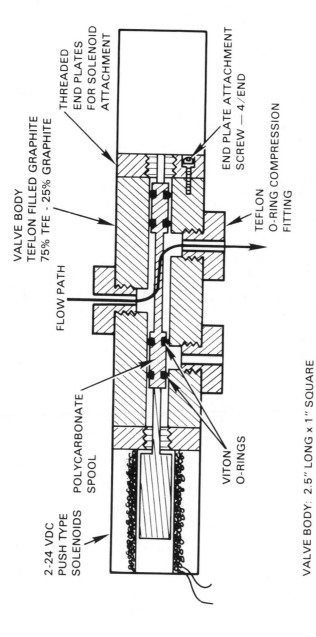

THREADED
END PLATES
FOR SOLENOID
ATTACHMENT

VALVE BODY
TEFLON FILLED GRAPHITE
75% TFE - 25% GRAPHITE

END PLATE ATTACHMENT
SCREW — 4/END

FLOW PATH

TEFLON
O-RING COMPRESSION
FITTING

POLYCARBONATE
SPOOL

VITON
O-RINGS

2-24 VDC
PUSH TYPE
SOLENOIDS

VALVE BODY: 2.5" LONG x 1" SQUARE

Figure 4.   A cross section of the AASV spool valve.

(CPU) board containing a Motorola 6802 microcomputer and support chips, a memory board containing both the random access memory (RAM) and the systems applications program on electrically programmable read-only memory (EPROMS), and an input/output (I/O) board that contains all interface circuitry. The control functions are implemented through eight peripheral adaptors with a total of 128 I/O lines, which can be programmed as inputs or outputs, and 32 interrupt lines, half of which can be programmed as inputs or outputs. Interfacing to the high-current components (motors, solenoids) is made through 8-A Darlington transistors. Two independent clocks are available: one for time of day and some timing sequences, and the other for a display of elapsed time. Data input to the computer is via a 16-key keyboard. Experimental parameters are entered in decimal form and converted to 16-bit integers for internal use. Computer outputs are presented on a 40-column alphanumeric printer. Cell currents are displayed on a 5-in. strip-chart recorder or on a 12-in. recorder external to the "dry" box.

The potentiostat functions are provided by a Princeton Applied Research Model 174 (PAR 174) potentiostat repackaged to reside within the instrument. During stripping, which is carried out in the differential pulse (DP) mode, data from the potentiostat are accessed by the microcomputer by digitizing the analog output of the PAR 174 with a 12-bit analog-to-digital converter with ± 5-mV resolution on a ± 10-V input range.

Software. The software developed consists of two major subroutines: one is the controlling program that carries out the experimental instructions entered via the keyboard; the second is a computing routine for calculating concentrations from peak currents. The controlling program "walks" the operator through the initialization of the system by the use of interactive questions and answers. The program currently has four different modes of operation, including one for medium exchange (13). Once the operator determines the proper mode and initializes the system, the start command is given and the analysis proceeds under computer control, without further operator intervention.

The computing software program performs the peak finding and identification tasks and determines the concentration of each metal in the sample at the end of a standard addition cycle. Concentration values for each metal, sampling time, date, and sequential sample number are printed out.

A typical printout of the questions is shown in Figure 5. Questions 0 to 5 set the experimental parameters. Delays in the sequencing of the valves (Questions 2 and 3) serve to minimize mixing. Question 6 determines the amount of standard to be added by timing the operation of the standard addition pump. Similarly, Questions 7 and 8 determine whether, and how much, acid is to be added to the sample. Question 11 asks whether the sample is to be recycled through the electrodes and the reservoir or simply put through once and discarded (4). Recycling takes a smaller volume of sample but risks internal contamination. However, test runs that measured the same sample consecutively from two to six times have shown no contamination problems with this system. The instrument has thus been used almost exclusively in this mode. Likewise, runs made with a sample followed by no sample, only Hg plate, have shown no carry-over from one sample to the next regardless of the metal concentrations. Because of the very short Hg plating time, contamination from recycling the Hg has not been observed.

Question 12 asks, in effect, which sample of a sequence is to be retained for standard addition. A standard addition is made to each sample when a 2 is input. An answer of n will make the machine do $n-1$ analyses before a sample is retained for standard addition. Questions 13–16 specify the metal concentrations of the standard, and Questions 18–23 provide the computing program with the values with which to bracket peaks and calculate peak areas. The Cu, Cd, and Pb concen-

```
ENTER YEAR                                19 PB LOWER LIMIT (IN VOLTS)=.62
        YEAR=1982
ENTER JULIAN DATE                         22 CU UPPER LIMIT (IN VOLTS)=.14
        JULIAN DATE=289
                                          21 CU LOWER LIMIT (IN VOLTS)=.45
SET REAL-TIME CLOCK TO CURRENT TIME
     THEN PUSH "GO"                       24 ZN UPPER LIMIT (IN VOLTS)=UNUSED

ENTER PARAM FILE (NEW=P, OLD=0)           23 ZN LOWER LIMIT (IN VOLTS)=UNUSED
        PARAM FILE (NEW=P, OLD=1=0
DO YOU WANT PARAM LIST PRINTED OUT?       26 RESV FILLING TIME (SEC)=25
ENTER YES=1, NO=0
        YES=1, NO=0                       25 INITIAL POTENTIAL=1.2
0 HG PLATTING TIME (MIN/SEC)=7
                                          FINAL POTENTIAL=-.1
1  SAMPLE PLATE TIME (MIN/SEC)=200
                                          DO YOU WANT TO LOAD THIS FILE?
2  VALVE DELAY1 (SEC)=1                    ENTER YES=1, NO=0
                                                 YES=1, NO=0=1
3  VALVE DELAY2 (SEC)=2                     FILE LOADED

4  SCAN TIME (MIN/SEC)=145                  ***********************************
                                           * IF THERE ARE NO CORRECTION AND  *
5  FLUSHING TIME (SEC)=X0                   * IF THE POTENTIOSTAT PARAMETERS  *
                                           * ARE SET, PUSH "GO" OTHERWISE     *
6  STD ADDITION TIME (SEC)=16              * PUSH "HALT" TO MAKE CHANGES      *
                                           * THEN PUSH "GO".                  *
7  ACID? (YES=1, NO=0)=0                    ***********************************

8  ACID ADDITION TIME (SEC)=UNUSED         DATA OUTPUT **CONCENTRATIONS IN PPB**

9  PURGE DELAY TIME (MIN/SEC)=200          *001   1982   209    1449:51
                                              CU    .54   PB    .40   CD   .18
10 ZINC ANALYSIS (YES=1, NO=0)=0
                                           *002   1982   209    1502:10
11 RECIRC MODE (YES=1, NO=0)=1                CU   1.38   PB    .10   CD   .18

12 CYCLE # FOR STD ADD =.2                 *003   1982   209    1512:59
                                              CU   1.78   PB    .05   CD   .28
13 CD STANDARD CONC - IN PPB=.4
                                           *004   1982   209    1523:50
14 PB STANDARD CONC - IN PPB=1.0              CU    .00   PB    .00   CD   .00

15 CU STANDARD CONC - IN PPB=2.0           *005   1982   209    1534:40
                                              CU   1.98   PB    .04   CD   .08
16 ZN STANDARD CONC - IN PPB=UNUSED

18 CD UPPER LIMIT (IN VOLTS)=.55

17 CD LOWER LIMIT (IN VOLTS)=.9

20 PR UPPER LIMIT (IN VOLTS)=.4
```

Figure 5. Printout of the initiation sequence and the final calculated concentrations.

trations calculated during a survey of San Diego Bay are also shown in Figure 5. Sample 4 shows no calculated concentrations, which means that the standard addition did not result in an increase in peak height. This observation is common in waters rich with organic matter (7). However it may also mean that the computing software failed to identify the peaks (vide infra).

   Procedure.   A typical procedure such as that carried out on the Naval Ocean Systems Center's (NOSC) instrumented survey platform is as follows. The main sampling pump is started, and a continuous source of seawater is maintained aboard the craft. (This pumping system, which supplies large volumes of seawater for many different kinds of analytical measurements, is built into the survey platform and not discussed here.) The ASV operator then places fresh Hg solution in the Hg reservoir, selects a prepolished electrode, inserts it into the electrode holder, and turns the CO_2 valve on. While O_2 is being removed, the system automatically starts to circulate the Hg through the electrodes at rest potential. At this point, the operator initializes the experiment by answering the questions and at the end pushing the GO button. The instrument now begins an analysis cycle by filling, flushing, and refilling the sample reservoir, and applying the preselected potential to the electrodes. Operator interaction is no longer required until the analyses have been completed and the instrument is manually shut down.

Results and Discussion

A flow-through ASV instrument needs flexibility to take advantage of new analytical procedures as they become available. The use of a microcomputer for both system control and data manipulation satisfies this need. Because programs can be easily modified or updated, changes can be made in the field to accommodate unforeseen problems. Similarly, by reprogramming, the hardware package can evolve fresh capabilities as new techniques, sensors, and equipment become available. For example, the addition of another reservoir and associated valves makes it possible to alter the medium present in the electrodes during the stripping cycle. Medium exchange may permit the analysis of metals other than Zn, Cd, Pb, and Cu. For example, by stripping into $HClO_4$ solution, Hg can be detected in seawater (15). Another benefit is that the computing power of the instrument permits data reduction in real time, which allows sampling and other operational decisions to be made in the field.

The instrument described here is in routine use on the NOSC survey platform and has seen service in the open ocean; in King's Bay, Georgia; Charleston, North Carolina; and San Diego Bay, California. Its operation by nonspecialists in trace-metal analysis, a notoriously difficult field, has been made possible by fully automating the procedure, including sampling, and by replacing the original wax-impregnated tubular mercury graphite electrodes (TMGE) with the isotropic carbon discs. The discs, although not as sensitive as the original tubular electrode, can be simply repolished to restore their sensitivity, thereby avoiding the difficult preparation of the TMGE (7).

Figure 6 presents values of Zn, Cd, and Pb concentrations measured in surface and bottom waters of King's Bay, Georgia, with the automated instrument. The measurements were conducted in the field, with the instrument on the survey platform. King's Bay is a shallow, sand- and mud-bottomed estuary lying just inshore of Cumberland Island on the south Georgia coast. Although the estuary is subjected to considerable fluvial runoff, the area is still relatively pristine with no obvious sources of anthropogenic metal inputs, except perhaps, the few ships anchored there. However, the area is subject to a considerable amount of both maintenance and construction dredging. The relatively similar concentrations found both at the surface and bottom attest to the vigorous tidal mixing present. The concentration levels of Zn and Cd are similar to those found in San Diego Bay (16). However, the Pb values are higher by a factor of 2–3. No attempt was

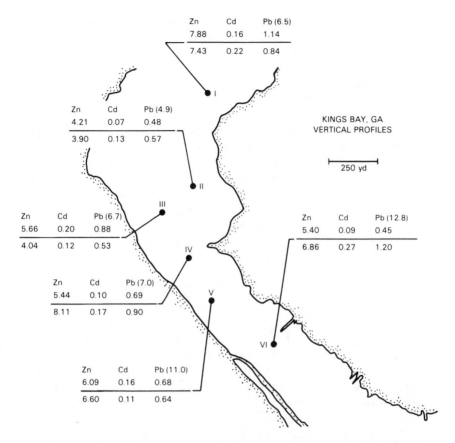

Figure 6. Concentrations (in μg/L) of Zn, Cd, and Pb in King's Bay, Georgia, during houseboat survey of October 28, 1982. Upper numbers are surface values; lower numbers bottom values. Depth to bottom is in parentheses.

made to titrate for possible organic ligands on this survey. The measurements represent that portion of the total-metal concentration termed "electroactive" or "labile" (7). Circumstantial support can be found in the observation that ASV-measured Cu concentrations correlate closely with potential values measured with a Cu ion-selective electrode (ISE) (17).

In a survey of San Diego Bay waters we used the computer-controlled instrument in conjunction with the Orion Cu(II) ISE to travel in and out of waters that were relatively high in Cu concentration (4). The potentiometric and voltametric data correlated very well. These data suggest that the combined use of the ASV instrument and the ISE may provide a way of obtaining quantified gradients of Cu activity in seawater.

The question of speciation and electrochemical availability is beyond the scope of this chapter. A detailed discussion of this topic can be found elsewhere (7).

In 1981, on the NOSC-organized *Varifront II* expedition, the computer-controlled instrument was used aboard the USNS R./V. *DeSteiguer* to measure Cu in surface waters of Baja, California. The voyage consisted of two separate "legs" to and from Acapulco, Mexico. Water for the underway experiments was collected from 1 m below the bow of the *DeSteiguer* by lowering a specially built conduit through the bow transducer well, which extends through the keel. A 70-m piece of Teflon-lined rubber hose was routed from the conduit inlet under the bow to the AASV instrument in the lab. Cu concentrations were analyzed within 30 min of sampling. Twelve samples from the Cabo Corrientes area were collected directly from the sampling stream before the water reached the AASV instrument for a direct comparison between AASV and the ammonium pyrrolidine dithiocarbamate (APDC) values (18).

Cu values (in micrograms per liter) obtained with the AASV instrument on Leg 1 are shown in Figure 7. The values range from a maximum of 0.61 μg/L off Point Eugenia to a minimum of 0.01 μg/L south of Cabo Corrientes. The distribution of values shows two groupings. Samples collected above 19° N latitude cluster about a mean of 0.20 + 0.16 (n = 31); samples collected from below 19° N average 0.08 + 0.05 (n = 24). The groupings coincide with a marked 3 °C temperature increase occurring south of the Baja Peninsula 200 miles west of Cabo Corrientes; therefore, the offshore waters south of Cabo Corrientes may be stratified and depleted in Cu as well as nutrients.

The average low value (south of 19° N latitude) is in good agreement with Cu values previously measured in this area (19), thus the sampling system was "clean." The values north of 19° N are, on the average, higher by at least a factor of 2. Although it is difficult to compare concentrations measured by different techniques (7), too many concentration "spikes" exist in our Cu data. These spikes may in fact point to a deficiency in the instrument rather than to natural phenomena. Although the current instrument operates well in estuarine waters, it is in fact operating at the limit of detection in oceanic waters. Figure 8 shows a recorder trace of the stripping scan made on Leg 1. At these low levels, the Cu "peak" is merely a shoulder on the Hg wave. This condition makes it difficult for the computing program to "find" the peak and calculate its area. Similarly, natural surfactants that tend to distort the baseline (7) can also cause large errors, particularly when the metal signal is low. This situation is also shown in Table I, which presents the results of the direct intercomparison between the AASV values and Cu concentrations obtained by cocrystallization. The data indicate that although seven of the AASV numbers are in the same range as the APDC values, four are at or below detection, without a similar

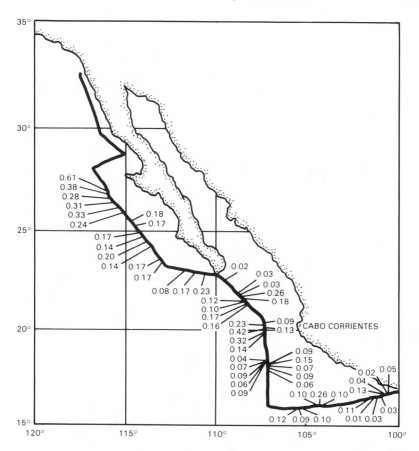

Figure 7. Cu concentrations (in μg/L) measured underway by computer-controlled ASV during Leg 1 of Varifront II expedition.

decrease in the APDC values. Therefore, the computing program could not draw a proper baseline under the original Cu signal.

Conclusion

The computer-controlled ASV instrument described is a more reliable and versatile device than previous instruments. Its reliability is supported by its long-term record in estuarine waters where it routinely measures trace-metal concentrations in near real time.

At present, the instrument is not sensitive enough to measure Cu in surface waters of the ocean with sufficient precision or accuracy. Nevertheless, because of the many experimental variables involved with ASV, the required sensitivity probably can be obtained with further effort.

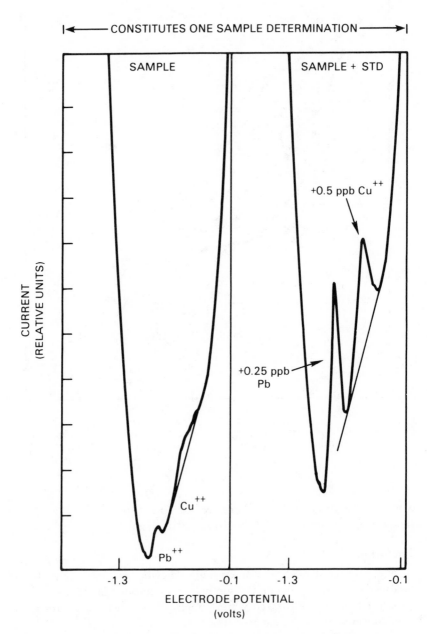

Figure 8. Recorder trace (scan rate, 10 mV/s) of the output of the analysis of a surface sample measured during Leg 1. Calculated concentrations are: Cu, 0.07 μg/L; and Pb, 0.02 μg/L.

Table I: Cu Concentration in Seawater by AASV and APDC

Sampling Date; Time	APDC	AASV
1/21; 22:56	0.06	0.07
1/21; 21:32	0.06	0.06
1/19; 20:38	0.08	0.09
1/16; 23:16	0.06	0.14
1/23; 18:42	0.06	0.11
1/23; 21:39	0.06	—
1/24; 00:03	0.09	0.22
1/25; 09:48	0.06	0.05
1/25; 15:42	0.07	—
1/25; 22:36	0.06	0.10
1/26; 01:47	0.06	0.01
1/26; 02:43	0.06	0.01

NOTE: All samples were taken in 1981. All results are in micrograms per liter.

Acknowledgments

We thank E. A. Boyle and his staff for carrying out Cu measurements in conjunction with ours.

Literature Cited

1. Zirino, A. *Cron. Chim.* 1973, March.
2. Lieberman, S. H.; Zirino, A. *Anal. Chem.* 1974, *46*, 20.
3. Zirino, A.; Lieberman, S. H. In "Analytical Methods in Oceanography," Gibb, T. R. P., Ed.; ADVANCES IN CHEMISTRY SERIES No. 147, American Chemical Society: Washington, D.C., 1975; p. 82.
4. Zirino, A.; Lieberman, S. H.; Clavell, C. *Environ. Sci. Technol.* 1978, *12*, 73.
5. Clavell, C. "Microcomputer Assisted Flow-Through ASV System"; In Proceedings of the Oceans 77 Conference, Los Angeles, Calif., Oct. 17, 1977.
6. Turner, D. R.; Whitfield, M. In "Marine Electrochemistry"; Whitfield, M.; Jagner, D., Eds.; Wiley-Interscience: Chichester, 1981; Chap. 2.
7. Zirino, A. In "Marine Electrochemistry"; Whitfield, M.; Jagner, D., Eds.; Wiley-Interscience: Chichester, 1981; Chap. 10.
8. Wang, J. *Environ. Sci. Technol.* 1982, *16*, 104A.
9. Batley, G. E. *Mar. Chem.* 1983, *12*, 107.
10. Brainina, K. Z. "Stripping Voltammetry in Chemical Analysis"; Wiley: New York, 1974.
11. Vydra, F.; Stulik, K.; Julakova, E. "Electrochemical Stripping Analysis"; Wiley: New York, 1976.
12. Wang, J. *Am. Lab.* 1983, July, 14–23.
13. Wang, J.; Ariel, M. *J. Electroanal. Chem.* 1977, *85*, 289–97.
14. Bokros, J.; LeGrange, C.; Schoen, F. J. In "Control of Structure of Carbon for Use in Bio-Engineering"; Walker, P. L.; Thrower, P. A., Eds.; Dekker: New York, 1972; Vol. 9, pp. 103–71.
15. Fukai, R.; Huynh-Ngoc, L. *Anal. Chim. Acta* 1975, *83*, 375.
16. Kenis, P. R.; Clavell, C.; Zirino, A. "Automated Anodic Stripping Voltammetry for the Analysis of Cu, Zn, Pb, and Cd for Environmental Monitoring," NOSC TR 243, June 1978.

17. Zirino, A.; Seligman, P. F. *Mar. Chem.* **1981**, *10*, 249–55.
18. Boyle, E. A.; Edmond, J. M. *Anal. Chim. Acta* **1977**, *91*, 189–97.
19. Boyle, E. A.; Huested, S. S.; Jones, S. P. *J. Geophys. Res.* **1981**, *86*, 9, 8048–66.

RECEIVED for review November 29, 1983. ACCEPTED May 7, 1984.

9

A Multiple-Unit Large-Volume In Situ Filtration System for Sampling Oceanic Particulate Matter in Mesoscale Environments

JAMES K. B. BISHOP, DANIEL SCHUPACK, ROBERT M. SHERRELL, and MAUREEN CONTE

Lamont–Doherty Geological Observatory of Columbia University, Palisades, NY 10964

The multiple-unit large-volume in situ filtration system (MULVFS) was developed and successfully deployed in the upper 1000 m in the northwest Atlantic for a study of warm core Gulf Stream Rings. The MULVFS is capable of simultaneously collecting 12 or more samples of size-fractionated >1-μm particulate matter, each filtered from 10,000–20,000 L of seawater. Total ship time required per station is 6–8 h. This chapter discusses advances in the designs of pump units, filter holders, and the electromechanical cable of the MULVFS. Improved methods of filter preparation and shipboard handling of samples are also described. The performance of the MULVFS is illustrated by particulate dry weight and Ca data from some of the 15 stations in warm core rings. Size-fractionated particulate Ca data collected from Warm Core Ring 82-B in April and June 1982 demonstrate unambiguously the aggregation of small particles in the upper 400 m. This aggregation, probably the result of zooplankton activities, resulted in a factor-of-three decrease in Ca concentration of small (1–5 μm), suspended particles in the 50–400 m deep thermostad waters of 82-B between the two cruises. The net loss of particulate Ca in 2 months resulted in a removal flux equivalent to 20–40% of the carbonate sediment accumulation rate in the slope-water region of the North Atlantic.

A CENTRAL PROBLEM IN MARINE GEOCHEMISTRY IS TO UNDERSTAND the global distribution, sources, sinks, and reactions of elements involved in the

0065-2393/85/0209-0155 $06.00/0
© 1985 American Chemical Society

biogeochemical cycle. Playing a major role in this cycle is particulate matter, consisting largely of organic and skeletal material of bacteria, phytoplankton, and zooplankton. In surface waters, active fixation of dissolved substances during phytoplankton growth converts many chemical elements into particulate matter, which is then transported as large-aggregrate particles to the deep ocean where remineralization and oxidation processes release the elements back into the dissolved state (1–5). Chemical species that are passively adsorbed, complexed, or coprecipitated in particulate matter also follow similar routes. Hence, understanding the chemistry, biology, and vertical flux of particulate matter provides a key to the understanding of the biogeochemical cycle in the ocean. In addition, comprehending the relationship between biogenic particulate matter and the physical and chemical characteristics of the water column in which it was formed may aid in the interpretation of the sedimentary record in paleoclimatic studies. Furthermore, knowledge of the behavior of biogeochemically cycled elements may elucidate pollutant dispersion processes in the ocean as well as constrain models of ocean circulation.

Concentrations of particulate matter in the ocean are typically 100–1000 $\mu g/kg$ in the upper 100 m, and decrease rapidly to 5–20 $\mu g/kg$ below 100 m (6). Hence, a major impediment to the study of particulate matter is one of adequate sampling. Only 150–600 μg of material is obtained from deeper waters with 30-L Niskin water samplers. This small quantity of material would restrict the analytical chemist to a reduced suite of elements and to using only the most sensitive techniques for their determination. This limitation was realized in the 1960s and prompted the development of the first in situ filtration systems capable of filtering 1000-L volumes of seawater (7, 8). The large-volume in situ filtration system (LVFS, 9, 10) evolved from the earlier concepts was capable of filtering up to 25,000 L of seawater through each of four filter holders containing a 53-μm mesh Nitex prefilter and a pair of 1-μm glass-fiber filters. Filtering a larger volume of seawater thus allowed the collection of sufficient material for precise analyses and ensured a more representative sampling of the size spectrum of particles in the water column (11). The 53-μm prefilter was used because most particles that dominate the vertical transport of material in the water column are bigger than 50 μm and, thus, can be separated on the prefilter from the smaller, slow-sinking material that comprises most of the suspended particulate mass. Flow-rate experiments using the LVFS have shown that large-aggregate material is filtered from the water column without significant fragmentation (12).

The LVFS has been successfully deployed on 11 cruises in the Atlantic and Pacific Oceans. Analyses of the samples have made known the distributions of over 20 elements and have documented the importance of particulate organic matter as a complexing and ion-exchange agent. Furthermore, the results have demonstrated that macroscopic aggregates (fecal matter)

(5) dominate the vertical transport of particulate elements over fecal pellets, and that a greater fraction of surface-produced particulate matter penetrates a given depth horizon in areas of greater primary production (5, 10, 13). However, the excessive mass of the LVFS (>500 kg) and its bulk (0.7 m^3) have limited operations to relatively calm seas. Also, 2 days of ship time is required to complete one 12-sample LVFS profile, thus LVFS sampling opportunities are restricted because of a shortage of abundant ship time. Furthermore, the long operation time required does not allow the study of many important biological and physical processes that alter particulate-matter distributions and sedimentation rates on time scales shorter than 2 days.

This chapter describes in detail the multiple-unit LVFS (MULVFS), which, like the LVFS, can filter large volumes (25,000 L) of seawater, but is small and lightweight, and takes only 6–8 h to complete one 12-sample profile. The MULVFS was developed as part of an interdisciplinary program (14) to relate particulate-matter chemistry, size distributions, and fluxes to a complete suite of simultaneous biological and physical measurements being made on the same water column within warm core Gulf Stream Rings. Some of our Rings data will illustrate the performance of the system and demonstrate one aspect of the application of large-volume in situ filtration sampling within the program. Furthermore, we will document improvements in filter-holder design, filter material, and sample-handling techniques that have allowed for lower detection limits and have expanded the suites of elements and compounds determinable.

Methods

The multiple-unit large-volume in situ filtration system consists of 12 pump–filter units, an electromechanical cable, a current monitoring and switching box interfaced to an Apple II control computer, a drum winch, and a level wind (Figure 1).

Each MULVFS pump–filter unit weighs only 36 kg in air and has minimal exposed surface area to reduce drag while deployed in rough seas (Figures 2 and 3). At the heart of the unit is a 0.5-hp, three-phase, 480-V AC-submersible well pump motor (Century Electric, Gettysburg, Ohio), mated to a well pump (Jacuzzi Brothers, Little Rock, Arkansas). Both the pump and motor had to be modified for application at sea. The direction of flow through the pump was reversed so that water would be sucked into rather than discharged from the threaded end of the pump. The pump pressure differential was reduced from 4 atm to less than 1 atm by removing impeller stages. This reduction resulted in a pressure versus flow rate curve that fell slightly below that of the LVFS (Figure 2B). The lower pressure differential was chosen because most filtering of particulate material for biological studies is carried out at less than 1 atm. The motor was fitted with an underwater electrical connector and sacrificial zinc anodes.

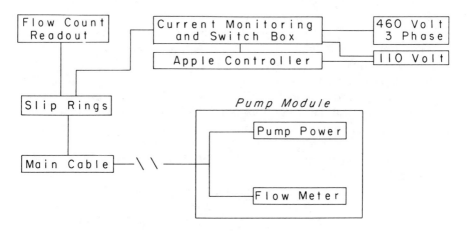

Figure 1. Block diagram of the multiple-unit large-volume in situ filtration
system (MULVFS).

During the operation of the pump, water is sucked through a filter-holder assembly into the pump and discharged through a telemetering flow meter located approximately 1 m below the filter holder. Contained within the filter holder is one 53-μm Nitex mesh prefilter and two 1-μm filters (254-mm effective filtration diameter) mounted in separate filter-holder stages. The top tubular baffle assembly of the filter holder is necessary to keep particles from being redistributed on the 53-μm filter during recovery operations. All MULVFS flow meters (Kent model C-700-B, Ocala, Florida) gave the same flow readings to within 0.3–0.6% (one standard deviation, $n = 15$) as determined by eight calibration experiments over flow rates ranging between 0.2 and 0.6 L/s.

The electromechanical (EM) cable was designed to permit the attachment and simultaneous operation of multiple pump units in the upper 1000 m of the ocean. Pump spacing ranged from 25 m near the surface to 100 m for depths deeper than 300 m (Figure 4). The EM cable is composed of a Hytrel (Du Pont, Wilmington, Delaware) jacketed Kevlar strength member (Wall Industries, Beverly, New Jersey) mated in parallel to an electrical cable. The electrical cable has three 12-gauge power conductors, a twisted shielded pair of 18-gauge conductors for data communication, and 15 electrical connection points ("pigtails") spaced along its 1000-m length (Blake Wire and Cable, Torrance, California). The strength member and electrical cable were mated at each pigtail and every 50 m with 1.2 cm wide nylon strapping and neoprene cement; they were further bound together at 2-m intervals by waterproof friction tape (American Bilt Rite Company, Mt. Laurel, New Jersey). Approximately 2% additional slack was allowed in the electrical cable when it was mated to the Kevlar mechanical cable to prevent mechanical loading of the conductors. This par-

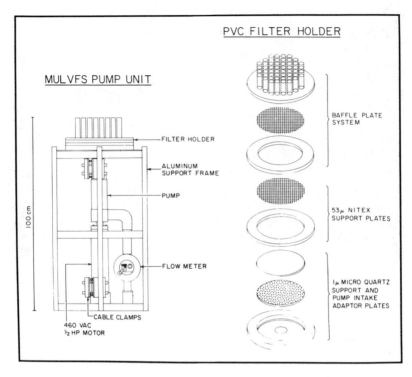

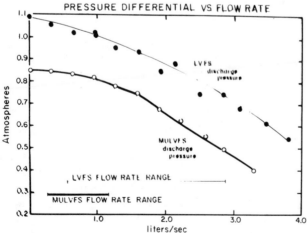

Figure 2. (Top) Detail of the MULVFS pump unit and filter-holder assembly. The filter holder consists of three stages. The upper two plates are baffles necessary to prevent particles from being redistributed on the filters during recovery. The next plate supports the 53-μm mesh Nitex prefilter, and the bottom plate supports the pair of 1-μm microquartz-fiber filters and seals to the pump intake.(Bottom) Pressure vs. flow rate curves for the MULVFS and LVFS pump units. Horizontal bars indicate operating ranges of the two systems during filtration.

Figure 3. One MULVFS pump unit during recovery (Oceanus 125). (Left)The open construction and low exposed surface area keeps drag forces minimal while deployed. The 36-kg weight of the unit makes it easy to handle on and off the wire. (Right)The filter holder is immediately covered with a polyethylene bag and excess water sucked from it before the pump is removed (with the aid of an air wrench) from the wire. Also shown on the right is a sediment trap used to passively collect material during MULVFS stations.

allel EM-cable configuration permitted mechanical and electrical attachment of multiple pump units to the separate cables while the pair is handled as a single entity. When a parallel cable configuration is used, repairs to the cable can be (and have been) carried out at sea. By using Kevlar as a strength member, the weight of the deployed cable is minimized.

Whereas the EM cable used with the LVFS was handled with a capstan and stored under low tension, the MULVFS cable had to be handled and stored under full tension. To minimize crushing loads on the cable, a large-diameter drum winch was used. Also a level wind was built, and a sheave was modified so that the 10–15 cm sized pigtail moldings would pass safely through them between the winch and the ocean. Slip rings (Meridian Lab, Middletown, Wisconsin) on the winch facilitated connection of the EM cable to the control electronics in the ship's laboratory.

During operation, the MULVFS is powered by 480-V three-phase AC power from the ship and is controlled by an Apple II computer programmed to disconnect the power in the event of a short circuit or imbalance in load. The computer also monitors the flow rate of one pump unit and records and graphically displays this information along with elec-

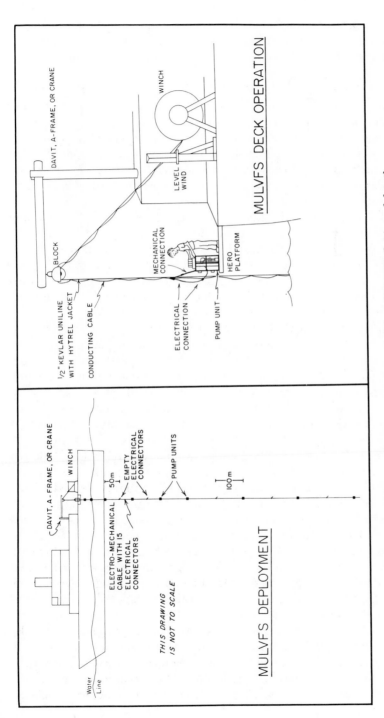

Figure 4. Shipboard deployment scheme for the MULVFS (left), and details of the MULVFS deck operation (right).

trical load data as a function of elapsed station time. Such information can be useful for determining the amount of ship time necessary to complete a station (discussed later) and for the evaluation of system performance.

Filter Cleaning, Handling, and Analysis

The need to process large numbers of samples at sea and in the laboratory required substantial improvement over previous filter-handling techniques. Furthermore, techniques had to be developed to allow fresh subsamples to be taken from the filters for organic analyses while allowing the remaining sample to be processed in normal fashion (5). All filter handling was performed with plastic or stainless steel tweezers and polyethylene gloves.

The filters used in the pump units generally consisted of one 305-mm diameter, 53-μm Nitex filter and two 280-mm diameter, 1-μm Gelman microquartz (MQ) filters (identical to Whatman type QM/A filters). The Nitex filters have been used for all LVFS and MULVFS samples; the MQ filters were chosen over Mead 935 BJ glass-fiber (GF) filters previously used in LVFS studies because they had much lower blank levels for many elements (e.g., Na, Ca, Mg, K, and Fe; Figure 5).

Both Nitex and MQ filters were cut from bulk material and processed according to the scheme illustrated in Figure 6. Batches of 100 MQ filters were stacked on a poly(vinyl chloride) (PVC) filtration stand with a 254-mm porous polyethylene insert above the level of the leaching solution in the filter-cleaning apparatus. During the filter-cleaning operation, a peristaltic pump sucked the leach solution from the bath and dispensed it onto the top of the filter stack at a rate exceeding the ability of the filters to absorb the solution. Leach solutions were 20-L volumes of filtered 18 Mohm deionized water [Milli-Q system (Millipore Corporation)] followed by two batches of 0.3 N HCl in deionized water. Following this 2-day procedure, the stack was rinsed with deionized water to remove the HCl and then separated into groups of 10 filters that were dried at 60 °C for 48 h on styrene racks in a polypropylene-lined drying oven. The filters were then transferred to a muffle furnance and, with groups separated by pyrex rods, were combusted at 450 °C for 3–4 h. After cooling and equilibrating with filtered laboratory air in a laminar-flow bench, the filters were repeatedly weighed to an accuracy of better than 1 mg. The weighing apparatus consisted of a hinged box draft-sealed to a top-loading balance (Sartorius Model 1206 MP) and a large-diameter stand. The stand supports the filters within the box while resting on the balance pan. After weighing, the MQ filters were bagged in pairs between acid-leached, 305-mm diameter, 149-μm mesh Nitex disks. This sandwich arrangement was necessary to provide support for the MQ filters during filtration. More accurate data can be obtained by ensuring that the filters of each pair were nearest neighbors

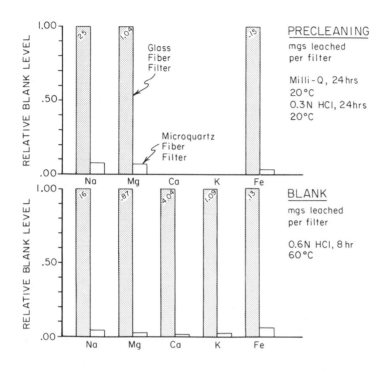

Figure 5. *The quantities of Na, Mg, Ca, K, and Fe leached during precleaning and during analysis are shown for MQ and Mead 931-BJ glass-fiber filters (used previously with the LVFS). These quantities demonstrate that the MQ filters are far cleaner by comparison. The MQ blank levels are low enough to measure accurately the trace elements Ba, Mn, Cd, and Co. Nitex blank levels have been described previously and are very low for most elements. (5).*

throughout all steps of the cleaning, combustion, and weighing procedures.

The 53-μm Nitex filters followed a similar but not identical processing scheme where 100-filter batches were placed in a large tank and leached with 0.1 N HCl in deionized water followed by deionized water. After additional rinsing in deionized water they were dried and weighed to the same precision as the MQ filters and stored with the MQ filter pairs in the polyethylene bags.

Prior to a station at sea (Figure 6), the MQ filter sandwich and the accompanying 53-μm Nitex prefilter were loaded into the filter holders and then covered with polyethylene bags to minimize contamination. The bags remained in place while the holders were bolted onto the pumps on deck and were removed just moments prior to submersion of the pumps. Upon

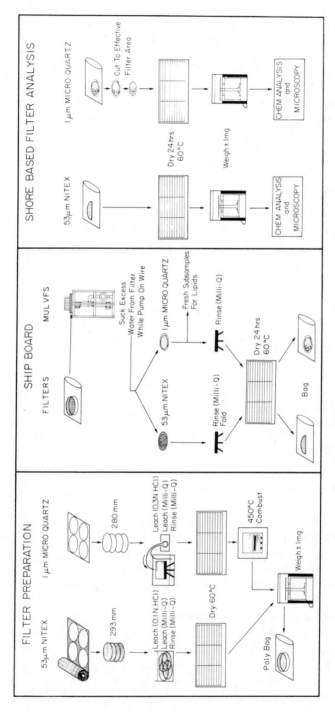

Figure 6. Schematic flow charts of filter preparation (left), shipboard handling (middle), and laboratory analysis schemes (right).

completion of sampling at depth, the pumps were recovered from the ocean. On emergence of each pump, its filter holder was immediately covered by a plastic bag to reduce contamination, and excess water was eliminated by applying suction through a vacuum line attached to the discharge side of the flowmeter. The filter holders were then unbolted and returned to the main laboratory for sample processing.

All collected particulate material was processed withing 3 h of the recovery of the last pump; the Nitex and MQ-filter sandwich (containing > 53 and < 53-μm particle-size fractions, respectively) were treated separately in a laminar-flow bench. Processing progressed in order of depth to minimize the degradation of the samples; material collected from the near-surface waters is most likely to degrade fastest.

The processing scheme for the 53-μm Nitex samples was similar to that described previously (*10*) where each filter was transferred to the PVC stand, rinsed with deionized water under suction to reduce sea salt, folded once onto itself to prevent loss of large particles, and then dried at 60 °C for > 24 h. Where samples larger than 53 μm have been required for lipid analysis, precombusted, 53-μm stainless steel mesh of identical specifications to the 53-μm Nitex was used. In this case a subsample was dried for the regular chemical analysis; the remaining sample was specially preserved.

The MQ-filter sandwich was similarly transferred to the filter-rinsing stand, the top 149-μm Nitex support was folded back, and three sharpened acrylic tubes were pressed into the filter pair. The MQ-filter area not enclosed by the acrylic tubes was rinsed with several aliquots of deionized water (totaling 200–300 mL) with suction alternately applied from below to reduce sea salt by an order of magnitude; this process was required for accurate dry-weight and major-ion determinations. The acrylic tubes were subsequently removed, and the fresh, unrinsed subsamples were preserved for lipid analysis. The remaining filter material was dried at 60 °C for > 24 h. Both the MQ and 53-μm Nitex samples were stored and transported flat in individual polyethylene bags to prevent loss of material.

Shore-based analysis (Figure 6) involved redrying the samples at 60 °C, equilibrating with the laboratory atmosphere, and weighing as just described. The only difference was in the treatment of the MQ filters; with the 149-μm supports removed, the 254-mm diameter area loaded with particulate matter was cut from the pair with an acrylic template and stainless steel scalpel. The dry weight of these samples was determined with the same method used for LVFS samples (*9*). Dried samples have been further subsampled for major and minor element analyses using atomic absorption spectroscopy, a carbon–hydrogen–nitrogen analyzer, and wet chemical techniques (*5*). Nitex samples also have been analyzed for the size distributions of different particle classes by light microscopy.

Results And Discussion

The MULVFS was deployed on three of the four cruises to sample warm core rings (WCR) of the Gulf Stream in 1982. Warm Core Ring 82-B formed in mid-February 1982, when a Gulf Stream meander pinched off and became isolated in the slope water (Figure 7) (*15, 16*). At that time, its core waters, derived from the Sargasso Sea and Gulf Stream, were warmer than 18 °C. By the time of the first WCR cruise in April (R.V. *Oceanus* 118), the core waters had cooled by surface heat loss and down mixing to

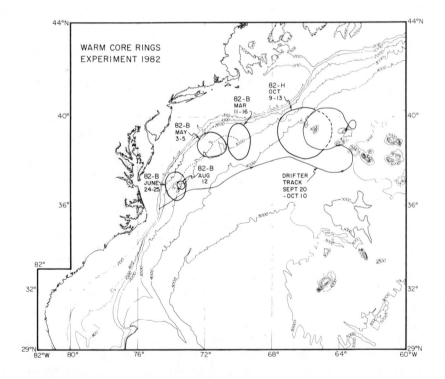

Figure 7. Locations of Warm Core Rings 82-B and 82-H during 1982 (15). The dashed line indicates the path of motion of the center of 82-B over time (provided by Don Olsen, Univ. of Miami, Rosensteil School for Mar. and Atmos. Sciences). The solid line indicates the path followed by a satellite tracked drifter released in the center of 82-B in August. This drifter stayed with the ring until the ring was reabsorbed by the Gulf Stream. It was then swept into the Gulf Stream and was transported rapidly to the area of formation of Ring 82-H by late September 1982.

produce a nearly isothermal (15.7 °C) layer extending from the surface to greater than 400-m depth; ring diameter was approximately 150 km. When Ring 82-B was revisited in June (R.V. *Oceanus* 121), the surface waters had become stratified from both solar heating and entrainment of lower salinity water from outside the ring into its surface layer. The isothermal 15.7 °C layer, *thermostad*, remained largely intact from below 30–50 to >400 m, and the diameter of the ring decreased slightly to approximately 140 km. By the time of the August cruise (R.V. *Oceanus* 125), Ring 82-B had decreased in size to <40 km and had lost 90% of its volume because of interactions with the Gulf Stream. Much of the water in the upper 100 m was from the Gulf Stream, and the thermostad was found only between 110 and 260 m. The final WCR cruise was made in late September 1982 (R.V. *Knorr* 98) when we witnessed the formation of a new 300-km diameter ring, Ring 82-H. This ring, by virtue of its larger diameter, contained substantial quantities of Sargasso Sea water in its core.

Ring 82-B was an area of high primary production (*17*). In April, primary production values, measured by the carbon-14 method, ranged between 300 and 1300 mg $C/m^2 \cdot day$. This rate was comparable to values measured at a station occupied in the Sargasso Sea but was lower than 2900 mg $C/m^2 \cdot day$ rates measured in the slope water. In June, the production values remained high, between 230 and 880 mg $C/m^2 \cdot day$, even though the production in both slope water and Sargasso Sea decreased significantly. This result was unanticipated and made WCR 82-B an even better site to study biogeochemical processes in the ocean.

MULVFS Performance.   The LVFS was used exclusively in April during *Oceanus* 118 because the MULVFS was still under development. On that cruise, reasonable spatial coverage of WCR 82-B was obtained with the LVFS at the cost of limiting most casts to shallower than 200 m because of ship-time constraints. The use of the LVFS also allowed us to compare MQ and GF filters and to test MULVFS control electronics. During *Oceanus* 121, in June 1982, the MULVFS worked only for half of one station before a flaw in an earlier design of the electromechanical cable resulted in mechanical loading and breakage of conductors in the electrical cable. The sampling program was completed with the LVFS, and a reoccupation of the single MULVFS station was made for comparison purposes. In August on *Oceanus* 125, the MULVFS, working most of the time, allowed us to collect samples from seven stations from Ring 82-B, the slope water, Gulf Stream, and Sargasso Sea. Collecting synoptic profiles with the LVFS in 82-B in August would have been nearly impossible because the translation speed of the ring center was approximately 20 km/day. During the 1982 *Knorr* 98 cruise in late September–early October, another eight MULVFS stations in Ring 82-H were occupied with complete success; some stations were in winds ranging from 30–40 knots and sea states up to 5 on the Beaufort scale. These three later cruises demonstrated that the MULVFS concept could be made to work at sea. However, the electromechanical cable, although mechanically functioning perfectly by the final WCR cruise, remained a continuing source of electrical problems. A combination of damage to the electrical cable during the *Oceanus* 121 and 125 cruises and repeated repairs to it have rendered it unreliable at this writing.

During the *Oceanus* 125 and *Knorr* 98 cruises the MULVFS took approximately 1 h to deploy to 1000 m, 4–6 h (discussed later) to complete sampling, and approximately 1 h for recovery; total station time required was 6–8 h. Total line tension on the electromechanical cable with 12 pumps attached was 500–700 kg, roughly half the line tension of the LVFS cable. Furthermore, when operating 12 pumps, the MULVFS consumed power at only three times the rate of the LVFS. This greater efficiency resulted from a better match among pump capacity, motor size, and flow rate of water through 1-μm filters.

We will present flow-rate, particulate dry weight, and particulate-calcium data from selected MULVFS and LVFS stations to (1) further document the sampling performance of the MULVFS and (2) demonstrate one application of large-volume in situ filtration to understanding the processes governing the distributions of particulate matter in a productive environment. Detailed presentation and interpretation of warm core ring data are outside the scope of this chapter.

Flow Rate.   The major factor governing flow rate of both LVFS an MULVFS was particle loading on the filters. Hence, the use of station time could be optimized by using flow-rate information acquired during MULVFS operation. Because only one pump could be monitored in real time, a deep pump, assumed to be in low-particle-concentration waters, was chosen. Data from two casts at the same station (Figure 8, Curve A) showed that flow rates decreased exponentially with time, the decrease being faster for a near-surface sample than for a deep sample as would be expected. From this observation, it was postulated that the volume of water filtered by each pump during a station could be used to indicate the systematics of the particle-concentration profile. The exponential decrease of flow rate (F), in liters per second, with time can be written as

$$F = F_o e^{-kt} \tag{1}$$

where F_o is approximately 1.2 L/s (Figure 8, Curve A), k is assumed to be proportional to particulate mass, and t is the elapsed time.

From Equation 1, k values could be calculated from the total volume of water filtered. This calculation was done for the MULVFS stations occupied on *Oceanus* 125 and *Knorr* 98 for each sample deeper than 100 m. The k values plotted against total dry-weight concentration for each sample (Figure 8, Curve B) show a linear relationship, thus confirming the use of k

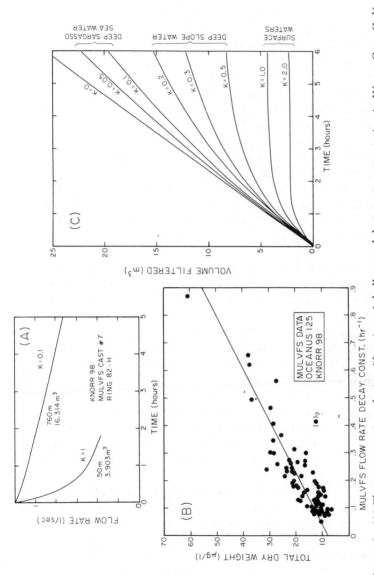

Figure 8. (A) Flow rate vs. time during filtration of shallow and deep water at a station in Warm Core Gulf Stream Ring 82-H. (B) Flow-rate decay constant vs. total particulate matter, dry-weight concentration for all samples deeper than 100 m. The linear fit to the data is $\mu g/L = 53k + 7.6$ with a correlation coefficient of 0.914; and $s(y) = 4\mu g/L$. Scatter in the data is attributed to additional factors such as size distribution or chemical differences among samples. (C) Filtration performance of the MULVFS in different environments.

to estimate dry-weight concentration. Secondary factors such as variability in particle-size distribution and organic carbon, carbonate, and siliceous contents are responsible for the scatter in the data.

The k value also can be used to predict how much water could be filtered in a given water mass (Figure 8, Curve C) for a given amount of ship time. For example, it would be possible to filter substantially more than 25,000 L of Sargasso Sea water deeper than 300 m in 5–6 h. In contrast, the point of diminishing returns is reached for the more particle-rich deep slope water in 3 h with just 10,000 L filtered; sampling of the near-surface waters is essentially complete within 1–2 h.

Particulate Calcium in WCR 82-B.   Calcium is a major component of particulate matter and occurs largely as calcite (5). Most particulate Ca is small and is predominantly present in 10–20-μm sized coccolithophores in surface waters and in 1–5-μm sized coccoliths below the euphotic zone (5). Individual coccoliths sink only fractions of a meter per day (2, 5). Hence, the <53-μm Ca concentration in waters below 100 m would be expected to show the least temporal variability of all particulate elements sampled. For this reason, Ca data from below 100 m was used for comparison of samples collected with GF and MQ filters and by the LVFS and MULVFS.

Comparison of the samples collected in April with either MQ or GF filters showed that there is no difference in <53 μm Ca concentration (Figure 9, Curve A). Comparison of the June LVFS and MULVFS profiles (Figure 9, Curve B) also showed excellent agreement. The factor-of-four difference in concentrations of the shallowest samples is attributable to the fact that the profiles were collected 1 week apart and at slightly different radial distances from the center of the ring. The <53 μm thermostad samples replicated within 1–2 nmol/L. This 5–10 % variability may be natural. Samples from the main thermocline were assumed to be best for comparison purposes by virtue of being deeper and in a more stratified region of the water column. Indeed, Ca values from 600 m deep LVFS and MULVFS samples agreed to within several percent. Therefore, instrumental and filter differences can be ignored when discussing data from different WCR cruises.

The LVFS samples from the center of WCR 82-B in April showed that <53 μm Ca concentration levels in the thermostad at the beginning of the cruise were nearly uniform at 69 nmol/kg. One week later, <53 μm Ca concentrations had increased in the upper 100 m but had decreased in deeper waters by 30 %. Accompanying the <53 μm Ca loss from deeper waters was a gain in >53 μm Ca concentration such that the relative contribution of the larger size class to total suspended Ca increased from 16 to 28 %.

Ring 82-B samples collected during the June cruise showed astonishing changes in particulate Ca distribution. Between April and June <53 μm

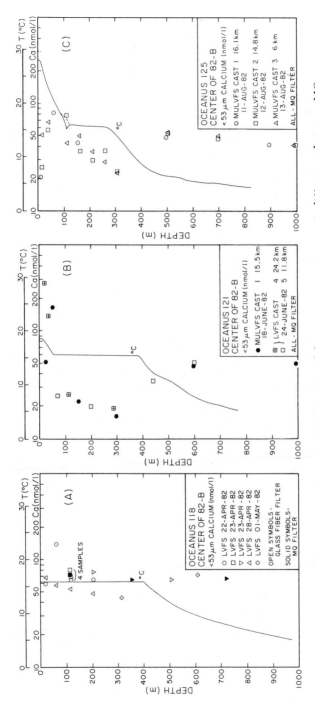

Figure 9. (A) LVFS data from the center of 82-B in April demonstrates no systematic differences between MQ (solid symbols) and GF filters in their ability to retain particulate Ca. (B) Profiles of <53 μm particulate Ca at stations near the center of Ring 82-B (June) occupied with LVFS and MULVFS systems. The agreement between the two systems is excellent, especially in deeper samples. (C) Three replicate casts from Ring 82-B (August) showing again excellent reproducibility of samples collected in the main thermocline. Solid lines de-note water-column temperature (°C).

Ca concentrations in the thermostad had dropped by a factor of three, to 20-nmol/kg levels, in spite of the concomitant increase in concentrations in the surface waters (Figure 9, Curve B). Also, in June, the <53 μm Ca concentrations in the nearly isothermal 15.7 °C thermostad layer of Ring 82-B varied as a function of distance from ring center; values rose from the 20-nmol/L levels at stations near the ring center (shown in the figure) to >40 nmol/L at the ring edge. In addition, the >53 μm Ca in ring center also had decreased significantly in the thermostad, contributing 18% to total particulate Ca, down from the 28% contribution in late April.

By August, Ring 82-B had lost much of its character. The surface layer of the ring had been almost entirely replaced by Gulf Stream water. Consequently, the near surface <53 μm Ca concentrations observed in August were lower than those observed in June. Interactions with the Gulf Stream also caused the thermostad layer to be severely eroded such that its volume had decreased by an order of magnitude. The August particulate Ca values in the thermostad were higher and more scattered relative to June data (Figure 9, Curve C). Because of the substantial loss of thermostad water and the June observation of a horizontal Ca gradient in the thermostad, the August data cannot be used to infer unambiguously net changes in particle abundances in the thermostad. The August data, however, do show that samples from three separate profiles replicate each other within 1–2% for waters in the main thermocline.

The particulate-Ca data in WCR 82-B show a dramatic loss of Ca from the thermostad waters in ring center between April and June. Theoretically, the mechanisms capable of removal of the particulate Ca are water-mass replacement, in situ dissolution, individual particle sinking, or aggregation of fine material into large, fast-sinking particles. The mechanism of water-mass replacement cannot explain this loss because the thermostad of Ring 82-B had uniquely defined hydrographic properties and showed little alteration between April and June. In situ dissolution is unlikely because most of the western North Atlantic water column is supersaturated with respect to calcite, the carbonate mineral phase of coccoliths (18). The slow-sinking rates (<1 m/day) of coccoliths, the dominant component of particulate Ca, rule out removal by direct sedimentation. Aggregation, the incorporation, probably by zooplankton, of small particles into larger, faster-sinking material, is the favored mechanism as evidenced by the increasing quantities of material in the >53-μm fraction observed in late April. Thus, our data provide the first unambiguous confirmation of this process in waters deeper than 100 m and can be used to further calculate the residence time of fine particles in the 100–400-m depth interval. The net loss of Ca from the <53-μm fraction between April and June corresponded to a mean removal flux of 150 μmol/m^2 day. In magnitude, this flux amounted to approximately 20–40% of the carbonate accumulation rates typical for the slope-water region (19, 20). Because the >53 μm Ca concentration had dropped significantly by June, most of the Ca removal probably had occurred in May. Our data therefore indicate that small particle abundances in the upper 400 m may be altered significantly on time scales of 1 month, especially in productive systems.

Summary

This chapter has documented the design and performance of the new multiple-unit large-volume in situ filtration system. The major features of the MULVFS are

1. The MULVFS pump unit weight has been reduced by a factor of 15 relative to the LVFS while maintaining filtration performance. The light weight and low exposed surface area of the MULVFS units make it practical to deploy the system in higher sea states and winds than heretofore possible.

2. The station time requirements for filtering 12 samples from the upper 1000 m by large volume in situ filtration have been reduced from 2 days to 6–8 h. This sixfold to eightfold increase in sampling efficiency has resulted in a better ability to sample moving mesoscale features the size of warm core Gulf Stream Rings.

3. Advances in filter preparation and the use of microquartz-fiber filter material has resulted in a lowering of blank levels for most elements by one to two orders of magnitude. This lowering has allowed the suite of elements determined in our samples to be expanded to include trace metals, such as Ba, Fe, Mn, Cd, and Co, that are of geochemical interest.

4. We have developed the methodology to permit subsampling of fresh collected material at sea. This methodology has enabled us to perform analyses of the lipid class compounds in our samples in addition to major and minor elements.

5. MULVFS flow-rate information can be used to optimize the use of station time because the dominant factor governing flow rate is particle loading on the filters. Therefore, the flow-rate data may be used also to infer the systematics of the particulate dry-weight profile at a station while at sea. This information is valuable for planning sampling strategy for the next station.

6. Analysis of particulate Ca in LVFS and MULVFS samples from WCR 82-B has shown that instrumental and filter differences may be ignored when comparing samples collected with either system.

An example of the usefulness of large-volume in situ filtration is illustrated by the particulate Ca data collected in the interdisciplinary Warm Core Rings program. Particulate Ca in the < 53-μm size fraction showed a factor-of-three drop in concentration in the thermostad of WCR 82-B between April and June 1982. In the environment of WCR 82-B, this drop is evidence for the process of reaggregation, probably by zooplankton, of the fine material into large fast-sinking particles. The net removal flux of Ca from the thermostad was 150 μmol of Ca/m^2 day, which amounted to between 20 and 40% of the accumulation rates of carbonate sediments in the slope-water region of the northwest Atlantic.

The MULVFS is a powerful sampling tool for elucidating element pathways in the biogeochemical cycle. It is ideally suited for sampling particulate matter in fast changing environments, in moving mesoscale features such as warm core Gulf Stream rings, and in the open ocean. In addition to oceanic sampling applications, a single filtration unit powered by a portable generator has been deployed with great success from bridges, docks, and small boats to sample rivers and lakes in the New York area. With currently available technology, the MULVFS can be improved to collect samples from greater depths and to acquire continuous sensor data (e.g., flow rate, pressure, and temperature) to pinpoint the environmental conditions of the samples.

Acknowledgments

We thank the officers, scientific parties, and crews of the research vessels *Oceanus* and *Knorr* for their help in deploying the MULVFS. Chief scientist, P. Wiebe, during all MULVFS and LVFS cruises, provided encouragement during difficult moments at sea. Members of the Warm Core Rings Project Office also helped to make our seagoing experiences a pleasure. Scott Weaver developed assembly language routines for our Apple computer. Jay Ardai and John Sindt (L-DGO), and R. D. Wilkes (Jacuzzi Pump) guided us through some very difficult stages in MULVFS development. Many of our friends spent their spare hours working on the system at Lamont. The manuscript was reviewed by P. E. Biscaye, Y. H. Li, and I. Fung. This research was supported by the National Science Foundation Grant, OCE 80-17468.

Literature Cited

1. Broecker, W. S.; Peng, T. H. "Tracers in the Sea"; Eldigio: Lamont–Doherty Geological Observatory, Palisades, New York, 1982; 690 pp.
2. Honjo, S. *Mar. Micropaleontology* 1976, *1*, 65.
3. Wiebe, P. H.; Boyd, S. H.; Winget, C. *J. Mar. Res.* 1976, *34*, 341.
4. Knauer, G. A.; Martin, J. H.; Bruland, K. W. *Deep-Sea Res.* 1979, *26*, 97.
5. Bishop, J. K. B.; Edmond, J. M.; Ketten, D. R.; Bacon, M. P., Silker, W. B. *Deep-Sea Res.* 1977, *24*, 511.
6. Brewer, P. G.; Spencer, D. W.; Biscaye, P. E.; Hanley, A.; Sachs, P. L.; Smith, C. L.; Kadar, S.; Fredericks, J. *Earth Planet Sci. Lett.* 1976, *32*, 393.
7. Laird, J. C.; Jones, D. P.; Yentsch, C. S. *Deep- Sea Res.* 1967, *14*, 251.
8. Spencer, D. W.; Sachs, P. L.; *Mar. Geol.* 1970, *9*, 117.
9. Bishop, J. K. B.; Edmond, J. M. *J. Mar. Res.* 1976, *34*, 181.
10. Bishop, J. K. B.; Collier, R. W.; Ketten, D. R.; Edmond, J. M. *Deep-Sea Res.* 1980, *27*, 615.

11. McCave, I. N. *Deep-Sea Res.* **1975**, *22*, 491.
12. Bishop, J. K. B., *EOS* **1982**, *63*, 46.
13. Bishop, J. K. B.; Ketten, D. R.; Edmond, J. M. *Deep-Sea Res.* **1978**, *25*, 1121.
14. The Warm Core Rings Executive Committee, *EOS* **1982**, *63*, 834.
15. Joyce, T. M.; Schmitt, R. W. *EOS* **1982**, *63*, 995.
16. Joyce, T. M.; Wiebe, P. H. *Oceanus* **1983**, *26*, 34.
17. Hitchcock, G. L.; Smayda, T. J. *EOS* **1982**, *63*, 995.
18. Broecker, W. S.; Takahashi, T. *Deep-Sea Res.* **1978**, *25*, 65.
19. Balsam, W. *Palaeogeogr. Palaeoclimatol. Palaeoecol.* **1981**, *35*, 215.
20. Balsam, W. *Science (Washington, D.C.)* **1982**, *217*, 929.

RECEIVED for review July 5, 1983. ACCEPTED January 3, 1984.

10

Oxygen Consumption in the Ocean: Measuring and Mapping with Enzyme Analysis

T. T. PACKARD

Bigelow Laboratory for Ocean Sciences, McKown Point, West Boothbay Harbor, ME 04575

The chemical basis of oceanic oxygen consumption originates with enzymatic reactions in plankton. This chapter discusses the measurement of the reaction rate of the enzyme systems that control 90% of the planktonic oxygen consumption. The reaction rate then gives the potential rate of oxygen consumption, from which oxygen consumption rates can be calculated in upwelled sewage, upwelling ecosystems, and abyssal waters. The use of an automated enzyme analyzer for mapping rates of oxygen consumption is also described. These maps show association between upwelled seawater and low oxygen consumption rates and between mature, seasoned upwelled seawater and high oxygen consumption rates. Deepwater oxygen consumption was associated with enhanced surface productivity and was used to calculate age and currents in the deep ocean. For waters 3000 m below the Peruvian Current, an age of 685 years and a current speed of 0.6 mm/s were calculated.

MAPPING CRITICAL WATER QUALITY INDICES IN REAL TIME is necessary for the development and verification of realistic mathematical models of the ocean. The advent of automated chemical analyses and computer mapping (1–4) has made real-time mapping of static chemical properties a reality. But these static properties, for example, nutrient salts, chlorophyll, and salinity, are not sufficient to describe the state of a system, nor can they be used to predict the recovery of a perturbed system. The dynamic properties, especially those that control the remineralization and oxidation of organic matter to CO_2 and NO_3, namely oxygen consumption, denitrification, and nitrification, must be measured (5–8). These processes are

0065-2393/85/0209-0177 $09.25/0
© 1985 American Chemical Society

normally measured by time-consuming incubation techniques that, although useful for experimental small-scale studies, are too slow to ensure the rapid information feedback required for mesoscale oceanographic studies. This chapter focuses on measuring and mapping the rate of oxygen consumption in surface and deep ocean waters, examines the biochemical basis of oxygen consumption, and demonstrates how knowledge of this biochemistry can be used to calculate and map oxygen consumption in the sea.

The Chemical Basis of Oxygen Consumption

Oxygen consumption, in contrast to oxygen production, occurs at all depths and regions of the ocean where oxygen and organisms are present. Without the organisms oxygen consumption would cease because nonenzymatic utilization of oxygen is negligible (9). Because marine microorganisms are ubiquitous, oxygen consumption always proceeds at some finite rate. This rate of consumption reflects their metabolism, which in turn reflects the oxygen consumption of many enzyme-catalyzed reactions. These reactions fall into two groups: those associated with oxidative phosphorylation—the production of adenosine triphosphate (ATP) (10, 11)—and those associated with nonphosphorylating oxygenases in the microsomes and cytoplasm (12, 13).

In the first group are the reactions coupling the oxidation of organic matter by glycolysis:

$$C_6H_{12}O_6 + 2NAD^+ + 2PO_3^{2-} + 2ADP \rightarrow$$
$$2CH_3COCO_2^- + 2NADH + 2H^+ + 2ATP + 2H_2O \quad (1)$$

and the tricarboxylic acid cycle:

$$CH_3COCO_2^- + 2H_2O + FAD + 4NAD^+ \rightarrow$$
$$3CO_2 + FADH_2 + 4NADH + 4H^+ \quad (2)$$

with the oxidative phosphorylation of the electron-transport system (ETS):

$$NADH + H^+ + FADH_2^+ + O_2 + 5ADP + 5PO_3^{2-} \rightarrow$$
$$NAD^+ + FAD^+ + 2H_2O + 5ATP \quad (3)$$

A reaction associated with nonphosphorylating oxygenases is shown in Reaction 4, an example of a minor oxygen-consuming reaction in biological systems. The oxygen consumption by cytochrome P-450 occurs in the microsomes of animals and the "microbodies" of plants.

$$RH + O_2 + NADPH + H^+ \rightarrow ROH + H_2O + NADP^+ \quad (4)$$

where R represents substrates such as natural lipids and foreign lipophilic substances.

Both groups of reactions are found in bacteria (*14*), all higher animals (*15*), and plants (*16*); however, oxidative phosphorylation is responsible for 90% of the oxygen consumed (*17*). Oxidative phosphorylation is driven by the respiratory electron-transport system that is embedded in the lipoprotein inner membrane of eukaryotic mitochondria and in the cell membrane of prokaryotes. It consists of four complexes (Scheme I). The first is composed of nicotinamide adenine dinucleotide (NADH) oxidase, flavin mononucleotide (FMN), and nonheme iron–sulfur proteins (*18, 19*), and it transfers electrons from NADH to ubiquinone. The second is composed of succinate dehydrogenase (SDH), flavin adenine dinucleotide (FAD), and nonheme iron–sulfur proteins (*20*), and it transfers electrons from succinate to ubiquinone (*21, 22*). The third is composed of cytochromes b and c, and nonheme iron–sulfur proteins (*23*), and it transfers electrons from ubiquinone (UQ) to cytochrome c (*24*). The fourth complex consists of cytochrome c oxidase [ferrocytochrome c:O_2 oxidoreductase; EC 1.9.3.1 (*25*)] which transfers electrons from cytochrome c to O_2 (*26, 27*).

These four complexes together are often referred to as the respiratory chain, but their organization is better described as a multienzyme system than as a chain (*28*). Accordingly, these four complexes will be called the electron-transport system (ETS) throughout this chapter. At the terminal end of the ETS, cytochrome oxidase is the ultimate mediator between intermediary metabolism and oxygen; however, its rate-limiting role has been preempted by the flavoprotein end of the ETS (*17*). Chance and Williams (*29–31*) demonstrated this preemption by calculating first-order rate constants for each step in the ETS. They showed that flavoprotein oxidation occurred at one-half the reaction rate of the cytochrome oxidase reaction. Therefore, the consumption of oxygen by mitochondria can only proceed as fast as the maximum velocity of the flavoenzymes, that is, the NADH–ubiquinone (EC 1.6.5.1) and the succinate–ubiquinone oxidoreductase (EC 1.3.99.1).

Oxygen Consumption: The Basis of Energy Production

The purpose of oxidative phosphorylation is to generate ATP. Because all living systems are in a nonequilibrium state, they require this energy to maintain themselves against the constant tendency to degrade and deteriorate according to the second law of thermodynamics. Furthermore, if these living systems are to grow, reproduce, or pursue an aggressive existence they require additional energy. Therefore, they first developed the substrate-level phosphorylation reactions, such as those in glycolysis, and second, they developed phosphorylating membranes that utilize the free energy of succinate and NADH oxidation to produce ATP. The oxidation is carried out by the four complexes of the ETS embedded in these membranes. Thus, oxygen consumption and ATP production are coupled. The four electron-transport complexes generate and maintain a proton differential (ΔpH) and an electric-charge differential (an electromotive force

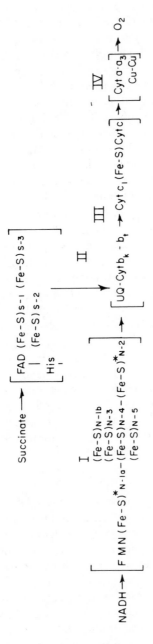

ETS according to Chance, (1977)

Scheme I. The electron flow pattern between the electron donors (succinate and NADH) and the electron acceptor (oxygen) in the mitochondrial respiratory ETS. The diagram shows the organization of the ETS into four isolatable enzyme complexes.

Abbreviations: FAD, flavin adenine dinucleotide; Fe-S, iron-sulfur proteins that can be identified in separate clusters by electron paramagnetic resonance analysis (the s-1, s-2 subscripts identify these iron-sulfur proteins as part of the succinate dehydrogenase complex); His, the histidine linkage between FAD and the large (70,000 daltons) protein moiety of the enzyme; FMN, flavin mononucleotide; N-1a, N-2 subscripts identify these iron-sulfur proteins as part of the NADH-dehydrogenase complex; UQ, ubiquinone; Cyt b_t and Cyt b_k, cytochrome b-566 and b-563, respectively.

symbolized as $\Delta\Psi$) across the mitochondrial inner membrane or across the bacterial plasmalemma. The sum of the pH and charge differentials produces the protonmotive force (P), which is the source of energy for mitochondrial ATP synthesis: $P = \Delta\Psi - 2.3[RT(\Delta pH/F)]$. The synthesis is actually carried out by the enzyme, ATP synthase, that discharges the proton differential and simultaneously phosphorylates ADP $(15, 17, 32, 33)$.

Topology of the Complexes in the Mitochondrial Membranes

The subunits of the respiratory electron-transport system can be isolated and studied independently of each other. Hackenbrock (28) presented data supporting the argument that these complexes $(see$ Scheme I, $34)$ are not arranged in linear sequence as in a chain, but are randomly distributed in the "plane" of the inner membrane. They are oriented with their long axis perpendicular to the plane of the membrane with 70–83% of their mass extending into cytosol and matrix space on either side of the membrane. Furthermore, Hackenbrock argues that they can migrate independently in the membrane plane in response to electrophoretic forces.

This conceptual model of the electron-transfer complexes migrating randomly in the plane of the inner membrane has been modified by Capaldi (35), who argues that the high-protein concentration (50%) in the mitochondrial matrix (36) would retard the lateral migration. In his modification, the electron-transfer complexes, as well as the ATP synthetase and the ATP–ADP translocase complexes, are fixed in the membrane rather than being free to migrate laterally (Figure 1). This change does not impair the capacity to transfer electrons because ubiquinone is left free to diffuse randomly and transfer electrons from Complexes 1 and 2 to Complex 3, and cytochrome c is left free to transfer electrons from Complex 3 to Complex 4 (Figure 1). Capaldi (35) calculates that cytochrome c can collide with both Complexes 3 and 4 within a single turnover time of the ETS (50–100 ms/cytochrome c oxidase).

Bacterial Oxygen Consumption

In bacteria, oxygen consumption associated with oxidative phosphorylation occurs in the cell membrane because these organisms do not have mitochondria $(37, 38)$. In this membrane the ETS is organized similarly to the ETS in eukaryotes (39), that is, it has the same basic iron–sulfur proteins, flavoproteins, quinones, and cytochromes to transfer electrons from the dehydrogenases to oxygen. However, some differences exist, the most striking of which occur at the terminal oxidase. The cytochrome c oxidase of mitochondria is often supplemented with simpler polypeptides in the form of cytochrome o (26), or cytochromes d, z_1, and c_{co} $(39, 40)$. Other differences are (1) the ubiquinone (coenzyme Q) of mitochondria is often supplemented or replaced by an analog, menaquinone; (2) cytochrome c is sometimes missing; (3) transhydrogenases at the dehydrogenase level are com-

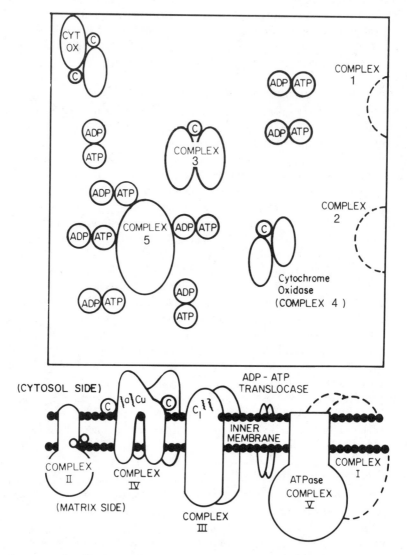

Figure 1. Distribution of the electron-transport complexes in the mitochondrial
inner membrane.

The upper diagram is a planar view; the lower one is a cross section through the membrane.
Each complex has been drawn to scale according to Capaldi (35); the distance between
Complex 1 and 3 is approximately 310 Å and between Complex 3 and 4, it is approximately
255 Å. The area around Complex 3 in the upper panel is about 200,000 Å. (Reproduced
with permission from Ref. 35.)

mon; and (4) the ETS often branches after the quinone level to accommodate two terminal oxidases (*39, 40*). Regardless of these differences, the respiratory capacity, and the molecular properties of the terminal oxidases, the bacterial ETS is not significantly changed.

Oxygen Consumption in Plants

Four types of oxygen consumption occur in plants; dark respiration, photorespiration, chloroplast respiration, and cyanide-resistant respiration. Other reviews of photorespiration have been published (*41, 42*); therefore it will not be discussed here. Also, because knowledge of respiratory ETS in plants is based on the higher plants in contrast to the algae, this section will be confined largely to the higher plant literature.

Chloroplast respiration is a novel form of respiration that has only recently been discovered. Chloroplast membranes exhibit a nonphosphorylation ETS that consumes oxygen (*48*). Ferredoxin and reduced nicotinamide adenine dinucleotide phosphate (NADPH) serve as electron carriers and glyceraldehyde-3-phosphate serves as the electron donor. This system probably originated in the respiratory ETS of the chloroplast's free-living ancestors, the cyanobacteria (*49*).

The most important form of plant respiration is housed in the mitochondrial inner membrane just as it is in mammals, and the system is similarly composed of NADH and succinate dehydrogenase, ubiquinone, and cytochromes b, c, c_1, a, and a_3. However, plant mitochondria have additional pathways to oxidize exogenous NADH, NADPH, malate, and other substrates as well as an alternative, cyanide-resistant, redox link between ubiquinone and oxygen (Figure 2). These pathways provide plants with a variety of responses to meet changes in metabolic and environmental conditions. Both the outer and inner mitochondrial membranes of plants contain NADH dehydrogenases, which give plant mitochondria the ability to oxidize exogenous NADH and sets them apart from mammalian mitochondria that normally only oxidize exogenous succinate (*43*).

Plant mitochondria also have the ability to oxidize exogenous NADPH without transhydrogenation to NADH. Oxidation is done via a rotenone-resistant pathway that bypasses Site I (Figure 2) and transfers electrons to ubiquinone (*44–45*). A third characteristic of plant mitochondria is the presence of an alternative oxidase that enables plants to consume oxygen in the presence of cyanide and other inhibitors of cytochrome oxidase (Figure 2). The physiological purpose of this alternative oxidase and its chemical pathways are unknown. It could serve as a heat-generating mechanism analogous to brown adipose tissue in mammals (*46*), or it could serve to oxidize potentially harmful fatty acid peroxy radicals (*47*). Enough research has not been done to strongly support either mechanism.

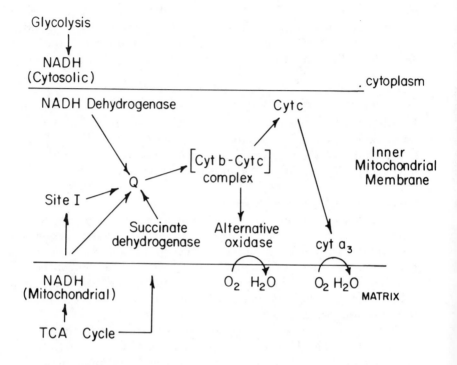

Figure 2. The respiratory ETS in plant mitochondria and the position of the "alternate oxidase." (Reproduced with permission from Ref. 43.)

Minor Oxygen-Consuming Systems

Although the oxygen consumed during oxidative phosphorylation accounts for 90% of the oxygen utilization of living organisms (at least for all the higher organisms, yeast, protozoans, and bacteria studied so far), many other enzymes require oxygen. Furthermore, nonphosphorylating electron-transport systems are found in the endoplasmic reticulum of higher organisms and in some bacteria. One ETS degrades fatty acids and uses NADH as the electron donor while transporting electrons via cytochrome b_5 to oxygen (*13*). A second ETS (Reaction 4) degrades steroid hormones and xenobiotics via NADPH, flavoproteins, and cytochrome P-450 (*50*). It also uses oxygen as the ultimate electron acceptor; the rate-limiting step occurs before the terminal oxygenase. In addition, other oxidases and oxygenases not associated with membrane preparations degrade a variety of amino acids and phenolic and other compounds. Ascorbate oxidase, tryptophan 5-monooxygenase, and lactate monooxygenase are examples (*51*).

However, because these systems play a minor role in the economics of oxygen consumption in living systems, they will not be discussed further.

Methods

In the research reported here we have assessed the oxygen-consumption rate at the sea surface, in the deep sea, and in vertical ocean profiles by employing the biochemistry of oxygen consumption. To a first approximation this biochemistry is nearly identical in bacteria, protists, metazoans, and plants (*15–17, 39*). Accordingly, we assume that this universalism extends to phytoplankton, zooplankton, and bacterioplankton, and we reasoned that by measuring the rate-limiting reaction of the oxygen consumption process in oceanic microbes we could calculate their oxygen consumption rate. Therefore, we used an enzyme assay that measures the combined activities of succinate–tetrazolium oxidoreductase (EC 1.3.99.1), NADH–tetrazolium oxidoreductase (EC 1.6.99.3), and NADPH–tetrazolium oxidoreductase (EC 1.6.99.1) in samples of oceanic particulate matter (*52–61*). The tetrazolium that we used is 2-(*p*-iodophenyl)-3-(*p*-nitrophenyl)-5-phenyltetrazolium chloride (INT). Reaction 5 shows the potential sites of INT reduction in beef heart mitochondria (*62, 64*).

$$\text{NADH} \atop \text{succinate} \Bigg\rangle \hspace{-1em} \longrightarrow \text{Fp} \xrightarrow{\text{INT}} \left\{ \text{CoQ} \atop \text{Cyt b} \right\} \xrightarrow{\text{INT}} \text{Cyt } c_1 \rightarrow \text{Cyt } c \rightarrow \text{Cyt } aa_3 \rightarrow O_2 \qquad (5)$$

Reaction 6 shows the range of sites over which INT is reduced by the ETS in artichoke mitochondria (*63, 64*).

$$\text{succinate} \rightarrow \text{Fp} \rightarrow \underbrace{\text{CoQ} \rightarrow \text{Cyt b} \rightarrow \text{Cyt } c \rightarrow \text{Cyt } aa_3} \rightarrow O_2 \qquad (6)$$

$$\text{INT} \longrightarrow \text{formazan}$$

The enzyme activities measured by this assay represent the maximum reaction rate (V) of succinate dehydrogenase (EC 1.3.99.1) and NADH–ubiquinone oxidoreductase (EC 1.6.99.3)(*see* Scheme I, Complexes I and II), which are the electron-donor systems to the cytochrome chain (Complexes III and IV). Because the electron-donor systems are rate limiting, the assay measures the capacity (V_{max}) of the respiratory ETS. In addition, the NADPH–tetrazolium oxidoreductase provides an estimate of the oxygen-consuming capacity of the microsomal ETS (Reaction 4) (*64*). The ETS assay was developed to rapidly assess the oxygen consumption rate in both the deep sea and in the surface waters (*52–58*). As developed, the assay measures the combined oxygen-consuming capacity of living microbial populations. By using the assay, the depth profiles of the in situ oxygen consumption, or as Craig (*86*) calls it, "the deep-metabolism," can be determined within hours to many thousands of meters. When automated and interfaced with a data-acquisition and computer-mapping system, the ETS assay can be used in surface water to provide maps of oxygen consumption within hours after an ocean survey is completed (65).

This chapter describes the use of this enzyme assay in investigations of upwelling off Oregon, Peru, and Mauritanea, in a survey of a sewage outfall off Los Angeles; and in an open ocean survey of the Costa Rica Dome. Only in the upwelling survey off Mauritanea was the automated version of the assay used (65). The manual methods were used for the deep water below the Costa Rica Dome (58), for the near-surface waters of the outfall off Los Angeles, and for the Oregon upwelling (56). These manual methods have been discussed elsewhere (66–68).

The Automated ETS Assay.   The automated ETS assay system was a laboratory-built (51), enzyme-analysis system dedicated to measurements of the respiratory ETS in cell-free homogenates (Figure 3). The system facilitated the processing of the large number of seawater samples that are required to map seawater oxygen consumption during ocean survey cruises. It ensured standardization of the timing and performance of all the manipulations and transfers in the chemical phase of the enzyme assay. This automated system has been used to map seawater oxygen consumption rates in the euphotic zones of upwelling and coastal waters (51); it has not been used to map the rates in oligotrophic or deep-sea waters.

The system was designed for computer interface in a manner similar to the nutrient Autoanalyzer array. It had a potentiometric output, 0–1 linear with 0–1 absorbance from the spectrophotometer. The starting switch flagged a data-acquisition system that took readings every 15 s for 3.5 min. The enzyme reaction was zero order and linear with time during the incubation period. Reaction rates were calculated by regression analysis of the time course. The device did not free the analyst of all hand manipulations. The cell-free homogenate was prepared by hand and injected into the mixing chamber; then the system proceeded automatically. All reagents were prepared according to Kenner and Ahmed (59). Fifteen seconds after starting, a pair of syringe pumps added the reagent mixtures (Figure 3). The reagents were preheated, mixed for 15 s in the mixing chamber with the homogenate, and then drawn into the temperature-controlled reaction chamber and flow cell where the mixture was monitored. Later the excess was removed, and the mixing chamber and flow cell were rinsed with distilled water. All of the subsystems returned to their base state quickly and were ready for immediate use. A complete cycle took 5 min; this rate enabled the apparatus to make 12 analyses per hour, a rate comparable to data-acquisition rate for seawater nutrient analysis.

The rate of reduction of INT to the insoluble formazan (64)(Reaction 7) in the flow cell was measured by the absorption of the reaction mixture at 490 nm (59). A blank was prepared identically except that homogeniza-

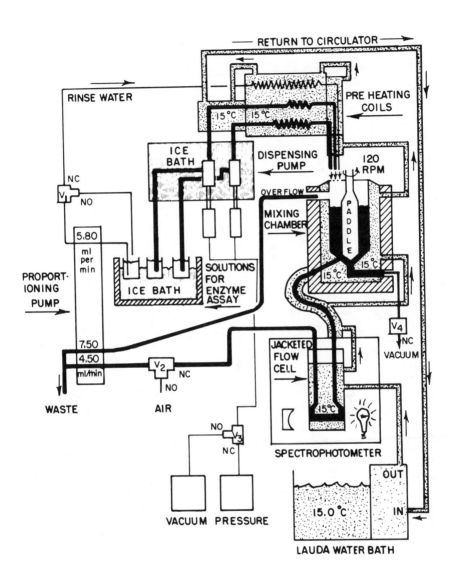

Figure 3. The electron-transport system.

Reactants are pumped from storage, then heated and injected into a mixing chamber (black flow lines). Sample is added to the mixing chamber, the mixture is injected into the flow cell, and the formazan production is monitored every 15 s for 3.5 min. The mixture is flushed, the system is rinsed, and the cycle begins again. The hatched regions are maintained at surface seawater temperatures. The four solenoid valves function most of the time in normally open (NO) or normally closed (NC) positions. A Technicon Autoanalyzer pump and a Bausch and Lomb Spectronic 88 spectrophotometer were used.

$$\text{INT} + 2H^+ + 2e^- \rightarrow HCl + \text{Formazan} \qquad (7)$$

tion buffer was substituted for homogenate in the reaction mixture. The average of several blanks was subtracted from the assay to obtain the rate of the enzymatically catalyzed tetrazolium reduction. The calculation of ETS activity accounts for the liters of seawater sampled (v) and the milliliters of the homogenate used (f), and the milliliters of the reaction mixture (S) (59). The following equation was used for calculating ETS activity (milliequivalents per hour per liter): $ETS_o = [7.54\, SH\,(A - B)]/fv$, where H is the volume of the uncentrifuged homogenate in milliliters, A is the change in absorbance of the 10-mm light path per minute, and B is the absorbance change in the blank. The units of $[SH\,(A - B)]/fv$ are absorbance per mole per liter. The coefficient, 7.54, incorporates the molar extinction coefficient of the INT formazan ($A_{490}^{1\,cm} = 15.9 \times 10^3/M$ cm), the two-electron stoichiometry of INT reduction (Reaction 7), and a time conversion. Because the ETS activity (ETS_o) is measured at a constant temperature (T_o) and the samples were drawn from seawater of varying temperature (T_i), a temperature correction must be made to calculate the ETS_o at in situ temperature (ETS_i). We used the following expressions based on the Arrhenius equation:

$$ETS_i = ETS_o \exp \frac{15.8}{R}\left(\frac{1}{T_o} - \frac{1}{T_i}\right)$$

where R is the gas constant (1.987 cal/mol deg), and 15.8 is the activation energy in kilocalories per mole (57). Both the incubation (T_o) and in situ (T_i) temperatures were expressed in absolute temperatures.

Interpretation of the ETS Measurement: Conversion of Oxygen Consumption. The measurement of ETS activity in oceanic particulate matter is a measure of the capacity or potential of that matter to consume oxygen. It is not a measurement of the in situ rate. The relationship between the in situ rate (r) and the potential rate (R) may be controlled either by the ADP concentration $(29-31, 69)$, by the ratio of ATP to ADP $(70-72)$, or by the ratio of ATP to ADP and inorganic phosphate (P_i) (73) by an expression analogous to the Michaelis–Menten expression, $r = SR/(K_M + S)$, where S equals either [ADP], [ADP]/[ATP], or [ADP]/([ATP] $\times$ [P_i]), and K_M equals the level of S at which $r = 1/2R$. However, because the respiratory control is not well understood $(69-74)$, an average value for $S/(K_M + S)$ has been determined experimentally in cultures and in the field and is used to calculate in situ oxygen-consumption rates from ETS measurements $(60, 61, 75, 76)$. A detailed discussion of these coefficients is found elsewhere $(66-75)$; a brief discussion follows.

For phytoplankton and phytoplankton-dominated euphotic zones, a coefficient of 0.15 is used with the Kenner and Ahmed (59) version of the ETS assay. This coefficient is the average ratio of respiration to ETS activity in eight species of phytoplankton, mostly diatoms (Table 1). In calculating this average, the results for *Ditylum brightwellii* were excluded because they appeared to be anomalous. The ratio was determined on organisms in their log growth phase, and no dinoflagellates were used in the experiments. Thus, although this ratio has been used in assessing respiration from ETS measurements in many euphotic zones (54), it should be most accurate in diatom blooms. At the end of a diatom bloom, when the inorganic nutri-

Table I. Ratios of Oxygen Consumption In Vivo to Respiratory Electron-Transport Activity In Vitro in Marine Phytoplankton

Phytoplankton	Respiration/ETS	N
Bacillariophyceae (Diatoms)		
Coscinodiscus angstii	0.137 ± 0.006	17
Cyclotella sp.	0.106 ± 0.006	10
Ditylum brightwellii	0.319 ± 0.010	13
Phaeodactylum tricornutum	0.161 ± 0.007	10
Skeletonema costatum	0.151 ± 0.008	11
Thalassiosira fluviatilis	0.135 ± 0.011	8
Chlorophyceae		
Dunaliella tertiolecta	0.169 ± 0.016	12
Coccolithophorids		
Cricosphaera carteri	0.159 ± 0.007	11
Haptophyceae (Isochrysids)		
Isochrysis sp.	0.177 ± 0.018	14

NOTE: ETS activity was converted to oxygen equivalent units (5,6 L of oxygen is reduced by 1 electron equivalent) so the respiration-ETS ratio is unitless. The number of experiments is given by N. Data are taken from Reference 60.

ent salts have been stripped from the seawater, a lower ratio should be used. Ratios as low as 0.06 have been found (60) for such nutrient-deficient conditions in culture.

In working with bacteria or zones in the ocean where bacteria dominate the biological community, a different coefficient should be used. In situations where the bacteria are growing and the ETS method of Packard et al. (55) is used to determine the oxygen consumption, a coefficient of 5.0 should be used. This value is based on 49 experiments with five species of marine bacteria (76). In situations where the bacteria are not growing, as in senescence, a coefficient of 0.43 should be used (Table II) (76). Calculations of deep-sea oxygen-consumption rates (68, 76) were made with the assumption that the deep water populations were dominated by bacteria in a senescent phase; therefore, the coefficient of 0.43 was used.

Table II. The Ratio of Respiratory Oxygen Consumption to ETS Activity in Five Species of Marine Bacteria in Three Different Stages of Batch Culture

Species	Phase[a]	Mean	Coefficient of Variation (%)	N
Vibrio adaptatus				
Peptone	growth	2.91	12	6
	senescent	0.492	9	24
Vibrio anguillarum[b]				
Peptone	growth	4.40	7	6
	senescent	0.468	6	6
Glucose	growth	5.90	41	7
	senescent	0.434	25	4
	starvation	0.110	28	4
Vibrio sp.				
SA774	growth	5.63	19	6
	senescent	0.445	14	18
Serratia marinorubra				
Peptone	growth	8.25	26	6
	senescent	0.491	19	18
Pseudomonas perfectomarinus				
Peptone 1 and Peptone 2	growth	4.78	35	9
	senescent	0.306	38	28
Glucose	growth	3.84	20	9
Weighted mean	growth	5.02	29	49
	senescent	0.426	20	98
	senescent	0.110	28	4

[a]Growth-phase data represent means of all samples prior to attainment of maximum populations; senescent-phase data represent means of samples taken following termination of growth in which stable, steady cellular levels of respiration and ETS had been attained; and the starvation data indicate the further depression of respiration–ETS following senescence.
[b]Following glucose depletion, V. anguillarum never attained steady levels of respiration per cell.
[c]The respiration–ETS of senescent P. perfectomarinus from experiments Peptone 1 and Peptone 2 were not significantly different ($P \leq 0.10$) and were pooled.

Direct determinations of oxygen consumption in surface seawater have become possible (77) so that ETS measurements can be directly calibrated under field conditions (Figure 4) (67). For research in the upper 50 m of the ocean, direct calibration will become the preferred procedure.

Observations and Results

Oxygen Consumption in Near-Surface Waters of Coastal Upwelling Areas.   Oxygen consumption in the surface waters of the California Bight off Los Angeles, the Oregon upwelling off Cascade Head, and the northwest African upwelling off Cape Blanc was mapped by measuring ETS activity. The maps show that oxygen consumption in the surface water cannot be treated as a constant in either the time or space domain.

CALIFORNIA BIGHT.   This experiment was designed to test the prediction that the effluent from a sewage outfall in an upwelling area would significantly enhance the respiratory oxygen-consumption rate in the surface waters by stimulating either a phytoplankton or a bacterioplankton bloom. In this test, the rate of oxygen consumption, chlorophyll, nutrient salts (PO_4^{3-}, NO_3^-, NH_4^+, and silicate), temperature, and salinity off Los Angeles in the vicinity of Long Point and Whites Point (Figure 5) were mapped. Upwelling occurred off Long Point, and industrial waste was discharged off Whites Point. The outfall discharged ammonium, phenols, and other noxious wastes into the ocean at the 50-m level through a diffuser designed to dilute the sewage so that it would not penetrate the thermocline and rise to the surface. If the system had worked according to design, the ammonium-rich effluent would not have risen into the euphotic zone; however, the natural upwelling in the region defeated the design. The currents at 25 m traveled northwest, the surface currents traveled southeast, and upwelling occurred near Long Point (78, 79). So, the sewage flowed at depth toward Long Point, was upwelled, and then moved back toward Whites Point in the surface flow (Figure 5). Because the sewage, when properly diluted, will set up a plankton bloom, high oxygen-consumption rates, and high algal or bacterial biomass off Long Point were expected. On all three dates of the mapping survey (March 31, April 1, and April 4) the highest levels of oxygen consumption and phytoplankton biomass lay inshore along the coast between Long Point and Whites Point. On April 4, a distinct plume was evident in both the oxygen-consumption rate and phytoplankton biomass (Figure 5). This plume occurred following a period of upwelling in which ammonium and nitrate rose into the euphotic zone (Figure 6). On April 2, April 3, and again late on April 4, high ammonium (> 4 μM) levels could be found at 10 m in the inshore region of the survey (Figure 6). This ammonium enrichment could easily have stimulated plankton growth that led to the development of the blooms on April 4 (Figure 5C). Therefore, the plume structure in the oxygen-consumption rate

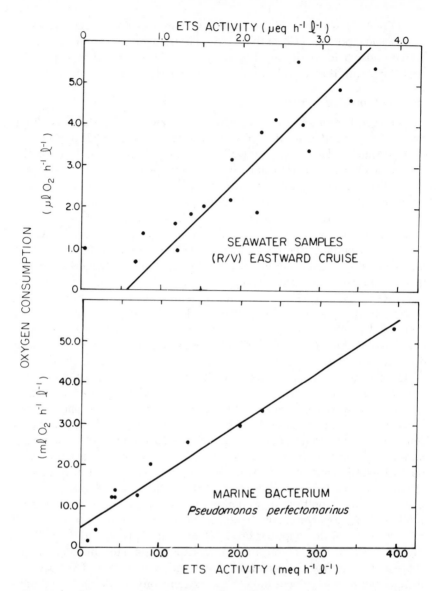

Figure 4. Two examples of the relationship between ETS activity and the oxygen-consumption rate in natural seawater samples (top) and in laboratory cultures of the marine bacterium, Pseudomonas perfectomarinus (bottom).

Top: the equation R = 1.92 ETS − 0.99, where R is the oxygen-consumption rate, describes the samples taken from the Georges Bank–Gulf of Maine cruise on the R.V. Eastward in July 1980. Bottom: the equation R = 1.29 ETS + 4.71 describes the ETS dependence of oxygen consumption of the bacterium, P. perfectomarinus. The differences in the equations may reflect the differences of the analysis of bacterial and phytoplankton ETS. In bacteria the ETS is housed in the cell wall, so crude homogenates must be used. In phytoplankton the ETS is housed in the mitochondria; thus, partially purified homogenates free of cell walls and nuclei are used. (Reproduced with permission from Ref. 67 and 75.)

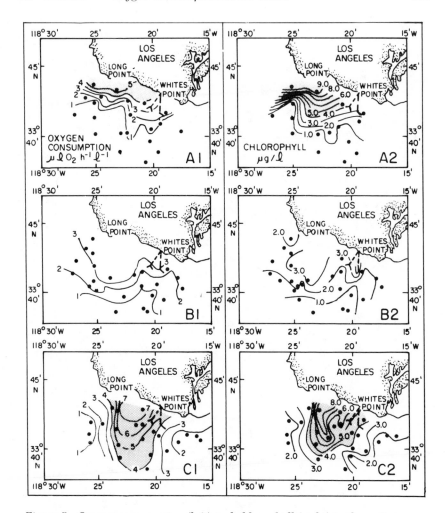

Figure 5. Oxygen consumption (left) and chlorophyll (right) in the surface waters (3 m) on three different dates off the Palos Verdes district of Los Angeles.

The three maps were made on March 31 (A), April 1 (B) and April 4 (C) during the Outfall I cruise of the R.V. Thomas G. Thompson. The ranges of nutrient and temperature conditions are summarized in Figure 6. Note the well developed plume on April 4. Oxygen consumption was calculated from ETS activity measurements. A manual version of the ETS assay (53) was used.

was probably caused by a phytoplankton bloom that, in turn, was fed by ammonium upwelling at Long Point (Figure 7) that originated at the Whites Point sewage outfall. Whitledge et al. (78) came to a similar conclusion on the basis of hydrographic data.

OREGON COAST.   Upwelling occurs along the Oregon coast every summer (Figure 8). Cold waters with temperatures as low as 6 °C upwell

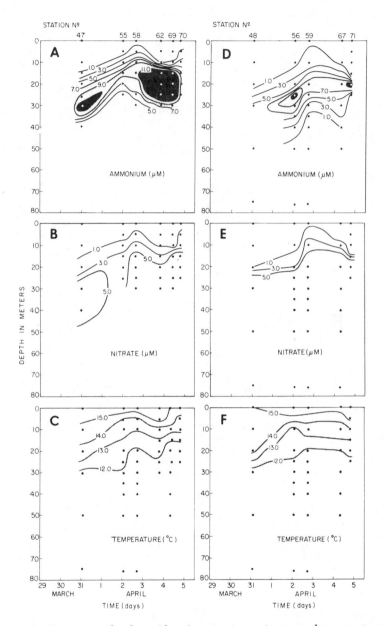

Figure 6. Time-series depth profiles of ammonium, nitrate, and temperature in the inshore region (Stations 47, 55, 58, 62, 69, and 70) and the offshore region (Stations 48, 56, 59, 67, and 71) between Long Point and Whites Point off Los Angeles.

Note the ammonium-rich subsurface waters rising between March 31 and April 3, and also the rising isotherms over the same period in the inshore stations. Warm nutrient-poor surface waters after April 3 suggest an incursion of offshore water. (The station locations can be found in Figure 1 of Reference 79.)

STATIONS

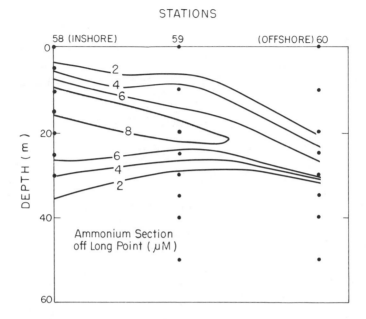

Figure 7. *The rising isopleths of seawater ammonium along an offshore–onshore*
section at Long Point, Los Angeles.

The ammonium serves as a tracer for the effluent from the White Point sewage outfall. Its
rise to the surface at the inshore station is driven by coastal upwelling. Once the ammonium
breaks into the euphotic zone it can stimulate phytoplankton growth.

to the surface and bring with them nitrate levels as high as 28 μM. A plankton bloom should be stimulated by this nutrient injection, and this bloom should stimulate the total plankton community to consume more oxygen. Thus, the rate of oxygen consumption should increase offshore as the plankton population increases. To test this prediction, the oxygen-consumption rate, temperature, salinity, nitrate, ammonium, nitrite, and chlorophyll were mapped in the upwelling region. On the first three passes cold water upwelled next to the coast, especially off Cascade Head. Nitrate levels decreased from 24–28 μM inshore to undetectable levels offshore (80). The chlorophyll and the oxygen-consumption rate displayed an inverse trend with respect to nitrate and temperature. Oxygen-consumption rates and chlorophyll were low inshore and high offshore (Figure 8). On the last survey the system had relaxed and the warm offshore water had moved in over the area like a blanket. The lowest temperature in the area was 14.6 °C where previously water of 8 °C could be found, and the rate of oxygen consumption, which previously had increased offshore, now decreased from 8 μL of O_2/h L inshore to less than 2 μL of O_2/h L offshore (Figure 9). Thus, two contrasting patterns of oxygen consumption in an

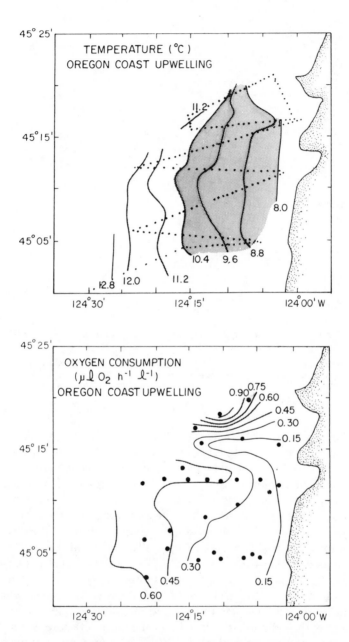

Figure 8. The offshore temperature gradient and offshore oxygen-consumption gradient along the Oregon coast when upwelling is strong (CUE II expedition). The highest rates of oxygen consumption occur offshore. These measurements and those in Figures 7 and 9 were made with the manual ETS assay.

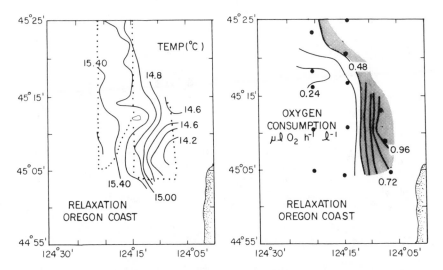

Figure 9. The offshore temperature (left) and oxygen-consumption (right) gradients along the Oregon coast when the upwelling relaxes (CUE II expedition). The highest rates of oxygen consumption occur inshore.

upwelling system emerged. The pattern for an upwelling event is low oxygen consumption in the freshly upwelled inshore water, increasing to high oxygen consumption offshore where the nutrient-rich water has stabilized in the surface waters and produced a plankton bloom. The pattern for an upwelling reversal or relaxation event is high oxygen consumption in the warm inshore water that has recently flowed onshore to replace the subsiding upwelling source water. The stable, plankton-rich offshore water moves inshore along the coast and oligotrophic oceanic water stands just offshore. This plankton-poor water is characterized by a low oxygen consumption rate.

NORTHWEST AFRICAN COAST.   The signature of a relaxation event was seen in the data from the *Joint I* expedition to the Mauritanean upwelling system in March and April 1974. During the mapping study off Cape Blanc (Figure 10), winds were conducive to upwelling until April 7 (*81, 82*). Shortly after that date the temperature began to rise inshore (*81*) and by April 9 the system had relaxed. As off Oregon, the relaxation off Cape Blanc could be seen in the maps of the oxygen-consumption rates because the highest values are found inshore (April 9 and 10, Figure 10), where as during the upwelling they occurred farther offshore.

The scenario of the upwelling that corresponds to maps of oxygen consumption in Figure 10 is described elsewhere (*82*). The upwelling along the coast was strong for the week preceding Grid 17 (Figure 10, March 30–April 5). During this upwelling event the density throughout the water column increased, the offshore near-surface transport increased, and the

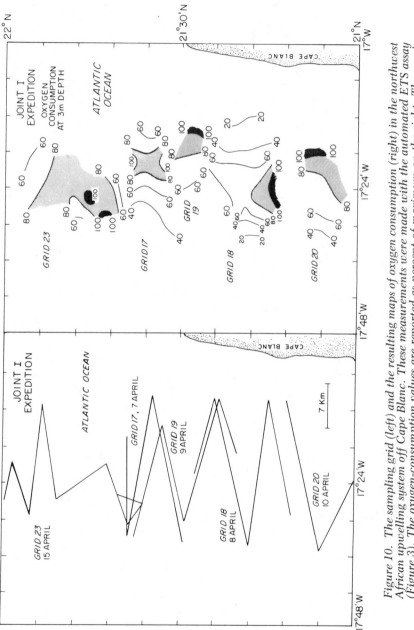

Figure 10. The sampling grid (left) and the resulting maps of oxygen consumption (right) in the northwest African upwelling system off Cape Blanc. These measurements were made with the automated ETS assay (Figure 3). The oxygen-consumption values are reported as percent of maximum on the right. The maximum values for the five grids were, from top to bottom, 1.25, 2.68, 3.21, 8.04, and 5.36 μeq/Lh.

temperature minimum and the nitrate and chlorophyll maxima moved to the edge of the continental shelf. The cross-shelf oxygen-consumption pattern that reflects the effect of this upwelling event on the metabolism of the plankton communities is shown in Grids 17 and 18 (Figure 10). The oxygen consumption peaks at the outer shelf and proceeds at slower rates inshore and seaward of this maximum as expected from the results of the Oregon experiment (Figures 8 and 9). At the end of the upwelling event the winds relaxed and then reversed (April 7–9). The cold nutrient-rich inshore water responded by subsiding, and the warm offshore water responded by moving in close to the coast (*81–83*). The pattern of oxygen consumption on April 9 and 10 reflects this relaxation (Grids 19 and 20). The zones with the higher rate of oxygen consumption were found inshore of their previous positions in Grids 17 and 18. The chlorophyll and photosynthesis maxima were likewise shifted inshore (*81, 83*). Between April 10 and 15 the winds blew toward the equator (upwelling favorable), and offshore transport resumed as during the March 30–April 5 period. These events caused coastal upwelling to reoccur and pushed the zone of high-oxygen-consumption rate offshore (Grid 23) near its shelf-edge location as before (Grids 17 and 18). Thus, the zone of enhanced oxygen consumption responded to upwelling off northwest Africa as it responded to upwelling off Oregon. This zone is found offshore between the upwelling source water and the oceanic oligotrophic water during an upwelling event; it is found inshore in front of the advancing oligotrophic water during a relaxation event.

Oxygen Consumption in the Deep Sea.   The rates of oxygen consumption in the deep sea are extremely low; 1 μL/L per year below 300 m is a value obtained by many approaches (*84–87*). Nevertheless, as advection–diffusion models show (*85, 86*), the oxygen consumption has a significant effect on the distribution of O_2, CO_2, $CaCO_3$, as well as NO_3^-, PO_4^{3-}, and silicate in the deep sea (*84, 85*). Before 1971 deep-sea oxygen utilization was calculated from advection–diffusion (*87, 88*) or nutrient regeneration (*85, 86*) models. Now calculations are based on microbial biomass (*84, 89*), ETS activity (*55, 57, 58, 68*), in situ incubations with ^{14}C-labeled substrates (*90, 91*), helium–tritium dating (*92*), or particle flux through the water column (*93*). From the models, from biomass, from ETS activities, from dating techniques, and from particle-flux measurements, the oxygen-consumption rates, between 1 and 4 km, fall in the 0.1–40 μL of O_2/L per year range. From the in situ incubations, the rates appear to be much higher as has been shown with biological oxygen demand (BOD)-type measurements (*94–96*). Rates ranging from 5 to 77 μL of O_2/L per year were found (*90, 91*) but such rates are likely to drive the deep-sea anoxic (*84*).

VERTICAL DISTRIBUTION.   The deep-sea oxygen consumption rates calculated from ETS activity fall in the 0.05–6 μL of O_2/L per year range for the waters below 1 km in the Peru Current and Costa Rica Dome region. From the sea surface to the bottom, the values diminish exponentially

with depth (68, 87, 93). An empirical model was developed (93) for predicting deep-sea rates of oxygen utilization from particle-settling velocities
and productivity in the euphotic zone. From this model, a family of depth
profiles was predicted for oxygen utilization in the water column (Figure
11). The ETS-derived values of oxygen utilization from the eastern tropical
Pacific fall around the predicted profile for a surface productivity of 100 g
of C/m^2 per year. Estimates of productivity for the waters where the ETS
measurements were made bracket this productivity value. Owen and Zeitschel (97) measured 70 g of C/m^2 per year, and Broenkow (98) calculated
140 g of C/m^2 per year. In agreement with this model (93), Ben-Yaakov
(99), Kroopnick (100), and Jenkins (92) showed that oxygen-utilization
rates decrease exponentially. For the Peru Current and the Costa Rica
Dome regions, Packard et al (68) found an exponential decrease of the
form: $R = Kz^{-a}$. In the Peru Current the coefficient K and the exponent a
were 1.15 and 1.0, respectively, and for the Costa Rica Dome they were
1.63 and 0.84, respectively.

HORIZONTAL DISTRIBUTION.   As the profiles just discussed show (93),
the deep-sea oxygen consumption rate should vary according to the surface
productivity. To demonstrate this covariation in the Costa Rica Dome region, ETS activity was measured in the deep sea, and chlorophyll and phytoplankton productivity were measured in the surface waters during the
PINTA expedition (58, 101). The results showed that under the statistical
center of the Costa Rica Dome (9° N, 89° W) (84) the deep-sea oxygen-

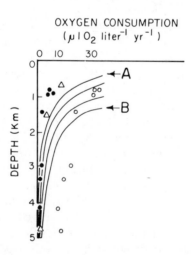

Figure 11. Depth profiles of oxygen consumption. The profile defined by the
solid circles (●) was calculated from ETS activity measurements (○) as reported
previously (55). The profile defined by the open triangles (△) is based on bacte-
riological studies (84). The family of continuous profile lines between A and B
was generated by Suess's (93) model of oxygen-consumption rates.

consumption rate is fivefold higher than comparable rates 160 km to the west (Figure 12). Therefore, surface water biological productivity has an observable effect on deep-water oxygen-consumption rates.

Discussion

Sea surface maps of static properties such as temperature, chlorophyll a, nutrient salts, and pH are readily available to oceanographic research (1). They can be constructed with data from ships (102, 103), airplanes (104), or satellites (105). Rarely, however, are sea surface maps constructed with the dynamic properties such as reaction rates, growth rates, or current velocities because their measurement is intrinsically slow and technologically difficult. The time scales for measuring most of the dynamic processes relevant to chemical oceanography range from hours to days; consequently, by conventional methods, real-time mapping is not feasible. However, by employing enzyme analysis, the potential or the capacity of many rate processes can be mapped in real time. The in situ rate can then be calculated from the potential rate and an empirically derived coefficient (66, 67), or from the potential rate and the kinetic control mechanism as described in the "Methods" section. Many chemical reactions in the ocean could be mapped by this procedure. The rate of CO_2 fixation could be mapped by measuring the activity of one of the enzymes associated with photosynthesis. Initial investigations with ribulose-1,5-bisphosphate carboxylase (EC 4.1.1.39) have not been fruitful (106), but other photosynthetic enzymes have not been tried. As an example, the enzyme, ferredoxin-NADP-reductase (EC 1.18.1.2), is theoretically a good candidate, because it serves as a major transmitter of reducing equivalents from the photosynthetic electron-transport system to a coenzyme ($NADP^+$) commonly involved in photosynthetic reactions. The rate of nitrate uptake and ammonium uptake could be mapped by measuring the activity of one or more of the enzymes associated with nitrogen assimilation. Nitrate reductase (EC 1.6.6.1) is a good index of recent nitrate-reducing activity (107), and glutamine synthetase (EC 6.3.1.2) may be a good index of ammonium uptake. Ammonium production can be predicted from glutamate dehydrogenase (EC 1.4.1.2) activity in zooplankton (108, 109). This chapter has described some investigations into the use of the enzymatic activity of the respiratory ETS in mapping the rate of oxygen consumption. The procedure was tried in the surface waters of upwelling areas off Oregon and Mauritanea, in the vicinity of a sewage outfall off Los Angeles, and in the deep water (3000 m) below the Costa Rica Dome and the Peru Current.

Surface-Water Oxygen Consumption.   Simpson and Zirino (103) observed that the pH of freshly upwelled seawater is distinctly lower than normal surface seawater; they used this characteristic to map the extent of freshly upwelled seawater off the coast of Peru. Other characteristics of freshly upwelled seawater are low chlorophyll, low temperature, low oxy-

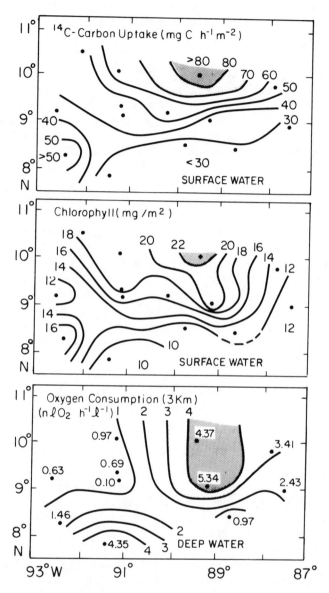

Figure 12. Deep-sea oxygen-consumption rates at 3000 m under the Costa Rica Dome and the overlying phytoplankton biomass and productivity in the surface waters. The maxima of all three properties are close to the center of the Dome (i.e., 9°N, 89°W). These measurements were made manually.

gen, and high-nutrient salts (*110*). From the results of the *CUE II* cruise to the Oregon upwelling, one can add to this list the low rate of oxygen consumption and its equivalent, low metabolism (Figure 8). Seawater downstream in an upwelling area is characterized by a high oxygen-consumption rate (high microbial metabolism), high chlorophyll a, high temperature, high pH, high oxygen, and low nutrients. Relaxed upwelling systems are characterized in the surface waters by the water types normally found downstream in an upwelling system, namely, high chlorophyll, high oxygen consumption rate, and temperature (Figure 9).

The maps off Mauritanea partially reveal three bands of oxygen consumption in an upwelling area (Figure 10). Bands of low oxygen consumption are discernable in the freshly upwelled water and in the distant offshore water; a band of high oxygen-consuming water lies in between. During a relaxation period the inshore band is eliminated. This banding and its shift with changing upwelling conditions would have been revealed much better if the maps had been made in exactly the same location each time instead of being made in different locations along the coast; however, when the sampling pattern was determined the short-term time-dependence of upwelling was not recognized. Very likely the current investigation of the California upwelling (Organization of Persistent Upwelling Structures, an NSF research program at the University of Southern California) will reveal, in detail, this type of banding now that the short-term temporal variability in upwelling systems is recognized and sampling is focussed on a single region, that is, Point Conception (*4*).

The maps of the ocean off Long Point (Figure 5) show the plume characterized by high oxygen consumption and high chlorophyll. Furthermore, they provide evidence that outfalls in the vicinity of upwelling must discharge their waste below the upwelling source water, which may be much deeper than the average thermocline depth.

Deep-Sea Oxygen Consumption. The deep-sea map of oxygen consumption under the Costa Rica Dome reveals for the first time that the deep-sea is not homogeneous with respect to metabolism and the oxygen-consumption rate. The rate appears higher under the productive upwelled water of the Dome (Figure 12). This observation suggests a link between oceanic surface waters and the deep sea below, and infers that deep-sea shadows of other oceanographic surface features such as equatorial upwelling zones or the Antarctic convergence also will be detectable by zones of enhanced oxygen consumption.

The vertical profiles of oxygen consumption (Figure 11) (*68*) show that, in spite of a connection between surface productivity and the deep water, most of the organic matter falling below the euphotic zone is oxidized in the upper 500 m. This zone serves as a great biological filter that protects the deep sea from receiving more organic matter than it can metabolize. The deep waters below the Peru Current are already seriously depleted of oxygen; only 2.4 mL/L of an original 7 mL/L remain. If more organic matter were to fall below 500 m the deep-water oxygen supply could be completely exhausted.

So far we have connected ETS activity to oxygen and the oxygen-consumption rate. In this context, it can also be used to estimate the speed of deep-sea currents and the age of the deep-sea water. Such estimates will be in error to the extent that advection and diffusion are not considered. Nevertheless, they serve as an exercise that scales the role of deep-sea biological oxygen consumption, and this exercise is rarely done. To make these estimates, prior knowledge of the current direction and the chemical properties of the water is needed. In the southeastern Pacific, the deep water east of the east Pacific rise at 2000 m has been observed to flow north (*111*) through passes in the ridge system that separate the various basins in this region (Figure 13). Furthermore, the deep-sea oxygen content decreases from basin to basin, starting with the southeastern Pacific Basin below the Antarctic circumpolar current, and ending with the Panama Basin. At the start of its journey in the southeast Pacific Basin the 2000-m water contains

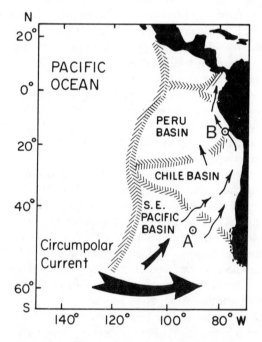

Figure 13. Deep-ocean circulation through the ridges and basins of the southeastern Pacific. The east Pacific rise at 120°W provides an effective barrier to deep water penetrating eastward from the central Pacific. The stations, A and B, are referred to in the text. The figure was drawn from Lonsdale's (111) description of the deep waters of the eastern south Pacific.

3.25 mL of O_2/L (*111*). By the time it reaches the southern part of the Peru Basin at 15° S it contains only 2.4 mL of O_2/L, having lost 0.85 mL of O_2/L along the way.

We measured ETS activity and calculated the oxygen-consumption rate for the water at 2000 m in the southern part of the Peru Basin (*68*). Because of the high productivity of the Peru current upwelling, this rate is likely to be higher than comparable deep-sea oxygen-consumption rates in other parts of the southeastern Pacific. Nevertheless, by assuming that these rates are roughly representative of the deep-sea rates in other basins and by assuming that advective and diffusive O_2 losses were smaller than respiratory losses, we calculated both the current speed and the time required for the water to flow from the southeast Pacific Basin to the Peru Basin. Assuming a deep oxygen-consumption rate of 3.8 μL of O_2/L per year (*68*), we calculate that it would take 224 years to consume 0.85 mL of oxygen. Because the oxygen minimum is strongest along the eastern boundary of the Pacific, diffusion and advection would act to replenish the deep-sea oxygen and minimize the observed depletion due to respiration along the trajectory of the current. Thus, 224 years is a lower limit. The distance between the study area and the starting point (Figure 13) is 4225 km. Thus the water must have been flowing 19 km/year or 0.6 mm/s if it traveled 4225 km in 224 years. This age only represents the time required for the deep water to flow from the northern part of the southeast Pacific Basin to the southern part of the Peru Basin. Because the oxygen at the starting point was already low (3.25 mL of O_2/L), significant oxygen consumption had already taken place in the water from the time it started traveling north from the depths of the Circumpolar Ocean to the time it reached the northern part of the southeast Pacific Basin. This oxygen consumption and the total age of the water can be calculated from knowledge of deep-sea oxygen distribution. At 2000 m under the Antarctic Convergence the water contains 5 mL of O_2/L (*112, 113*). Thus 1.75 mL of O_2/L was lost in transit to the southeast Pacific Basin at 40° S. At R = 3.8 μL of O_2/L per year, the transit time must have been 460 years. The total transit time from 70° S to 15° S would be 685 years. Similar ages (i.e., 750 and 900 years) were calculated for the 3000-m level under the Costa Rica Dome (*54, 57*). As mentioned before, such calculations in eastern boundary regions will yield low estimates of water-mass age because advection and diffusion have been ignored.

Acknowledgments

This work was supported by NSF Grants OCE–75–23718 AO1, OCE 77–18668, OCE 78–00610, and by ONR Contract N00014–76–C–0271. I thank Peg Colby and Pat Boisvert for enduring the many revisions of the manuscript, and J. Rollins for drafting all the figures. D. Harmon worked around the clock with me to map the Whites Point outfall and the Oregon upwelling. J. Abrahamson constructed the automated system shown in Figure 3 and used it in mapping the Mauritanean upwelling. A. Westhagen and J. Rix interfaced the automated ETS system to the IRIS computer system (Coastal Upwelling Ecosystems Analysis program) in the northwest Africa experiment. J. Kelley was chief scientist and coordinated the mapping programs in both the Oregon upwelling (CUE II) and the northwest Africa (JOINT I) upwelling experiments. J. H. Vosjan and two reviewers made helpful suggestions that improved the manuscript between drafts, and D. Corson made thoughtful modifications that improved the final version. I am indebted to them all for their cooperation and assistance in this work. I also thank E. Green, the director of Office of Naval Research's chemical oceanography program, for supporting and encouraging the application of enzymology to the mapping of rate processes in the ocean. This is contribution No. 84023 from the Bigelow Laboratory for Ocean Sciences.

Literature Cited

1. Zirino, A.; Clavell, C., Jr.; Seligman, P. F. Nav. Res. Rev. 1978, 31(6), 26–38.
2. Armstrong, F. A. J.; Stearns, C. R.; Strickland, J. D. H. Deep-Sea Res. 1967, 14, 381–89.
3. Dugdale, R. C. Geoforum 1972, 11, 47–61.
4. Jones, B. H.; Dugdale, R. C.; Brink, K. H.; Stuart, D. W.; Van Leer, J. C.; Blasco, D.; Kelley, J. C. in "Coastal Upwelling, Pt. A"; Suess, E.; Thiede, J., Eds.; Plenum Press: New York, 1983; pp. 37–60.
5. Arden, E.; Lockett, W. T. J. Soc. Chem. Ind., London 1914, 33, 523, 1122–24.
6. Metcalf, L.; Eddy, H. D. "Sewerage and Sewage Disposal"; McGraw-Hill: New York, 1930; p. 283.
7. Mann, K. H. Adv. Ecol. Res. 1969, 6, 1–71.
8. Platt, T.; Mann, K. H.; Ulanowicz, R. E. "Mathematical Models in Biological Oceanography"; Unesco Press: Paris, 1981; p. 156.
9. Richards, F. A. in "Treatise on Marine Ecology and Paleontology"; Hedgpeth, J. H., Ed.; Geol. Soc. Amer. Mem. No. 67, New York, 1957; pp. 185–235.
10. Green, D. E. in "Comprehensive Biochemistry"; Florkin, M.; Stotz, E. H., Eds.; Elsevier: New York, 1966; pp. 309–26.
11. Ernster, L.; Schatz, G. J. Cell Biol. 1981, 91, 227S–255S.
12. Hayaishi, O. in "Biological Oxidations"; Singer, T. P., Ed.; Interscience Publishers: New York, 1968; pp. 581–601.
13. Siekevitz, P. in "Mitochondria and Microsomes"; Lee, C. P.; Shatz, G.; Dallner, G., Eds.; Addison-Wesley: Massachusetts, 1981; pp. 541–62.
14. Jones, G. F. J. Gen. Microbiol. 1979, 115, 27–35.
15. Wehrle, J. P. Curr. Top. Membr. Transp. 1982, 16, 407–30.
16. Ducet, G.; Lance, C. "Plant Mitochondria"; Elsevier: New York, 1978; p. 454.
17. Tzagoloff, A. "Mitochondria"; Siekevitz, P., Ed.; Plenum Press: New York, 1982; p. 342.
18. Slater, E. C. in "Electron Transfer Chains and Oxidative Phosphorylation", Quagliariello, E.; Papa, S.; Palmieri, F.; Slater, E. C.; Siliprandi, N., Eds.; North-Holland: Amsterdam, 1975, pp. 3–14.
19. Ragan, I. C.; Galante, Y. M.; Hatefi, Y.; Ohnishi, T. Biochemistry 1982, 21, 590–94.

20. Weiss, H.; Wingfield, P.; Leonara, K. in "Membrane Bioenergetics"; Lee, C. D.; Schatz, G.; Ernster, L., Eds.; Addison-Wesley: Massachusetts, 1979; pp. 119–32.
21. Singer, T. P.; Ramsay, R. R.; Paech, C. in "Mitochondria and Microsomes"; Lee, C. P.; Schatz, G.: Dallner, G., Eds.; Addison-Wesley: Massachusetts, 1981; pp. 155–90.
22. Singer, T. P. in "Oxidases and Related Redox Systems"; King, T. E.; Mason, H. S.; Morrison, M., Eds.; John Wiley: New York, 1965; pp. 448–81.
23. Trumpower, B. L. *J. Bioenerg. Biomembr.* 1981, *12(3–4)*, 151–64.
24. Gutman, M. *Biochim. Biophys. Acta* 1980, *594*, 53.
25. International Union of Biochemistry, "Enzyme Nomenclature, 1978"; Academic Press: New York, 1979; p. 606.
26. Chance, B. in "Oxygen and Living Processes"; Gilbert, D. L., Ed.; Springer-Verlag: New York, 1981; pp. 200–209.
27. Wikström, M.; Krab, K.; Saraste, M. "Cytochrome Oxidase, A Synthesis"; Academic Press: New York, 1981; p. 198.
28. Hackenbrock, C. R. *Trends Biochem. Sci. (Pers. Ed.)* 1981, *6*, 151–54.
29. Chance, B.; Williams, G. R. *J. Biol. Chem.* 1955, *217*, 409–27.
30. Chance, B.; Williams, G. R. *J. Biol. Chem.* 1955, *217*, 429–38.
31. Chance, B.; Williams, G. R. *Adv. Enzymol. Relat. subj. Biochem.* 1956, *17*, 65–134.
32. Mitchell, P. *Ann. N.Y. Acad. Sci.* 1982, *341*, 564–84.
33. Papa, S. *J. Bioenerg. Biomembr.* 1982, *14(2)*, 69–86.
34. Chance, B. *Annu. Rev. Biochem.* 1977, *46*, 955–1026.
35. Capaldi, R. A. *Biochim. Biophys. Acta* 1982, *694*, 291–306.
36. Srere, P. A. *Trends Biochem. Sci. (Pers. Ed.)* 1980, *5*, 120–21.
37. Imagawa, T.; Nakamura, T. *J. Biochem. (Tokyo)* 1978, *84*, 547–57.
38. Bergsma, J.; Strijker, R.; Alkema, J. Y. E.; Seijen, H. G.; Konings, W. N. *Eur. J. Biochem.* 1981, *120*, 599–606.
39. Jones, C. W. in "International Review of Biochemistry, Microbial Biochemistry"; Quayle, J. R., Ed.; University Park Press: Baltimore, 1979; Vol. 21, p. 381.
40. Ludwig, B. *Biochim. Biophys. Acta* 1980, *594*, 177–89.
41. Burris, J. E. in "Primary Productivity in the Sea"; Falkowski, P. G., Ed.; Plenum Press: New York, 1980; pp. 411–32.
42. Raven, J. A.; Beardall, J. *Can. Bull. Fish. Aquat. Sci.* 1981, *210*, 55–82.
43. Moore, A. L.; Rich, P. R. *Trends Biochem. Sci. (Pers. Ed.)* 1980, *5*, 284–88.
44. Palmer, J. M.; Moller, I. M. *Trends Biochem. Sci. (Pers. Ed.)* 1982, *7*, 258–61.
45. Nash, D.; Wiskich, J. T. *Plant Physiol.* 1983, *71*, 627–34.
46. Kelly, G. J. *Trends Biochem. Sci. (Pers. Ed.)* 1982, *7*, 233.
47. Rustin, P.; Dupont, J.; Lance, C. *Trends Biochem. Sci. (Pers. Ed.)* 1983, *8*, 155–57.
48. Kow, Y. W.; Erbes, D. L.; Gibbs, M. *Plant Physiol.* 1982, *69*, 442–47.
49. Peschek, G. A. *Biochem J.* 1983, *210*, 269–72.
50. Coon, M. J.; Black, S. D.; Koop, D. R.; Morgan, E. T.; Persson, A. V. in "Mitochondria and Microsomes"; Lee, C. P.; Schatz, G.; Dallner, G., Eds.; Addison-Wesley: Massachusetts, 1981; pp. 707–28.
51. Hayaishi, O.; Yoshida, R. in "Mitochondria and Microsomes"; Lee, C. P.; Schatz, G.; Dallner, G., Eds.; Addison-Wesley: Massachusetts, 1981; pp. 611–28.
52. Packard, T. T., Ph. D. Thesis, University of Washington, Seattle, 1969.
53. Packard, T. T. *J. Mar. Res.* 1971, *29*, 235–44.
54. Packard, T. T. *J. Mar. Res.* 1979, *37(4)*, 711–42.
55. Packard, T. T.; Healy, M. L.; Richard, F. A. *Limnol. Oceanogr.* 1971, *16*, 60–70.
56. Packard, T. T.; Harmon, D.; Boucher, J. *Tethys* 1974, *6(1–2)*, 213–22.
57. Packard, T. T.; Devol, A. H.; King, F. D. *Deep-Sea Res.* 1975, *22*, 237–49.

58. Packard, T. T.; Minas, H. J.; Owens, T.; Devol, A. in "Ocean Sound Scatter Prediction"; Andersen, N. R.; Zahuranec, B. J., Eds.; Plenum Press: New York, 1977; p. 859.
59. Kenner, R. A.; Ahmed, S. I. *Mar. Biol.* **1975**, *33*, 119.
60. Kenner, R. A.; Ahmed, S. I. *Mar. Biol.* **1975**, *33*, 129.
61. Owens, T.; King, F. D. *Mar. Biol.* **1975**, *30*, 27.
62. Lester, R. L.; Smith, A. L. *Biochim. Biophys. Acta* **1961**, *47*, 475–96.
63. Kalina, M.; Palmer, J. M. *Histochemie* **1968**, *14*, 366–74.
64. Altman, F. D. *Histochem. Cytochem.* **1976**, *9(3)*, 1–56.
65. Abrahamson, J.; Setchell, F.; Jones, V.; Packard, T. T. CUEA Technical Report No. 48, 1980, p. 54.
66. Christensen, J. P.; Packard, T. T. *Limnol. Oceanogr.* **1979**, *24(3)*, 576–83.
67. Packard, T. T.; Williams, P. J. LeB. *Oceanol. Acta* **1981**, *4(3)*, 351–58.
68. Packard, T. T.; Garfield, P. C.; Codispoti, L. A. in "Coastal Upwelling, Pt. A."; Suess, E.; Thiede, J., Eds.; Plenum Publishing: New York, 1983; pp. 147–73.
69. Jacobus, W. E.; Moreadith, R. W.; Vandegaer, K. M. *J. Biol. Chem.* **1982**, *257*, 2397–402.
70. Davis, E. J.; Lumeng, L. *Febs Lett.* **1974**, *48*, 250–52.
71. Davis, E. J.; Lumeng, L. *J. Biol. Chem.* **1975**, *250*, 2275–82.
72. Küster, U.; Letko, G.; Kunz, W.; Duszynsky, K.; Bogucka, K.; Wojtczak, L. *Biochim. Biophys. Acta* **1981**, *636*, 32–38.
73. Erecínska, M.; Wilson, D. F. *J. Membr. Biol.* **1982**, *70*, 1–14.
74. Erecínska, M.; Davis, J. S.; Wilson, D. F. *Arch. Biochem. Biophys.* **1979**, *197(2)*, 463–69.
75. Packard, T. T.; Garfield, P. C.; Martinez, R. *Deep-Sea Res.* **1983**, *30(3A)*, 227–43.
76. Christensen, J. P.; Owens, T. G.; Devol, A. H.; Packard, T. T. *Mar. Biol.* **1980**, *55*, 267–76.
77. Williams, P. J. LeB.; Jenkinson, N. W. *Limnol. Oceanogr.* **1983**, 576–84.
78. Whitledge, T. E.; Dugdale, R. C.; Hopkins, T. S. *Journ. Etud. Poll. CIESM, Abstract* **1972**, 115–17.
79. MacIsaac, J. J.; Dugdale, R. C.; Huntsman, S. A.; Conway, H. L. *J. Mar. Res.* **1979**, *37*, 51–66.
80. Jones, V.; Garfield, P.; Packard, T. T. Bigelow Laboratory for Ocean Sciences Technical Report No. 19. 1981, p. 72.
81. Barton, E. D.; Huyer, A.; Smith, R. L. *Deep-Sea Res.* **1977**, *24*, 7–23.
82. Jones, B. H.; Halpern, D. *Deep-Sea Res.* **1981**, *28A*, 71–81.
83. Jones, B. CUEA Technical Report No. 37. 1978, p. 263.
84. Williams, P. M.; Carlucci, A. F. *Nature (London)* **1976**, *262*, 810.
85. Munk, W. H. *Deep-Sea Res.* **1966**, *13*, 707.
86. Craig, H. *J. Geophys. Res.* **1971**, *76(21)*, 5078–86.
87. Riley, G. A. *Bull. Bingham Oceanogr. Collect.* **1951**, *13*, 1–126.
88. Wyrtki, K. *Int. J. Oceanol. Limnol.* **1967**, *1(2)*, 117–47.
89. Holm-Hansen, O.; Pearl, H. W. in "Memorie dell' Instituto Italiano di Idrobiologia"; Melchiorri-Santolini, U.; Hopton, J. W., Eds.; 1972, *29*, 151–68.
90. Jannasch, H. W.; Eimhjellen, K.; Wirsen, C. D.; Farmanfarmaian, A. *Science (Washington, D.C.)* **1971**, *171*, 672–75.
91. Sorokin, Y. I. in "The Changing Chemistry of the Oceans"; Dryssen, D.; Jagner, D., Eds.; Almquist and Wiksell: Stockholm, 1972; pp. 189–204.
92. Jenkins, W. J. *J. Mar. Res.* **1980**, *38*, 533–69.
93. Suess, E. *Nature (London)* **1980**, *288*, 260–63.
94. Seiwell, H. R. *Nature (London)* **1937**, *140*, 506–7.
95. Rakestraw, N. W. *J. Mar. Res.* **1947**, *6*, 259–63.
96. Skopintsev, B. A. *Oceanol.* **1976**, *15*, 556–60.
97. Owen, R. W.; Zeitschel, B. *Mar. Biol.* **1970**, *7*, 32–36.
98. Broenkow, W. W. *Limnol. Oceanogr.* **1965**, *10*, 40–52.
99. Ben-Yaakov, S. *Mar. Chem.* **1972**, *1(1)*, 3–26.

100. Kroopnick, P. M. *Deep-Sea Res.* **1974**, *21*, 211–27.
101. Kuntz, D.; Packard, T. T.; Devol, A.; Anderson, J. Technical Report 321, Dept. Oceanogr., University of Washington, 1975.
102. Kelley, J. C. in "The Ecology of the Seas"; Cushing, D. H.; Walsh, J. J., Eds.; Saunders Co.: Philadelphia, 1976; pp. 361–87.
103. Simpson, J. J.; Zirino, A. *Deep-Sea Res.* **1980**, 27, 733–44.
104. Stuart, D. S. in "Coastal Upwelling"; Richards, F. A., Ed.; American Geophysical Union: Washington, D. C., 1980; pp. 32–43.
105. Yentsch, C. S. *Nature (London)* **1973**, *244*, 307–8.
106. Glover, H. E.; Morris, I. *Limnol. Oceanogr.* **1979**, *24*, 510–19.
107. Blasco, D.; Packard, T. T.; MacIsaac, J. J.; Dugdale, R. C. *Limnol. Oceanogr.* **1984**, *29*, 275–86.
108. Bidigare, R. R.; King, F. D.; Biggs, D. C. *J. Plankton Res.* **1982**, *4*, 895–911.
109. Bidigare, R. R.; King, F. D. *Comp. Biochem. Physiol.* **1981**, *70B*, 409–13.
110. Minas, H. J.; Packard, T. T.; Minas, M.; Coste, B. *J. Mar. Res.* **1982**, *40(3)*, 615–41.
111. Lonsdale, P. *J. Geophys. Res.* **1976**, *81*, 1827–41.
112. Duedall, I. W.; Coote, A. R. *J. Geophys. Res.* **1972**, *77(12)*, 2201–3.
113. Reid, J. L. *John Hopkins Oceanogr. Stud.* **1965**, *2*, 1–85.

RECEIVED for review September 6, 1983. ACCEPTED February 27, 1984.

11

Bioluminescence: A New Tool for Oceanography

JON LOSEE, DAVID LAPOTA, and STEPHEN H. LIEBERMAN
Naval Ocean Systems Center, San Diego, CA 92152

Because bioluminescence in marine surface waters (upper 100 m) is primarily due to small plankton, it can be successfully characterized by relatively simple photometer systems. The two basic types of bioluminescence detectors are an open type that views directly out into the seawater and a closed type that views a closed volume through which seawater is pumped. The bioluminescence variability is an interdependent phenomenon often associated with changes in physical and chemical parameters. For example, ocean frontal regions are almost always associated with enhanced levels of bioluminescence. Bioluminescence spectral content and signal kinetics often indicate the type of organisms present.

$\mathbf{B}$IOLUMINESCENCE IS LIGHT EMITTED BY LIVING ORGANISMS; it has been observed in the marine environment in all trophic levels from bacterial decomposers to predator zooplankton and fish. The many observations of marine bioluminescence (1–7) indicate that bioluminescence is a widespread phenomenon and serves a crucial role in the general ecology of marine plants and animals.

Although certain marine organisms produce light when mechanically perturbed, this phenomenon has been used little, and the distribution and activities of the bioluminescent organisms in situ are little known. A few concerted efforts (3–9) showed that bioluminescence is observed almost everywhere ships go; that is, the phenomenon is nearly ubiquitous and highly variable.

The instrumentation used to measure in situ marine bioluminescence fits into three basic categories. These categories include, first, a closed system in which seawater is pulled into a light-tight volume "viewed" by a detector (usually a photomultiplier tube) (7, 10–12) that measures bioluminescence stimulated by the turbulently flowing seawater. Second, open detectors view directly out into the seawater (3, 6) and measure stimulated bioluminescence. Open detectors might measure spontaneous bioluminescence if the platform vehicle is quiescent. A variation of the open-type system uses two detectors in coincidence to define a common volume of seawater for spontaneous bioluminescence (13, 14). Last, a remote detector (usually a sensitive video camera) has been used to view the ocean from an aircraft (15, 16) for bioluminescence stimulated from wind, waves, and fish schools.

Because biological and chemical variations in the upper ocean are not random, but rather interdependent phenomena often associated with specific physical events (e.g., upwelling, divergence, convergence, and stratification), correlations should exist between the occurrence of bioluminescence and other physical and chemical parameters. Our prime objective was to determine if correlations could be established between the occurrence of bioluminescence and the distribution of other oceanographic variables. Specifically, we wanted to physically characterize bioluminescence in the marine environment, determine how bioluminescence can be used to characterize planktonic communities in situ, and determine the relationships between the spatial and temporal distribution of planktonic bioluminescence with physical, chemical, and biological variables in the open ocean.

Since 1979, we have measured marine bioluminescence in diverse ocean waters ranging from tropical waters off St. Croix, V.I. to 73° N latitude to observations made under the pack ice of the Beaufort Sea, and finally from the submersibles *Alvin* and *Johnson Sea Link* to depths of 3650 m. This chapter describes the instrumentation developed for this purpose and presents examples of our measurements to date.

Experimental

Our primary measurement technique was to pull seawater through a 25-mL volume chamber with a pump. The organisms emit light when stimulated by the turbulently flowing seawater. This light is viewed by a photomultiplier tube (PMT). Two in situ measurement systems were used on surface ships. The on-board bioluminescence detector pulls seawater from below the ship's hull for continuous real-time measurements of surface bioluminescence; a bathyphotometer was used on station to depths of 100 m. An additional laboratory system was used to measure bioluminescence flashes from individual plankters isolated from plankton tows and pumped collections.

On-Board Bioluminescence Detector. Seawater was pumped from a 5-cm inside diameter (ID) stainless steel pipe that is put through the ship's sea chest to a point about 40 cm below the ship's hull. (A sea chest is a large opening in the hull backed by a pipe and valve.) With this arrangement, the seawater intake was at a depth of 3 m. From the 5-cm ID pipe at the sea chest, the seawater flowed through 16 m of 2.5-cm ID hose to the on-board photometer system (Figure 1). (For a discussion of the perturbation due to the hose, *see* the Appendix.) The pump, a self-priming Jabsco model 11810, with a flow rate of 0.56 L/s, was placed in line downstream of the viewing chamber. Two RCA 8575 PMTs with an S-20 response, used in photon count mode, were symmetrically mounted on opposite sides of the 25.5-

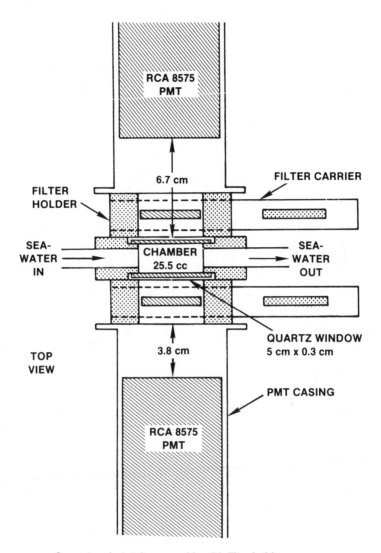

Seawater phototube assembly with filter holder.

Figure 1. On-board bioluminescence detector used for surface measurements. (Reproduced with permission from Ref. 7.)

mL viewing chamber. These PMTs viewed bioluminescence emanating through the quartz windows from the turbulently flowing seawater. Spectral data were obtained by inserting narrow-band optical filters (10-nm width) into the filter carrier in front of each PMT. (The narrow-band filters were checked for their response to light at angles other than normal incidence. A shift of approximately 20 nm in the direction of shorter wavelengths was observed for each filter for light incident at 30° to the normal; 40° is the maximum acceptance angle for the geometry used.)

One photomultiplier tube measured the total (broad-band, spectrally) bioluminescence, and it was used as reference to compare the signals through the various
filters. All signal counting was preceded by the dark-current counts of the PMT,
averaged for the subsequent signal count time, and subtracted. The net signal
count when filters were used was also corrected for the total transmittance of each
filter and for the quantum efficiency of the phototube (*see* Appendix).

Bathyphotometer Detector.   The submersible system, which employs the
same basic approach of pumping seawater past a PMT, also consisted of an RCA
8575 PMT used in the photon count mode, a filter-wheel disc with a capacity for
carrying 20 optical filters, a temperature thermistor, a beam transmissometer, and
a depth sensor. The central wavelengths of the optical filters used were 360, 370,
380, 390, 400, 420, 440, 460, 480, 500, 520, 540, 650, 580, 600, 620, and 640 nm.
Neutral-density filters of varying attenuation powers complemented the filter set,
and all components were contained in an aluminum pressure housing (Figure 2).

A ⅓-hp submersible pump (Gould Model 10 EJ) with a flow rate of 1.1 L/s was
mounted on the exterior of the detector to pull seawater through a viewing chamber fitted with a UV-transmitting (UVT) acrylic pressure window. The filter-
wheel disc was remotely rotated so that various filters could be inserted between
the UVT pressure window and the PMT. A light-tight 10-cm ID S-shaped plastic
black pipe was used for conveying seawater into the detector, which abruptly nar-
rowed down to a 16-mm ID input to the viewing chamber. The sudden reduction
in pipe size introduced turbulence and maintained a constant stimulus to the
plankton being viewed. The detector was deployable by the ship's hydrographic
winch to a maximum depth of 120 m. Two electrical cables, strapped together,
supplied power to the detector and pump and returned the PMT, depth, temperature, and transmissometer signals to the shipboard electronics.

The signal pulses of the PMTs from both systems and the laboratory plankton
test chamber (LPTC) used fast nuclear counting techniques. Neutral-density optical filters were used with all detectors when the signal must be limited so that the
electronics were not saturated. The signals from the PMTs were fast pulses (10–60-
ns widths); the number of pulses was directly proportional to the number of photons incident on the photocathode of the tube. Signal pulses were handled in several ways. Average signal levels were determined by several 100-s counts obtained
with Ortec Model 776 or 874 scalers. High time-resolution data were obtained by
binning the PMT signal pulses into time channels ranging from 10 μs to 1 s on two
Davidson multichannel analyzers (10- or 40-ms channels were normally used). The
analyzers were used synchronously, each with a 1023-channel capacity. Output
from the PMTs was also converted to a voltage proportional to PMT count rate and
was normally displayed on a strip-chart recorder that allows for a long-term continuous record of average bioluminescence.

Laboratory Plankton Test Chamber.   A major problem that marine scientists
face when conducting bioluminescence measurements at sea is correlating collected phytoplankton and zooplankton samples with either measured or visually
observed bioluminescence. Determination of which of the collected organisms are
luminescent has either been made visually while on board ship or by reviewing
published reports on these species. Both methods appeared to be inadequate because some luminous species may not flash or may flash at a level not visually observable when agitated following collection procedures. Literature accounts of luminescent species are quite incomplete. To compound this problem, some
bioluminescent forms, such as dinoflagellates, collected from one geographic area
have not been observed to be luminescent when collected from other areas (*17–19*).

To facilitate these field measurements, the Naval Ocean Systems Center's
(NOSC) laboratory plankton test chamber (LPTC) was built to measure biolumi-

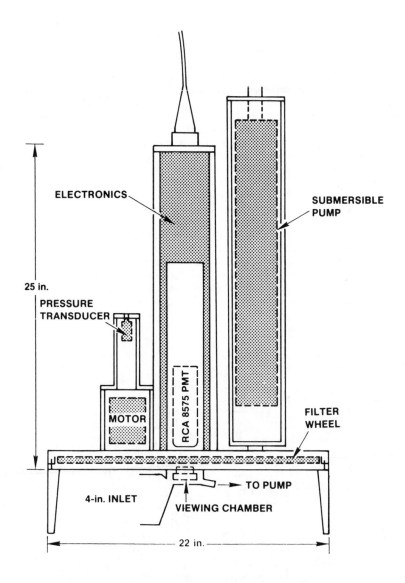

Figure 2. *Submersible bathyphotometer used for vertical profiling of biolumi-nescence. (Reproduced with permission from Ref. 20.)*

nescence from individual dinoflagellates and zooplankters recently isolated from fresh collections (*20*). Its usage complements the on-board system and bathyphoto-meter used for measuring bioluminescence in surface waters (upper 100 m). Imme-diate advantages for using the LPTC are that (1) a relatively large number, 40–80, of isolated organisms can be tested daily either during photophase or scotophase; (2) organisms recovered from the pump effluent of either system can be tested individually for assessment of "probable luminescence" contributing to the measured surface bioluminescence; and (3) confirmation of bioluminescence in known luminous organisms can be made as well as observation of bioluminescence in plankters not previously known to be luminescent.

To use the laboratory plankton test chamber, freshly collected samples were diluted with filtered seawater (filtered to 0.45 μm) to facilitate isolation of the desired test organism. The isolated organisms were then placed into quartz sample holders with a pipette and washed with approximately 6 mL of filtered seawater. The lucite bases of the sample holders were drilled to accommodate water removal by the applied vacuum. Whatman GF/C filter discs line the bottoms of the sample holders to retain the test organism following water removal (Figure 3). An onboard filtered seawater system was used, when possible, to obtain a continuous supply of fresh filtered (0.45 μm) seawater for live maintenance of plankton. The isolates were placed in 6-mL quartz sample holders that were fitted with 20-μm mesh lids and then placed in the filtered seawater bath for time periods of hours to a week. During this period, 12-h day–night cycles were maintained with fluorescent lights. After a period of time in the bath, the organisms were tested for bioluminescence with the LPTC system.

To observe the luminescence from the vacuum-stimulated organism, two RCA 8575 PMTs with S-20 response were mounted onto the test-chamber ports. One PMT measured the broad-band (350–650-nm wavelengths) from the induced flash; the other PMT measured intensity from select spectral regions (often the UV

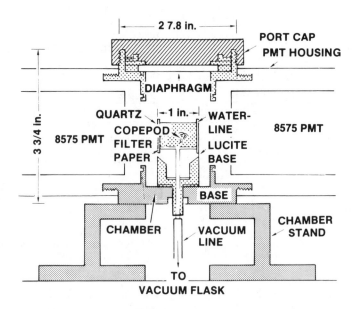

Figure 3. Laboratory plankton test chamber used for measuring bioluminescence from isolated phytoplankters and zooplankters. (Reproduced with permission from Ref. 20.)

region, wavelengths less than 390 nm). The PMT signals were handled by the two Davidson analyzers as previously discussed.

Recovery of the luminous plankter was accomplished by washing the contents of the sample holder several times with filtered seawater into a beaker and then preserving the contents in a buffered 5% formaldehyde solution. The entire sample was examined with a Bausch and Lomb dissecting microscope (magnification 10–30) for initial identification of the plankter (21–28) and observation of any contaminating organisms that may have been carried along in the isolation procedure and may have contributed to the luminous response. In instances where more than one plankter was found in the recovered samples, the entire high time-resolution scan was considered suspect and disregarded.

Calibration of Bioluminescence Detectors. All three systems were calibrated with a solution of the bioluminescent bacterium *Photobacterium phosphoreum* (29). The correct geometry was obtained by filling the detector chamber (or sample holder) with the bacterial solution. Also the measured optical spectra of the bacterial luminescence are very near those measured in situ. A sample of the bacterial solution was measured with a calibrated instrument (supplied by University of Georgia) tested against a secondary standard (radioactive scintillant combination), which gives the solution source strength in units of photons per second per milliliter for the entire bioluminescence spectrum (broad band). Our detectors gave an average count rate (PMT counts per second) when the solution was introduced into the chamber. Thus, for the on-board and bathyphotometer systems, we have PMT counts per second converted to photons per second per milliliter (the conversion factor is about 50). For the laboratory system, a conversion factor of 719 was obtained from PMT counts per second to photons per second.

Bioluminescence measured in this way gives a source strength (photons per second per milliliter) in previously unperturbed in situ seawater. Previous reports (3, 30) give bioluminescence measurements in units of microwatts per square centimeter. This unit is inappropriate for describing in situ bioluminescence. Units of radiance intensity (microwatts per square centimeter) are obtained from integrating the bioluminescence source strength (photons per second per milliliter) over a given turbulent volume by using the attenuation to the detector for the optical wavelengths involved. Therefore, units of microwatts per square centimeter not only involve the bioluminescence but also the turbulent volume used, the detector's geometry to that volume, and the attenuation of light in seawater. For example, in an infinite homogeneous sea that is turbulent throughout, the radiance intensity in any direction (microwatts per square centimeter) is proportional to the bioluminescence source strength (photons per second per milliliter) divided by the diffuse attenuation coefficient (reciprocal centimeters).

Bioluminescence Detectors Used on Submersibles. The detectors used on the submersibles were in pressure housings and were used either open (viewing out into the seawater) or closed (viewing in a chamber through which seawater is pulled by a pump). For operations using the *Johnson Sea Link* submersible (to depths of 600 m), both an open- and a closed type bathyphotometer were installed on the vehicle. The turbulent volume viewed in the closed system was 66 mL while a flow rate of 1 L/s was maintained. Both used 3 cm thick UVT acrylic pressure windows. The PMT (an RCA 8575) and electronics were contained in cylindrical pressure housings [12.7 cm outside diameter (OD) × 56 cm long] and pressure tested to 6800 m of seawater. The open detector was mounted clear of obstructions and downward viewing. When the submersible *Alvin* was used to depths of 3600 m, both detectors were used as open types and both were mounted free of obstructions for downward viewing. One detector was used with a Schott UG-11 optical filter. Thus, both the broad-band and UV-spectral regions of the bioluminescence signal were measured.

Other Parameters Measured Simultaneously with Bioluminescence.   Measurements of several other parameters were obtained from the seawater after it had traversed the bioluminescence detector. When working on station with the bathyphotometer, which was equipped with a pressure transducer, temperature and beam transmittance were measured at depth while seawater was pumped by the submersible pump at depth to shipboard with 110 m of 2.54-cm ID hose. Sea surface temperature was obtained continuously from a probe at the intake near the sea chest. The seawater, obtained from either the sea chest or the bathyphotometer, was pumped through a Turner Designs fluorometer to measure chlorophyll fluorescence, and past a pH probe (*31*) and a conductivity cell when available. Samples of seawater were frozen for subsequent nutrient analysis (NO^{3-}, NH_4OH, PO_4^{2-}, and NO_2). Plankton filtrates from 20 to 100 L (depending on plankton abundance) of seawater were collected from a 100-L effluent tank fitted with plankton net collection cups of 20-μm mesh porosity. The filtrate was split, filtered onto Whatman GF/C 4.25-cm filter discs, and frozen for subsequent carbon and nitrogen determinations. The other half of the sample was preserved in 5% buffered formaldehyde solution for taxonomic analysis.

Results and Discussion

On-Board System: Examples of Ocean Frontal Regions.   In December 1981 (*Varifront III* expedition), in the light-rich, nutrient-limited Gulf of California, on-board system measurements indicated that enhanced regions of near-surface bioluminescence are not randomly distributed but are almost always associated with thermal fronts (surface spatial gradients). In Figure 4, bioluminescence, temperature, and chlorophyll fluorescence are shown for a transect of about 20 nautical miles. These parameters correlate well and show an example of enhancement at a thermal front. The correlation of bioluminescence with thermal gradients occurred about 70% of the time in this region; both enhanced bioluminescence and chlorophyll fluorescence occurred on the cold sides of these fronts (*32*). Plankton analysis indicated that the enhanced bioluminescence was attributed to increased numbers of dinoflagellates. A frontal region from the Barents Sea is shown in Figure 5. In addition to the parameters named, pH is shown. In this light-limited, nutrient-rich area, these parameters show enhancements on the warm side of the thermal front.

The average open ocean (noncoastal) nighttime near-surface bioluminescence, as measured by the on-board system, is on the order of 2×10^5 PMT counts/s or 1×10^7 photons/s/mL broad band for turbulent seawater. In Figure 6, the relative bioluminescence signal is shown for a transect from Kodiak, Alaska to Pearl Harbor, Hawaii in October 1982. The overall signal north of about 43° N latitude (subarctic front) (about 5×10^7 photons/s/mL) is higher than that measured south of the front (near average). Observations at about 30° N latitude (subtropical front) indicate both low intensity and variability with the signal south of this front very low (about 5×10^5 photons/s/mL), which is typical of the Pacific Central Gyre (*33*). Both frontal regions were marked by changes in temperature and salinity. Furthermore, the diurnal variations in near-surface bioluminescence are

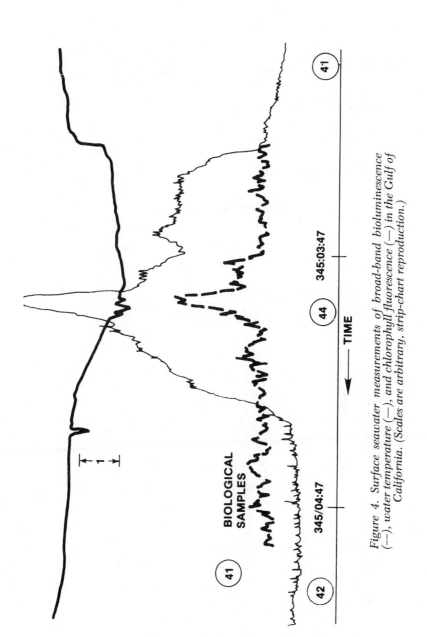

Figure 4. Surface seawater measurements of broad-band bioluminescence (--), water temperature (—), and chlorophyll fluorescence (—) in the Gulf of California. (Scales are arbitrary, strip-chart reproduction.)

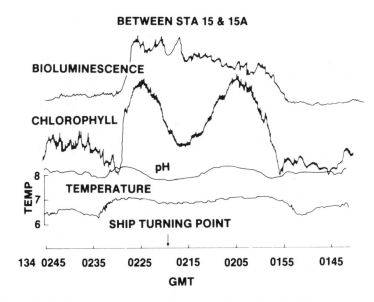

Figure 5. *Surface seawater measurements of broad-band bioluminescence, chlorophyll fluorescence, pH, and water temperature in the Barents Sea. (Scales are arbitrary, strip-chart reproduction.)*

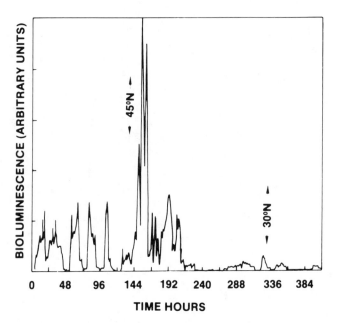

Figure 6. *Integrated hourly averages of bioluminescence intensity at the sub-arctic and subtropical fronts in the Pacific.*

pronounced north of the subarctic front. This variation is usually indicative of a dinoflagellate contribution to the bioluminescent signal (supported by biological samples). South of the front, little diurnal variation occurs, a condition indicative of zooplankton bioluminescence. Thus, measurements made at midnight are used when comparing bioluminescence levels for different regions.

High Time-Resolution Data.   Multichannel analyzer scans at 10-ms resolution obtained off Kodiak, Alaska are contrasted with those obtained near Hawaii in Figure 7. The scan near Kodiak shows many small events and many large events; the scan near Hawaii shows only a few large events. This difference indicates the change in type and number of organisms in these very different areas. Similar observations made in the Gulf of California for both broad-band and UV-spectral regions (Figure 8 and 9) showed marked differences from one area to another. Taxonomic determinations show that the distinct signal of Figure 9 is most likely due to *Pyrocystis pseudonoctiluca*, whereas the signal in Figure 8 is produced mainly by luminous zooplankton (copepods, nauplii, etc.).

Bathyphotometer Measurements.   Depth profiles to 100 m were routinely recorded on all expeditions. For example, the vertical bioluminescence profile in Figure 10 shows a distinct layer, centered at 30 m, that is often observed when a well-defined thermocline exists. Bioluminescence signal along with transmissometry, temperature, and chlorophyll fluorescence data are plotted from a station in the Gulf of Alaska on September 9, 1982 in Figure 11. The bioluminescence signal usually correlates well with transmittance and temperature, but not necessarily with chlorophyll fluoresence. Large-area correlations between chlorophyll fluoresence and bioluminescence have been observed by averaging both signals in the upper 50 m for stations in the Pacific during the *Varifront II* cruise in January 1981 from San Diego, California to Acapulco, Mexico (Figure 12).

Optical spectra measured at various locations with the bathyphotometer show differences particularly for wavelengths in the 360–460-nm region. Spectral measurements of bioluminescence in the Beaufort Sea, under the ice pack, and in the Norwegian Sea (Figure 13) indicate differences not only in spectral content, but in plankton organisms collected. Plankton analysis indicated a mix of dinoflagellates and crustaceans in the Beaufort Sea, and predominantly a crustacean population in the Norwegian Sea *(34)*. Additionally, the average bioluminescence signal was approximately 100-fold greater in the Beaufort Sea than that in the Norwegian Sea at the stations shown.

Bioluminescence Observations in Isolated Plankters.   The use of plankton chambers for the photoelectric recording of flash responses from luminescent dinoflagellates *(35, 36)*, calanoid copepods, and other zooplankters *(8, 37, 38)* is not novel. Although artificial stimuli (electrical and condenser shocks or vacuum and formaldehyde-solution stimulation) were

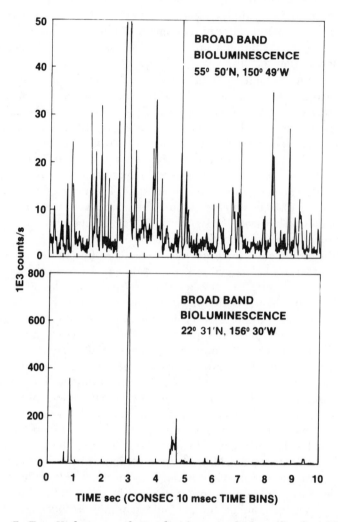

Figure 7. Top: *High time-resolution data [scan, multichannel scaling (MCS) mode] recorded by the on-board bioluminescence detector in the Gulf of Alaska. The signal intensity has been attenuated with a neutral-density 3 filter (1000× attenuation) inserted between the viewing chamber window and face of the photocathode of the PMT.*

Bottom: *High time-resolution data (scan, MCS mode) recorded by the on-board bioluminescence detector north of Hawaii. The signal intensity has been attenuated with a neutral-density 1 filter (10× attenuation) inserted between the viewing chamber window and face of the photocathode of the PMT.*

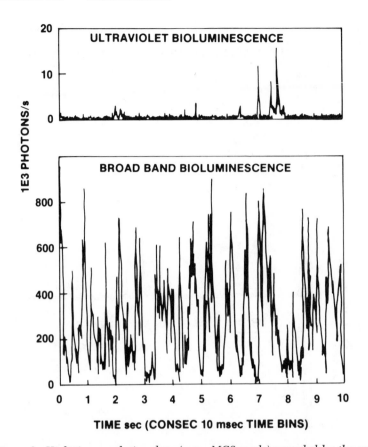

Figure 8. High time-resolution data (scan, MCS mode) recorded by the on-board bioluminescence detector in both the broad-band and UV portions of the bioluminescence spectrum from the Gulf of California. The signal is pro-duced mainly by luminous zooplankton.

The broad-band signal intensity has been attenuated with a neutral-density 2 filter (100× attenuation) inserted between the viewing chamber window and face of the photo-cathode of the PMT. The UV signal was observed through a Schott UG-11, 2.5 mm thick filter.

used to evoke a flash response from isolated organisms, the stimuli were nevertheless effective for identifying luminous species. Vacuum-triggered or induced flash responses have also been observed (38), but this method of stimulation was not investigated in any great detail.

By using the NOSC LPTC, new observations of bioluminescence were made on larval stages of the euphausiid shrimp *Nyctiphanes simplex* Hansen (Calyptopis II, Furcilia I, II, and III, and juveniles) and *Euphausia eximia* Hansen (Calyptopis I); the Calanoida copepods *Centropages furcatus* Dana, *Paracalanus indicus* Wolfenden, and *Acrocalanus longicornis* Giesbrecht; the Cyclopoda copepods *Corycaeus (Corycaeus)* sp. Dana and

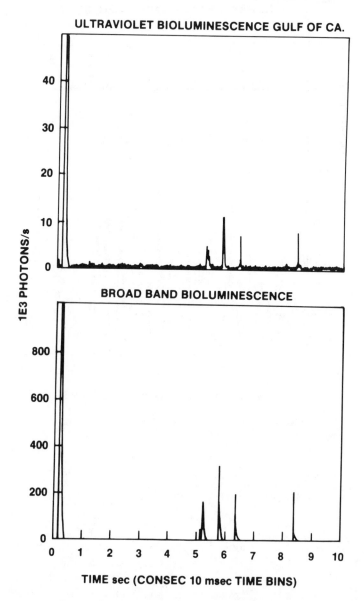

Figure 9. High time-resolution data (scan, MCS mode) recorded by the on-board bioluminescence detector in both the broad-band and UV portions of the bioluminescence spectrum from the Gulf of California. The distinct signal is probably due to Pyrocystis pseudonoctiluca. *The conditions are the same as in Figure 8. (Reproduced with permission from Ref. 20.)*

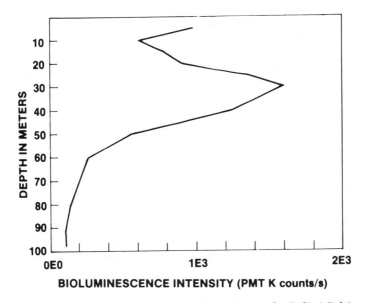

Figure 10. Vertical bioluminescence profile to 100 m in the Gulf of California (22:00–23:50, 12/7/81).

Corycaeus (Onychocorycaeus) latus Dana; and the dinoflagellates *Ceratium breve* Ostenfeld and Schmidt, *Ceratium horridum* Gran, and *Ceratium gibberum* Gourret. These observations indicate the important contribution of some of the small copepods and larval euphausiids to surface bioluminescence. Examples of recorded vacuum-induced flashes from the dinoflagellates *Ceratium breve* and *C. horridum* are shown in Figure 14 and those of *Centropages furcatus*, *Paracalanus indicus*, *Corycaeus speciousus*, and *C. latus* are shown in Figure 15 (20).

Discussion

The on-board system data presented, as well as our more extensive unpublished data, show that ocean frontal and patch features are indicated from the bioluminescence signal obtained from continuous measurements of the surface seawater obtained through the ship's sea chest. Surface seawater can be characterized by the average bioluminescence signal level, the bioluminescence signal patterns from the high time-resolution scans, and the optical spectral content of the signal. These data, in conjunction with chlorophyll fluorescence, temperature, pH, and salinity, provide an effective means of characterizing, both biologically and chemically, oceanic waters

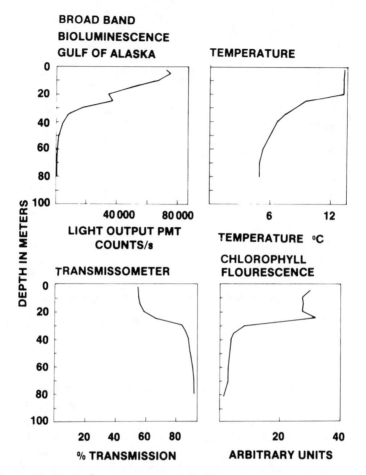

Figure 11. Vertical measurements of bioluminescence intensity, transmissometry, seawater temperature, and chlorophyll fluorescence from a station in the Gulf of Alaska (00:40–01:50, 9/9/82).

in near real time. Variations in these parameters also indicate the type of bioluminescent organisms present. For example, a patch of enhanced bioluminescence without a change in chlorophyll fluorescence can probably be attributed to zooplankton bioluminescence. If, in addition, little diurnal variation occurs in the bioluminescence signal patterns, the flash rise times are greater than 40 ms, and no signal at wavelengths shorter than 420 nm is observed, then the signal is very likely crustacean in origin.

Data obtained with the bathyphotometer as a function of depth generally show higher signals in the surface-mixed layer followed by an exponential-like decrease, with occasional maxima at deeper stratifications. The maximum signal almost always occurs above 50 m. Often a

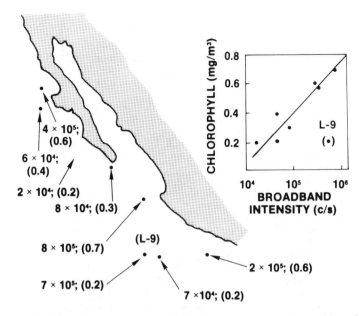

Figure 12. *Correlation of surface bioluminescence and surface chlorophyll fluorescence in coastal waters off Mexico.*

layer of bioluminescent organisms lies just above or just below the main thermocline.

Data obtained from the pumped (closed) deep bathyphotometer and the open bathyphotometer mounted on the *Alvin* submersible show that skylight penetrates to 200 m on a moonless night and to 500–700 m in daylight conditions. Therefore, bioluminescent measurements at depths to about 200 m at night and 500 m in the daylight can reasonably be done only with a closed system. (These depths can vary considerably with the turbidity of the seawater.)

Enough data now have been obtained with the measurement techniques described in this chapter to be confident of the reproducibility of the measurements. The bioluminescence measurements also often correlate to some other oceanographic parameters and generally show planktonic distributions that are consistent with overall concepts of plankton activity in the ocean. However, with all of the bioluminescence sensors, as with all planktonic sampling schemes, the question of avoidance of larger mobile organisms can be raised. Our on-board and bathyphotometer systems rarely sample adult euphausiids, which are well known for their mobility and bioluminescence. Many larger organisms, such as luminous fish and larger gelatinous forms, are not measured by our systems. An accurate esti-

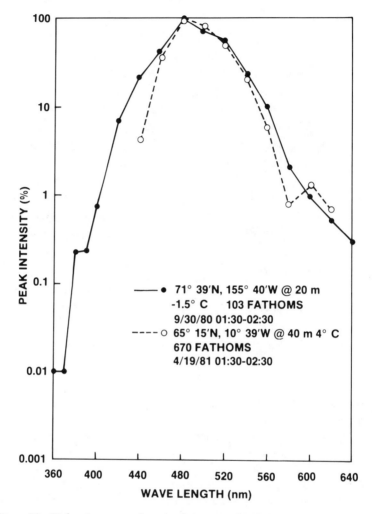

Figure 13. Bioluminescent color spectra measured in the Beaufort and Norwe-gian Seas.

mate of the bioluminescent organisms not measured because of avoidance is difficult and unknown at this time. However, some evidence does exist. Swift et al (9) estimate this bioluminescence at about 5% and our visual observations from the *Johnson Sea Link* submersible indicate that the concentration of large bioluminescent organisms is low compared to that of the smaller plankters in the upper 100 m. A measurement using the *Dolphin* submarine, which was equipped with both a silicon intensifier tube (SIT) video camera and an open photometer, showed only four or five large bioluminescent events per 6-h nighttime dive; a steady glow was observed around a metal

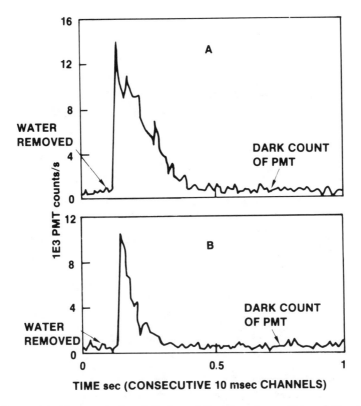

TIME sec (CONSECUTIVE 10 msec CHANNELS)

Figure 14. Vacuum-induced flashes of Ceratium breve *(A) and* Ceratium horridum *(B). (Reproduced with permission from Ref. 20.)*

rod that protruded into the flow stream to provide turbulence (39). The glow was presumably due to the many small bioluminescent plankters.

Conclusion

The techniques discussed herein provide new ways of characterizing planktonic communities in the oceans. They are essentially instrumental and experimentally simple and can provide real-time information about in situ populations of small (<5 mm) organisms. The spatial patterns of bioluminescence measured underway coincide with known chemical and physical phenomena; thus, the techniques provide a novel way of characterizing the surface ocean on mesoscales and larger scales. Similarly, vertical profiles of bioluminescence covary with other commonly measured oceanographic variables, thereby providing additional information on the vertical and temporal distributions of organisms. Finally, time-series and optical data vary among different populations of organisms. These data suggest that

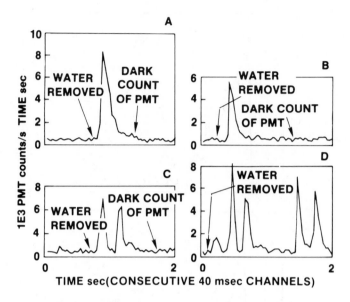

Figure 15. Vacuum-induced flashes from the copepods Centropages furcatus *(A)*, Paracalanus indicus *(B)*, Corycaeus speciousus *(C), and* Corycaeus latus *(D). (Reproduced with permission from Ref. 20.)*

bioluminescence measurements are useful for characterizing organisms. Because of the interrelationships existing between the chemical and physical environment and biological communities, it may be possible to characterize the marine environment via bioluminescence.

Appendix

The on-board and bathyphotometer systems both have cylindrical "viewing" chambers of 2-cm radius and 1.9-cm height. The inlet and outlet ports are on opposite sides perpendicular to the cylindrical axis (*see* Figures 1 and 2). The inlets, either a 2.54-cm ID hose for the on-board system or a 10-cm ID pipe for the bathyphotometer, reduce to 1-cm ID to jet the seawater into the chamber. The jet flares in the chamber and part of it bounces off the opposite side and creates an eddy circulation on both sides of the central flow. This circulation was checked by injecting ink into the hose. The flow rate used is usually 0.6–0.7 L/s, and thus the average residual time in the chamber is approximately 40 ms. The residence time in the eddies can be approximately 300 ms. Thus, the bioluminescence signal is an ensemble of flash events—some complete flashes and some part flashes. Even with this ensemble of flash events, the signals from one region with its group of organisms can be quite distinct from that of another (*34*).

The average residence time (approximately 40 ms) is shorter than typical flash decay times for the plankton organisms that are being measured

(3, 34). This basic assumption means that little fatigue is associated with the organisms in the chamber, thus, an initial stimulated response for unperturbed seawater is being measured. Therefore, the units are of a rate: photons per second per milliliter.

An alternative to this measurement (40) is to have a residence time in the viewing chamber long with respect to the flash decay times (approximately 1 s or longer). This basic assumption means that the organisms in the chamber are exhausted (in the sense of their ability to luminesce) before they leave the field of view. The signal is then the total stimulateable light from a volume of seawater and, therefore, the units used are photons per milliliter. This approach can be considered an integration in time of the rate, starting with what we measure as an initial rate and ending when the rate is zero. Thus, these two measurements are consistent with one another.

Our on-board system was checked for bioluminescence signal robustness of chamber design. An acrylic cylinder of similar design to those used in the Naval Oceanographic Office's (NAVOCEANO) systems was fitted with various inlet and outlet apertures (Figure 16). A PMT was fitted to view the center of the acrylic cylinder perpendicular to the cylindrical axis. San Diego Bay water was pulled through both the on-board system and the acrylic cylinder with two separate hoses. The hose inlets in the bay were tied together to ensure similar water in both systems. The ratios of the bioluminescence signals from the acrylic cylinder using various combinations of inlet and outlet apertures to that of the on-board system is shown in Fig-

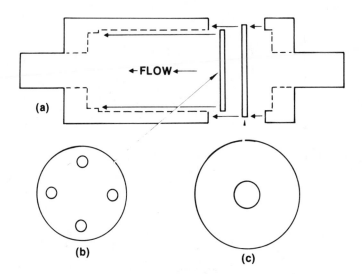

Figure 16. Schematic of test chamber that shows input/output plates (a). The output plate (b) has four holes. The input plate (c) has one hole.

ure 17. The combinations of apertures are denoted by a pair of numbers
that represent the number of holes in each aperture. The first number signifies the input aperture; a zero means that no aperture is present (open
2.54-cm ID hose), and prop signifies a free-spinning acrylic propeller. The
volume of bay water viewed in the acrylic cylinder was approximately
twice (50 mL) that of the on-board system, thus the expected ratio is about
2. (Differences in geometry and PMT efficiencies contribute little here and
are ignored.)

A comparison of (0,1) to (1,0) of Figure 17 shows that the input aperture has the largest effect on the bioluminescence signal. Although the various combinations changed the signal in a consistent way, the total change

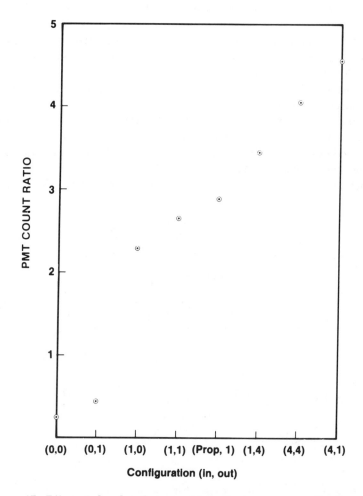

*Figure 17. Effect of chamber design on bioluminescence counts. Configura-
tion designations (in,out) are explained in the text.*

was only about twofold. The conclusion is that the details of chamber design are not very significant to the measurement technique.

The on-board system was also checked for signal changes as a function of flow rate. In the range of 0.1–1.0 L/s, the bioluminescence signal is directly proportional to the flow rate raised to the 1.6 power. This proportion is most likely due to the turbulent stimulus. Small variations in flow rate (approximately 0.2 L/s) are not very significant to the measurement technique.)

The effect of the 2.54-cm ID hose in transporting the seawater from the ship's sea chest to the on-board system (typically approximately 16-m length) has been investigated for both total signal, optical spectral content, and the statistical probability density function (PDF) of the signal rates (7). Even though the flow in the hose is fully turbulent (Reynolds number ~ 30,000), the data show little measurable difference in any of these properties (about a 30% loss in average signal with no significant spectral shifts or changes in the signal statistics). However, tests run with a stimulus at the intake of the hose showed that the properties could be easily altered. For example, a 7.6-cm ID pipe with stainless steel shavings (pot scrubbers) inside was placed on the hose intake and produced a significant loss in the optical spectral region from 440 nm to shorter wavelengths, and the high rates in the PDF were lost. (That is, the brightest flashes were eliminated from the bioluminescence signal.) The reason that this stimulus changes those properties but the hose turbulence does not is not known.

An intake test was done off Mexico in the Pacific for the bathyphotometer. An S bend of 5-cm ID pipe and one of 20-cm ID were interchanged to determine if these intakes changed the bioluminescence signals. A 2.54-cm ID hose was attached to the bathyphotometer and run up to the on-board system so that a normalization was provided for the seawater during the bathyphotometer measurements (made at 20-m depth). The data showed that the bioluminescence signal was about 10% higher for the 5-cm ID intake compared to that of the 20-cm ID intake. This result was considered insignificant and subsequently, a 10-cm ID intake has been used. During this same test, a comparison between bathyphotometer and on-board bioluminescence signals was available. The on-board system was pumping from 20-m depth through 30 m of 2.54-cm ID hose. The ratio of signals from bathyphotometer to on-board system (account taken of the neutral-density filters used in both systems) was 0.7. This result again indicates that prestimulus in the hose is not a significant detriment to the on-board system.

Literature Cited

1. Harvey, E. N. "Bioluminescence"; Academic Press: New York, 1952; 649 pp.
2. Turner, R. J. National Institute of Oceanography, N.I.O. Internal Report No. B4, July 1965, 30 pp.
3. Bityukov, E. P. *Oceanology (Engl. Transl.)* **1971,** *11,* 103–8.
4. Lynch, R. V., III *NRL Memo. Rep.* **1981,** 8475.
5. Gitel'zon, I. I.; Baklanov, O. G.; Filimonov, V. S.; Artemken, A. S.; Sha-

tokhin, V. F. *Works of the Moscow Society of Naturalists, Bioluminescence* **1968**, *21*, 147–55.
6. Losee, J. R.; Lapota, D.; Geiger, M.; Lynch, R., III "Progress Report Abstracts—Oceanic Biology Program," Office of Naval Research Annual Report, 1982, 29–30.
7. Losee, J. R.; Lapota, D. "Bioluminescence: Current Perspectives"; Nealson, K. H., Ed.; Burgess Publishing Co.: 1981; 143–52.
8. Rudjakov, J. A.; Voronina, N. M. *Oceanology* **1967**, *7*, 838–48.
9. Swift, E.; Biggley, W. H.; Lessard, E. J. in "Symposium on Chemical Oceanography: Analytics of Mesoscale and Macroscale Processes"; ACS: Washington, D.C., 1984.
10. Backus, R. H.; Yentsch, C. S.; Wing, A. *Nature (London)* **1961**, *192*, 518–21.
11. Seliger, H. H.; Fastie, W. G.; Taylor, W. R.; McElroy, W. D. *J. Gen. Physiol.* **1962**, *45*, 1003–17.
12. Clark, G. L.; Kelly, M. G. *Limnol. Oceanogr.* **1965**, *10*, (supplement), R54–R66.
13. Boden, B. P.; Kampa, E. M.; Snodgrass, J. M. *Nature (London)* **1965**, *208 (5015)*, 1078–80.
14. Boden, B. P. *J. Mar. Biol. Assoc. U.K.* **1969**, *49*, 669–82.
15. Cram, D. L. "Proceedings of the 9th International Symposium on Remote Sensing of the Environment"; Willow Run Laboratories: Ann Arbor, 1974; pp. 1043–49.
16. Cram, D. L.; Schulein, F. H. *J. Cons., Cons. Int. Explor. Mer* **1974**, *35(3)*, 272–75.
17. Nordli, E. *Oikos* **1957**, *8*, 200–65.
18. Sweeney, B. M. *Biol. Bull.* **1963**, *125*, 177–81.
19. Tett, P. B. "Scottish Marine Biological Association, Eunstaffinage Marine Research Laboratory"; Final Report on Agreement No. AT/992/03/RL, 1969; 71 pp.
20. Lapota, D.; Losee, J. R., *J. Exp. Mar. Biol. Ecol.* **1984**, *77*, 209–40.
21. Boden, B. P. *Proc. Zool. Soc. London* **1951**, *121*, 515–27.
22. Brodsky, K. A. *Tabl. Anal. Faune URSS, Zool. Inst., Acad. Sci.* **1950**, *35*, 1–442.
23. Einarsson, H. "Euphausiacea I, Northern Atlantic Species"; Dana Report, 1945 No. 27, 1–85.
24. Graham, H. W.; Bronikovsky, N. *Carnegie Inst. Washington Publ. Biol.* **1944**, *5*, 565, p. 209.
25. Lebour, M. V. *Plymouth Marine Biological Laboratory* **1925**, 250 pp.
26. Mayer, A. G. *Carnegie Inst. Washington Publ.* **1912**, *162*, 58 pp.
27. Motoda, S. *Publ. Seto Mar. Biol. Lab.* **1963**, *11*, 2, 209–62.
28. Owre, N. B.; Foyo, M. "Copepods of the Florida Current. Fauna Caribea Number 1, Crustacia, Part I: Copepoda"; Institute of Marine Sciences, University of Miami, 1967; 137 pp.
29. Lee, John, University of Georgia, developed the calibration technique (unpublished) and supplied the bacterium.
30. Clark, G. L.; Hubbard, C. J. *Limnol. Oceanogr.* **1959**, *4(2)*, 163–80.
31. Zirino, A.; Clavell, C.; Seligman, P. F.; Barber, R. T. *Mar. Chem.* **1983**, *12*, 25–42.
32. Lieberman, S. H.; Lapota, D.; Losee, J. R.; Zirino, A., unpublished data.
33. "Naval Oceanographic Office Data Across the Pacific," unpublished.
34. Losee, J. R.; Lapota, D., unpublished data.
35. Hardy, A. C.; Kay, R. H. *J. Mar. Biol. Assoc. U.K.* **1964**, *44*, 435–86.
36. Tett, P. B. *J. Mar. Biol. Assoc. U.K.* **1971**, *44*, 435–84.
37. David, C. N.; Conover, R. J. *Biol. Bull.* **1961**, *123*, 92–107.
38. Clark, G. L.; Conover, R. J.; David, D. N.; Nicol, J. A. C. *J. Mar. Biol. Assoc. U.K.* **1962**, *42*, 541–64.
39. Howarth, R. F., personal communication, NOSC code 8114.
40. Swift, E., this volume, 1984.

RECEIVED for review November 29, 1983. ACCEPTED May 23, 1984.

Distributions of Epipelagic Bioluminescence in the Sargasso and Caribbean Seas

ELIJAH SWIFT

Graduate School of Oceanography and Department of Botany, University of Rhode Island, Kingston, RI 02881

W. H. BIGGLEY

Biology Department, The Johns Hopkins University, Baltimore, MD 21218

EVELYN J. LESSARD

Graduate School of Oceanography, University of Rhode Island, Kingston, RI 02881

The nature of oceanic bioluminescence and some of the problems associated with its measurement are discussed in this chapter. No present instrument accurately measures the complete bioluminescence potential of the open ocean environment. At night we measured the bioluminescence stimulated mechanically in a pump-through bathyphotometer capable of capturing weakly swimming organisms. Using this instrument, we were able to describe the bioluminescence in the upper 120 m of the northern and southern Sargasso Sea and the Caribbean Sea. In the late summer, more bioluminescent organisms and more stimulated bioluminescence were observed in the northern Sargasso Sea than in the southern part. In the fall, the station at the southern edge of the southern Sargasso Sea near the Antilles and the station in the Caribbean Sea showed the highest flashing rates (i.e., concentrations of bioluminescent organisms). The bioluminescence in the Caribbean was comparable to that in the northern Sargasso Sea. These patterns reflect some of the general oceanographic features of the Sargasso Sea.

0065-2393/85/0209-0235$07.00/0

ON MACROSCALES AND MESOSCALES IN THE OCEAN, the communities of organisms change in parallel with changes in hydrographic features (1, 2). Different bioluminescent species are associated with the different communities. In the first part of this chapter, we explain how this diversity makes it difficult for any one instrument to measure all the actual or potential bioluminescence present. In the second part of the chapter, we will describe the instrument used for measuring bioluminescence, how it was used, and some characteristics of the bioluminescence measured in three regions: the Gulf Stream and northern Sargasso Sea, the southern Sargasso Sea, and the Caribbean Sea.

In nature, the ability to bioluminesce confers a number of presumed advantages, including camouflage, startling predators or grazers, luring prey, illuminating prey (e.g., flashlight fish), or attracting mates (3). These bioluminescent displays are carried out at a natural rate that is only rarely observed. For example, two light detectors directed at a common remote volume were used (4) to examine the frequency of flashing above, below, and at the depth of the deep scattering layers. A similar apparatus was used (5) to examine the flashing of small organisms in Scottish coastal waters. The natural rates of flashing are usually quite low. Further, these rates may tell more about the presence of predators or grazers than about the bioluminescent organisms themselves. For example, schools of fish like menhaden, which are not themselves bioluminescent, can be detected from airplanes with image intensifiers because they stimulate the bioluminescent organisms present. Thus, the concentrations of bioluminescent organisms can usually be observed by using mechanical stirring to elicit bioluminescence. The total bioluminescence that can be stimulated by vigorous mechanical stirring is termed the "bioluminescence potential" of the water sample (6). The stimulated bioluminescence of marine organisms usually produces orders of magnitude more light than the "natural" bioluminescence (7). Although the stimulated bioluminescence bears little relationship to the natural rates of the production of bioluminescence, it does allow the presence of bioluminescent organisms to be detected and their distributions to be mapped.

Both laboratory and field measurements (8–10) have demonstrated that most epipelagic organisms do not bioluminesce when mechanically stimulated during the day. Bioluminescent individuals from both epipelagic and mesopelagic species may migrate vertically over a daily cycle, usually concentrating near the surface at night. Thus, the vertical distributions of bioluminescent organisms change throughout the day and night. To take vertical migration into account, when comparing data taken at different times of the night, the measured values per unit volume can be integrated as a function of depth, and the bioluminescence potential can then be expressed on a water column (per square meter basis).

Mechanically stimulated bioluminescence can be produced by organisms present in a wide variety of concentrations. The ubiquitous bioluminescent marine bacteria living outside symbiotic relationships are not considered here because they do not produce bioluminescence unless associated with a rich organic food source. Mechanically stimulated bioluminescence can be produced by other organisms present in a wide range of concentrations (several hundred per liter to one per hundreds of cubic meters), a wide range of bioluminescence capacity per individual (10^7 to more than 10^{17} photons per individual), a wide range of sizes (lengths of 30 μm to more than 2 m), a wide range of swimming speeds (tens of centimeters per hour to several kilometers per hour), and finally, a variety of positions in the food web (*11–14*). All the variations in the taxonomy, size, rarity, elusiveness, stimulateability, as well as the color, time course, and intensity of the bioluminescence confound the measurement of the bioluminescence potential. Currently, no single instrument can capture, stimulate, and record all the bioluminescence potential from all the organisms in a marine community.

Bathyphotometers that measure stimulated bioluminescence are of several types. Early instruments used by Clarke et al. (*15–20*) to determine ambient light in the ocean also detected bioluminescent flashing. Higher rates of bioluminescent flashing occurred when the ship rolled. This action suggested that some of the light as recorded was mechanically stimulated. Kremer (*21*) took advantage of the interaction of the bathyphotometer with the organisms in mapping the distribution of ctenophores in Narragansett Bay with a towed bathyphotometer. The ctenophores bioluminesced when they struck a transparent plastic dome that was a fixed distance ahead of the bathyphotometer. Bioluminescent dinoflagellate distributions in bays were mapped by using a towable photometer (*7*). Underway bioluminescent flashing rates were determined (*8*) by sampling water from a through-hull opening. Underway sampling was improved (*22*) to investigate bioluminescence, chlorophyll, pH, salinity, temperature, and other factors concurrently. (*See also* other chapters in this volume.) Gitel'zon et al. (*23*) determined vertical profiles of bioluminescence by repeatedly lowering and raising a bathyphotometer inside a light shield with open ends. The raising and lowering produced a seawater flow inside the light shield. The flow past a constriction near a photomultiplier window stimulated organisms to bioluminesce. An instrument was developed (*6*) that enclosed a known volume of water before the water was agitated with a hydrodynamically defined mechanical stimulus. In a final type of instrument mentioned here, a known volume of water is pumped through a light-shielded chamber, and the bioluminescence is stimulated by the action of the pump impeller (*7, 8, 24–27*). We used this type of instrument.

The measurement of bioluminescence potential with a pump-through bathyphotometer presents several problems (*11*). The first problem is avoidance: Large organisms with good swimming ability can avoid the intake of the bathyphotometer. In some pelagic communities, they might represent a substantial part of the bioluminescence capacity. For example, in the Antarctic, the dominant bioluminescent krill *Euphausia superba*, which forms dense swarms, would probably be able to avoid the intake of most currently used pump-through bathyphotometers.

Second, the residence time of bioluminescent organisms in the sample chamber of the bathyphotometer may be inappropriate. If the numbers of bioluminescent organisms are being counted, then the residence time should be short so the bioluminescent signals of individual organisms can be separated. On the other hand, if the residence time is too short, then it will be difficult to establish the amount of light produced by individuals. When the bioluminescent display persists longer than the residence time of the organism in the sample chamber, the full bioluminescence potential of the organism will not be recorded (7). This effect is particularly noticeable with organisms that produce clouds of bioluminescence with slow decay times like the ostracods.

Third, in both pump-through and towed bathyphotometers, the distance between the bioluminescing organism and the photomultipier varies to some extent. This variation makes the true intensity of the bioluminescence difficult to determine.

Fourth, no ideal way to calibrate the photometers has been found. Seliger et al. (25) calibrated their pump-through bathyphotometer with dilute solutions of luminous marine bacteria. They assumed that the average spatial distribution of stimulated bioluminescence inside the impeller housing of their bathyphotometer was the same as that of the continuously luminous bacteria. They noted

> The major advantages of using luminous bacteria are (a) the light emission of a light solution is essentially continuous and constant, not requiring external stimulation and (b) the spectral emission of the particular strain of bacteria used (*A. fisheri*) very closely matches that of the dinoflagellates, obviating the necessity for knowledge of the phototube spectral sensitivity.

However, oceanic bioluminescence in some regions is not due to dinoflagellates but to a number of zooplankton groups (28). The color of the bioluminescence from other organisms varies with taxonomic group (29–32). For an accurate calibration, both the phototube spectral sensitivity and the types of organisms being stimulated must be known.

A fifth problem is that organisms may be prestimulated and lose some of their bioluminescence before they reach the sample chamber.

Sixth, the degree of stimulation in the sample chamber is not hydrodynamically defined. Thus, comparing the degree of mechanical stimulation from instrument to instrument is not usually possible. The degree of stimulation may affect the amount of the signal recorded. Then, the amount of bioluminescence stimulated and recorded may vary from bathyphotometer to bathyphotometer and among organism groups in a yet unknown way.

Seventh, if the rare but comparatively bright organisms are important, large volumes should be sampled to obtain statistically significant estimates of bioluminescent potential. Most current bathyphotometers do not pump large volumes of water. These factors suggest that more bioluminescence potential is in the sea than is currently measured with pump-through bathyphotometers.

Nekton and micronekton are not usually considered in measurements of bioluminescence. However, large volumes as vigorously stirred as those in a ship's wake may include them. In the region we will describe, their contribution does not appear to be very important, less than a few percent of the total (unpublished data). To estimate nekton contribution to the bioluminescence potential, they must be collected and concentrated first with large, rapidly moving trawls. The bioluminescence potential of the organisms can be measured in a shipboard photometer. Multiplying the concentrations of these organisms by their bioluminescence potential produces their potential contribution per cubic meter. Unfortunately, many large organisms do not survive being captured in a trawl. Thus, only estimates based on the characteristics of the robust species can be made. Keeping organisms alive and in a condition approximating the natural one is a problem common to all the organisms collected in nets for determinations of their bioluminescent potential.

Marine Bioluminescence Measurements: Methods

Most of the bioluminescence measurements we describe were obtained with a bathyphotometer that had a relatively low pumping rate (15 L/min) because it was designed to measure vertical distributions of bioluminescent dinoflagellates. Thus, it efficiently sampled bioluminescent members of the protozoans, dinoflagellates, and most copepods, larvaceans, and ostracods. Efficient sampling was not possible for strong swimmers or rare organisms such as euphausiids, micronekton, colonial tunicates, squid, and fish. On the basis of the bioluminescence measured, we were able to examine three regions of the Sargasso Sea (described in the "Results" section).

On R.V. *Endeavor* Cruise 054 and the R.V. *C.O. Iselin* cruise in 1976, only small amounts of bioluminescence were detected during the day because of either the photoinhibition of the organisms present (8) or the organisms migrating out of surface waters during the day (24, 25). However, the measurements reported here were made starting at least 2 h after sundown by using the bathyphotometer system similar to that shown in Figure 1. Organisms were drawn into the entry chamber through a circular hole. The organisms were then drawn down through a 2.0-cm

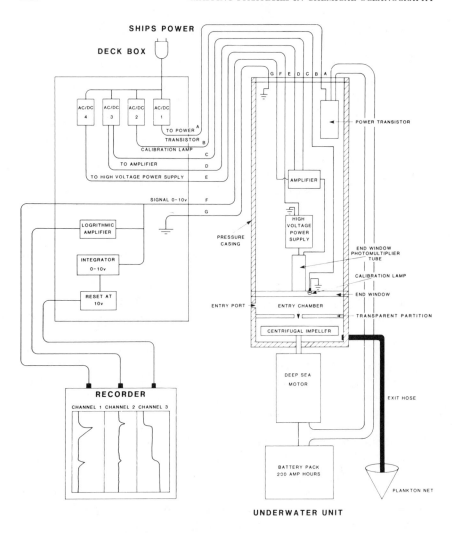

Figure 1. A schematic diagram of the bathyphotometer used.

diameter orifice in the clear plastic floor of the entry chamber. The orifice
was centered over a centrifugal pump impeller. Organisms left the impeller
chamber through a section of 1.9-cm diameter hose and were collected with
a plankton net with 25-μm aperture mesh netting. The residence time of
water in the entry and impeller chambers, where bioluminescence could be
detected by the photomultiplier, was ~1 s. For the few times the
bathyphotometer was lowered during the day, a light trap over the entry
chamber entrance was a 10 cm long, right-angled section of 1.4-cm inside
diameter (ID) pipe with the end pointing downward.

Organisms were stimulated to bioluminesce in the entry and impeller chambers by the agitation caused by the pump impeller. The bioluminescence was detected with a bialkali photomultiplier tube [Electrical Musical Industries, Ltd. (EMI) 9824A]. The photomultiplier was mounted behind a polyacrylate window in a watertight chamber with an electrometer amplifier and a high-voltage power supply. The signal due to stimulateable bioluminescence was recorded as a voltage pulse (log or linear response) and was also integrated electronically as a function of time to indicate the total light content of each flash. The flashes, which represented distinct bioluminescent organisms, were separated by at least 1 s (the residence time of water in the detection chamber); for our bathyphotometer, this value meant that concentrations less than ~ 2000 bioluminescent organisms per cubic meter were measured. With organism concentrations higher than that, the question of whether one organism produced more than one distinct group of flashes could not be objectively resolved; thus, reported concentrations represent estimates. The signals were recorded with a Gould-Brush Model 220 recorder with full-scale response of 100 Hz. The deck unit supplied DC voltage to the high-voltage power supply, amplifier, and calibration lamp, and to the relays that initiated underwater battery power to the pump motor.

The signals from the bathyphotometer were calibrated at sea by immersing the bathyphotometer in a tank of seawater and pumping through a known number of previously isolated dark-adapted cells of the bioluminescent dinoflagellate *Pyrocystis noctiluca* Murray ex Haeckel, 1885. *Pyrocystis noctiluca* was used because it is large enough (300–600 μm in diameter) and common enough in tropical waters to make it easy to pick out of plankton samples. It is also robust enough to survive being collected in a net. *Pyrocystis noctiluca* is the brightest photosynthetic dinoflagellate commonly found in these waters (Figure 2). Thus, by using cells collected at sea, the signals can be sorted into *Pyrocystis*-sized signals and smaller. The smaller signals could be from dinoflagellates and dim zooplankton; the larger signals could be from zooplankton. Factors that change the photon output of the *Pyrocystis* are discussed elsewhere (9, 33, 34). Table I shows the photon yields for *P. noctiluca* collected in all regions of the Sargasso Sea over 6 years. The values in the table come from batches of 10 dark-adapted cells stimulated to exhaustion in a laboratory photometer. For all of the measurements, the dinoflagellate produced a rather consistent photon yield of ~ 2–5 $\times$ 10^{10} photons per cell. An average value of 5 $\times$ 10^{10} photons per cell was used when we calculated the calibration factor for the bathyphotometer with R.V. *Endeavor* Cruise 054 data. The R.V. *C.O. Iselin* cruise was exceptional in that photon yields were low by one order of magnitude at the stations where bathyphotometer calibrations were performed (Table II). An average photon yield of 4.7 $\times$ 10^9 photons per cell was used for calculating the calibration factor for the bathyphotometer on the R.V.

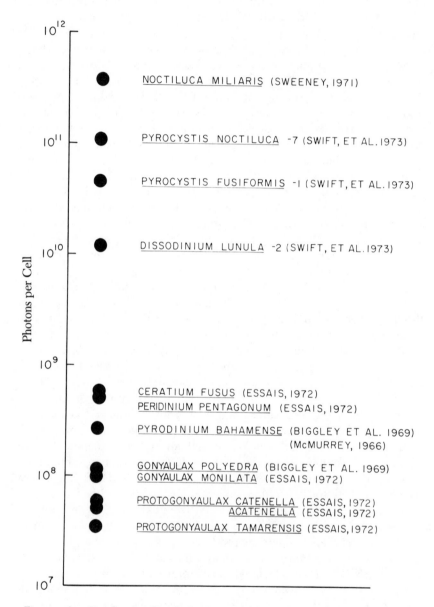

Figure 2. Total stimulated bioluminescence produced by different dinoflagellates.

Table I. Photon Yields Per Cell for *Pyrocystis noctiluca*

Region	Cruise, Date	Number of Stations	Photons Per Cell ($\times 10^{10}$)
Anegada	*Iselin*, October 1976	5	1.6 ± 0.8[a]
Passage	*Endeavor* 18, February 1978	2	2.4 ± 5.5
Regional mean			2.0
Southern	*Iselin*, October 1976	5	0.55 ± 0.5[b]
Sargasso	*Endeavor* 18, February 1978	5	5.9 ± 1.4
Regional mean			3.2
Northern	*Trident* 169, July 1975	4	4.3 ± 0.9
Sargasso	*Endeavor* 54, August 1980	1	4.3
	Endeavor 73, August 1981	5	5.1 ± 1.4
Regional mean			4.6

NOTE: The samples were collected near the surface, isolated, and stimulated to exhaustion by a laboratory photometer. The mean values (with ± 95 % confidence limits) indicate the magnitude of the year-to-year, region-to-region, and cruise-to-cruise variability in measurements of photon yield of *P. noctiluca*.

[a] Values for the R.V. *C.O. Iselin* cruise for Anegada Passage are higher than reported for Station Z in Table II as they include higher values obtained before bathyphotometer calibrations started.

[b] Cells collected during this cruise were unusually small; size may explain the small values.

Table II. Calibrations Using *Pyrocystis noctiluca* During R.V.
C.O. Iselin Cruise, October 1976

Station, Date	Laboratory Photometer (10^9 photons/cell)	Bathyphotometer (10^{10} photons/V)
Z, 10/17	4.7 ± 1.6	23.5[a]
		13.4[a]
Y, 10/19	4.6 ± 0.8	4.2
		4.0
Y, 10/20	7.7 ± 0.6	3.5
X, 10/21	1.7 ± 0.1	2.8
Mean for all stations	4.7 ± 2.9	3.6 ± 0.6

NOTE: To determine the photon yield of their bioluminescence, batches of 10 *P. noctiluca* cells were stimulated mechanically to exhaustion in a laboratory photometer. The batches of *P. noctiluca* cells were pumped through the bathyphotometer, and the values given assume that the cells were stimulated to exhaustion in the photometer. The mean for all values, stations, and dates is given with its 95 % confidence limit.

[a] These calibrations were excluded from the calculations of the cruise average. The batch of cells used produced much less light than in the other calibrations.

Iselin cruise. On all cruises, cells for calibration were collected in near-surface waters so that cells would have high levels of bioluminescence (34). The dinoflagellates were collected late in the afternoon by net tows, isolated by pipette into filtered seawater, collected from the same depth, and held in 10-mL pipettes with enlarged tips. After several hours in the dark, and at the time of natural darkness, they were gently introduced a few at a time into the sample chamber intake of the bathyphotometer with the pump running. Batches of the same cells were assayed in the laboratory photometer after being isolated into 3 mL of filtered seawater, held several hours in darkness, and stimulated mechanically to exhaustion.

To determine the fraction of the total mechanically stimulatable luminescence (9), which was stimulated and recorded as *P. noctiluca* or any other organism passed through the bathyphotometer, we first had to calibrate the photomultiplier in the bathyphotometer. To do this, we added an aliquot of a culture of the bioluminescent dinoflagellate *Pyrocystis lunula* to the sample chamber and released the dinoflagellate's bioluminescent potential by injecting acetic acid to drop the pH in the chamber from 8.2 to less than 5. In parallel with the bathyphotometer measurements, aliquots of the same culture were added and were stimulated with a similar pH drop in an absolutely calibrated laboratory photometer (36). Cultures of bioluminescent bacteria have been used (25), and the use of aqueous luminol in the bathyphotometer chamber has been suggested (35) as a method of calibration. To determine the fraction of bioluminescence potential released by single cells of scotophasic *P. noctiluca*, we compared the signals when the cells passed through the bathyphotometer with the signals from the laboratory photometer where cells were mechanically stimulated to exhaustion.

About one-tenth of all of the bioluminescent potential of *P. noctiluca* was detected as it passed through the bathyphotometer. As with other pump-through bathyphotometers, the stimulation may not have been intense enough in the sample chamber to release the total bioluminescence before the cells passed through the sample chamber. Also, the release of the bioluminescence may have been too slow to occur completely while the cell was in the sample chamber (7). If the bioluminescence potential of other organisms is underestimated to the same extent as *P. noctiluca*, then the total bioluminescence potential calculated for the water column and reported here and elsewhere (28) is correct. The average bioluminescence produced by *Pyrocystis* and the other important bioluminescent organisms in this region—copepods, larvaceans, and ostracods—seems to be in the same ratio in the bathyphotometer as in a laboratory photometer where the organisms are stimulated to exhaustion. This similarity suggests that they do give up similar fractions of their bioluminescence in the bathyphotometer. The ratios were relative to *P. noctiluca*: one- to threefold for copepods, one- to threefold for larvaceans, and three- to fivefold for ostracods (unpublished data). However, more extensive measurements are needed of what fraction of their total bioluminescent potential organisms like dino-flagellates, larvaceans, copepods, and ostracods actually produce in our bathyphotometer. This knowledge would allow more accurate estimates to be made of the bioluminescence potential in the water column.

Profiles taken in the Sargasso Sea near the same location one to several nights apart were not very different. To illustrate this, we have compared values of stimulateable bioluminescence per square meter of sea surface for a number of instances where data were taken at the same general location. As Table III indicates, the mean value of the coefficient of variation between such stations is about 15%. Some of this variability may have been due to "patchiness" or true differences in the concentrations or taxonomic composition of the bioluminescent organisms. However, some of it is due to a few large and very bright organisms that contribute a large part of the stimulated bioluminescence. Sampling variability (whether or not these large organisms are sampled) adds to variability from patchiness. The coefficient of variation calculated from an entire night's bioluminescence pro-

Table III. Comparisons at Closely Spaced Stations of Measurements of Epipelagic-Stimulated Bioluminescence Per Square Meter of Sea Surface

Cruise, Station	Date	Latitude, Longitude	Photons Per Square Meter ($\times 10^{15}$)	Coefficient of Variation (%)
Endeavor 54, 4[a]	Sept. 4, 1980	37°11′ N, 68°18′ W	8.8	
	Sept. 5, 1980	38°08′ N, 68°38′ W	10	
		Mean value	9.4	12
Endeavor 54, 2[b]	Aug. 24, 1980	28°46′ N, 68°16′ W	4.1	
	Aug. 26, 1980	28°43′ N, 68°21′ W	2.8	
		Mean value	3.5	34
Endeavor 49, 2[c]	March 6, 1980	23°08′ N, 66°24′ W	2.7	
	March 8, 1980	23°09′ N, 66°30′ W	3.0	
	March 9, 1980	23°13′ N, 66°31′ W	3.4	
		Mean value	3.0	13
C.O. Iselin, Y[d]	Oct. 19, 1976	19°39′ N, 67°32′ W	7.3	
	Oct. 20, 1976	19°43′ N, 67°23′ W	6.3	
		Mean value	6.8	15
C.O. Iselin, W[e]	Oct. 23, 1976	24°09′ N, 75°36′ W	4.0	
	Oct. 24, 1976	24°15′ N, 75°41′ W	4.1	
		Mean value	4.0	2
Mean value for all stations				15

NOTE: These measurements were determined on different nights at similar locations with the pump-through bathyphotometer. The coefficient of variation is the standard deviation from the mean value divided by the mean value.

[a] Measurements were taken 29 km apart.

[b] Measurements were taken 6 km apart.

[c] Measurements were taken 6 km apart on March 6 and March 8, 1980. Measurements were taken 4 km apart on March 8 and March 9, 1980.

[d] Measurements were taken 9 km apart.

[e] Measurements were taken 8 km apart.

files taken 5–7 min apart in the tropical Pacific Ocean from a slowly drifting ship (23) was largest near the surface where the bioluminescence was at a minimum. Coefficients of variation near the surface were as large as 56% and were smallest at the depth of the bioluminescent maximum with values of 12–14% (23). Sampling variability may be responsible for the higher variability found near the surface than in the maximum. The coefficient of variation between the total light in the tropical Pacific Ocean profiles between the surface and 200 m was ~4% (23). This value suggests a consistency throughout the night of epipelagic bioluminescence measurements when expressed on a water column or per square meter of sea surface basis.

For hydrographic information, we used 12–14 bottle casts from the surface to 200 m with 5-L poly(vinyl chloride) sample bottles (Niskin). Temperature from the paired reversing thermometers (± 0.01 °C) was used to correct the trace of temperature with depth from an expendable bathythermograph (XBT-Sippican Corp.). Analyses run on the water samples included salinity with an inductive salinometer (Plessy Model 6230). Extracted chlorophyll was measured in a fluorometer (Turner Model 111) (46), calibrated with a pure chlorophyll a standard (Sigma Chemical Company). The usual methods were used for analyses of phytoplankton nutrients on ship: ammonium (37), inorganic phosphate free from arsenic interference (38), and hydrolyzable organic phosphate (38, 39, 40). Samples for nitrite and nitrate were filtered, frozen, and analyzed after the cruise with the method of Wood et al. (41).

Results

The Caribbean Sea is represented by a station in Anegada Passage (Station Z) and a station about 90 mi north of Puerto Rico in the southern Sargasso Sea (Station Y). Station Y is influenced by water flow from the Caribbean Sea. Typical stations were selected to form a north–south (N–S) section from Stations 4 and 1 in the northern Sargasso Sea, to Stations 2 and 3 in the southern Sargasso Sea, to Stations Y and Z as Caribbean stations (Figure 3). Figure 4 shows the numbers of epipelagic (0–120 m) bioluminescent organisms on a per square meter of sea surface basis (summing up discrete flashes per unit volume with depth). As seen in the profiles of bioluminescence with depth, most of the bioluminescent signal is observed in this epipelagic layer. As shown by the frequency of flashing, the number of bioluminescent organisms decreased from north to south about twofold in the Sargasso Sea in August. Profiles made at the southern edge of the region in October indicated that the concentrations of epipelagic bioluminescent organisms increased over the values found in the southern Sargasso Sea in August (about ninefold at Station Z). The observed flashes were separated into those that produced more light than a typical *P. noctiluca* (presumably zooplankton) and those that produced less light (presumably dinoflagellates and small zooplankton). The numbers of organisms producing more light than a *P. noctiluca* were relatively greater at stations with high concentrations of

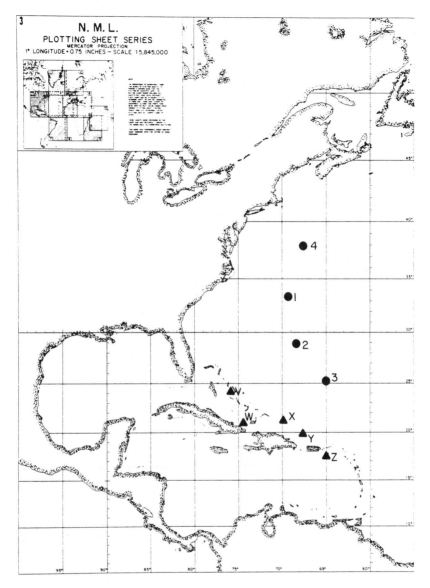

Figure 3. The location of stations for the Iselin (▲) *and* Endeavor 54 (●) *cruises.*

bioluminescent organisms than at those with low concentrations. Thus, they are numerically more important at Stations 4 and 1 and Stations Y and Z than at Stations 2 and 3.

A similar N–S trend in the estimated value of the total stimulated bioluminescence per square meter was recorded by the bathyphotometer (Figure 5). About twice as much light was produced per square meter in the

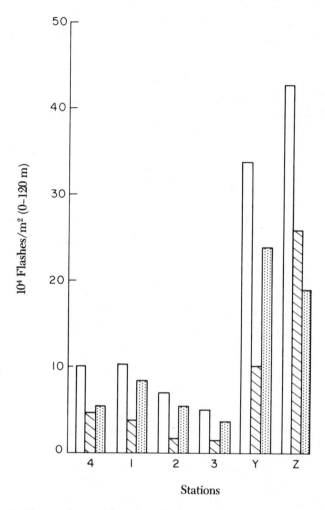

Figure 4. The numbers of epipelagic bioluminescent organisms per square meter of sea surface along a N–S section in the Sargasso Sea from the Gulf Stream to the Greater Antilles (see Figure 1). Flashing organisms are divided into three catagories: total bioluminescent organisms (□), bioluminescent organisms producing more light than P. noctiluca (∖∖) and bioluminescent organisms producing less light than P. noctiluca (▨).

northern Sargasso Sea as in the southern Sargasso Sea. In the Caribbean Sea, the stimulated bioluminescence per square meter was more than in the southern Sargasso Sea (threefold more at Station Z than at Station 3). Figure 4 shows that small flashes were generally the most frequent at all stations; Figure 5 shows that large flashes (more light than a P. *noctiluca*) produced most of the total light. This latter result is more evident at the stations with high values of total bioluminescence. Subsurface peaks in bioluminescence

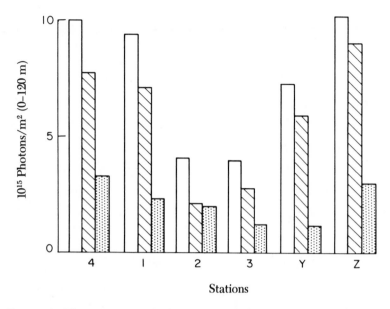

Figure 5. *The amount of bioluminescent light produced by epipelagic organisms per square meter of sea surface along a N–S section in the Sargasso Sea from the Gulf Stream to the Greater Antilles. Bioluminescent organisms are classed three ways: total organisms (□), organisms producing more light than* P. noctiluca *(▧), and organisms producing less light than* P. noctiluca *(▨).*

were more pronounced in the stations with high values of bioluminescence than in the ones with low values, as will be seen in the station profiles discussed now.

Northern Sargasso Sea. As summarized in Figures 4 and 5, this region, including the Gulf Stream, is one of higher bioluminescence activity than in the southern Sargasso Sea. A station in the northern Sargasso Sea (Station 1) is shown in Figure 6. A feature that distinguishes profiles in this region is the distinct subsurface peak in stimulated bioluminescence potential. In this profile, the peak at ~80 m produced ~1.3×10^{14} photons/m^3. *Pyrocystis noctiluca* is in greater numbers than further south, and it produces more bioluminescence per cell than at the R.V. *C.O. Iselin* stations. At Station 1 it accounted for ~15% of the total light production in the bioluminescence peak. Peak flashing rates (~1700/m^3) were about twice as high as at the stations in the southern Sargasso Sea. The chlorophyll maximum is deeper than the bioluminescence maximum.

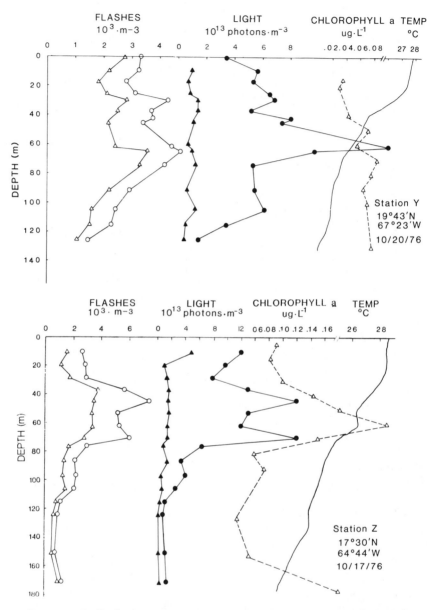

Figure 6. Bathyphotometer profiles of the concentration of bioluminescent organisms and the bioluminescence they produced at Stations 1 and 3. Organisms producing less light than P. noctiluca (—△—) produce most of the total flashes (○) at any depth, but the organisms producing dim flashes (▲) do not contribute much to the total bioluminescence (●). The profile of the chlorophyll peak is shown for comparison (--△--).

Southern Sargasso Sea.   Station 3 at 25° N is typical of the Sargasso Sea south of the position of the subtropical fronts (Figure 6). The maximum bioluminescence was ~6 × 10^{13} photons/m³. This value is about one-tenth of the values in Anegada Passage at Station Z and one-half of the values in the northern Sargasso Sea. Here, the contribution to light production by bright (>5 × 10^{10} photons) organisms is relatively smaller (Figure 6). The *Pyrocystis* found in the profile would produce about one-tenth of the bioluminescence in the water column. The depth of the maximum in both flashing rate and light production is above the depth of the chlorophyll maximum (Figure 6). In this region, the chlorophyll maximum may be below the biomass maximum because cells may tend to adapt to low light intensities at depth by increasing their pigment per cell (*42*).

Caribbean Region.   At Station Z, a pronounced peak occurred in bioluminescence between 50 and 80 m (Figure 7). This peak contained the most bioluminescent water in our N–S section with ~2 × 10^{14} photons/m³. The bioluminescence peak differed from other stations because it was associated with the depth of the chlorophyll peak at about 70 m. This depth was also the depth for the population peak of *P. noctiluca* (Figure 8), but the ~60 cells/m³ at the depth of the peak would produce <0.04 % of the observed bioluminescence. Further north at Station Y (Figure 7) some features suggest elements of Caribbean influence. A pronounced peak of bioluminescence is still seen at 60 m. This peak represents ~1.3 × 10^{14} photons/m³, which is almost entirely associated with organisms producing more light than *P. noctiluca*. At Station Y, *P. noctiluca* were present at about 10/m³ (Figure 8), so <0.12 % of the bioluminescence was apparently due to that species. Nutrient concentrations were considerably lower and the NO_3 nutricline is much deeper at this station than at Station Z (Figure 9).

Discussion

The large-scale patterns in bioluminescence that are observed appear to be related to the hydrography and productivity of the regions examined, at least on a macroscale level. For example, in the Caribbean region in October, when our bathyphotometer profiles were made in Anegada Passage (Station Z) and at the southern edge of the Sargasso Sea (Station Y), nutrients, chlorophyll, and bioluminescence profiles suggested more eutrophic conditions than farther north in the southern Sargasso Sea (Figures 9 and 10).

Water from the Caribbean Sea is advected to a varying extent north into the Sargasso Sea at the surface where it mixes with less productive water of the type found in the southern Sargasso Sea. This idea is supported by Wüst's analysis (*43*), which predicts a general surface flow of 17–35 km/day in a northeasterly direction across the shallow banks of the lesser Antilles and out of Anegada Passage into the southern Sargasso Sea. This predicted flow would carry low-salinity water, originating primarily with

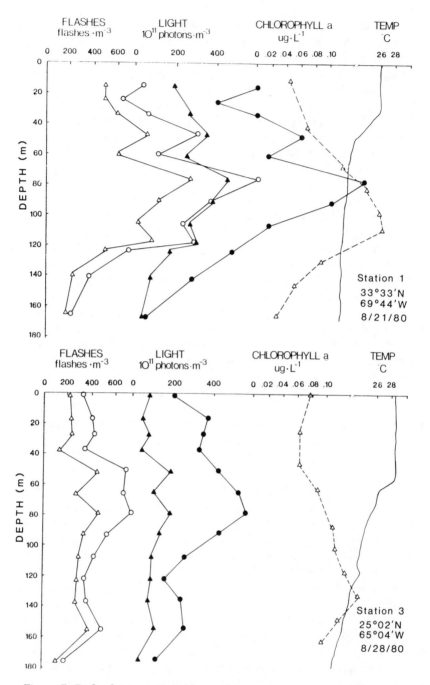

Figure 7. Bathyphotometer profiles of the concentration of bioluminescent organisms and the bioluminescence they produced at Stations Y and Z. Symbols are the same as in Figure 6.

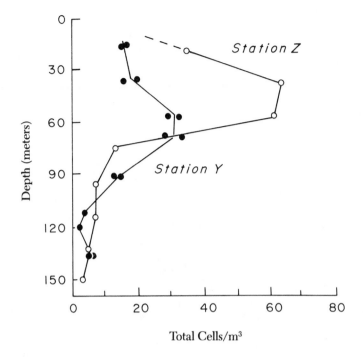

Figure 8. The concentrations of P. noctiluca *with depth at Stations Y and Z.*

the Amazon and Orinoco Rivers, as well as with fresh water due to high-surface precipitation in the vicinity of Puerto Rico (Plates 5 and 6, Ref. 43). This surface water is underlain with a salinity maximum due to subtropical underwater, which appears to originate to the northwest (Plate 13, Ref. 43). The resulting halocline produces a stable water column with a shallow mixed layer. The countercurrent flow (43) of surface and subtropical underwater may produce a "nutrient trap" in this region. Relatively high rates of nitrogen fixation may occur in Anegada Passage (44). The presence of shallow banks may contribute to local eutrophication of surface waters (44). All these factors may explain the high phytoplankton biomass (suggested by the chlorophyll values) and high concentration of bioluminescent organisms.

The southern Sargasso Sea is difficult to separate from the northern region on the basis of the water-mass concept of biogeography (2). However, these two regions have different fauna (45). The east–west (E–W)-tending thermal fronts at about 28° N are an important feature. North of these fronts, primary production is higher (47, 48), nitrogen fixation and *Oscillatoria* populations are lower (44), and the communities of mesopelagic fish (45) and acantharians (49) exhibit differences. Surface water south of this

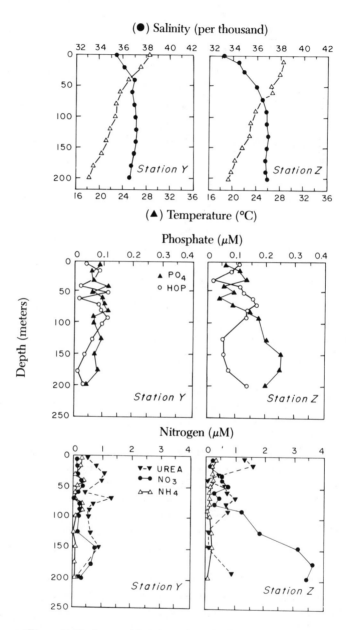

Figure 9. *Hydrographic information for Stations Y and Z.*

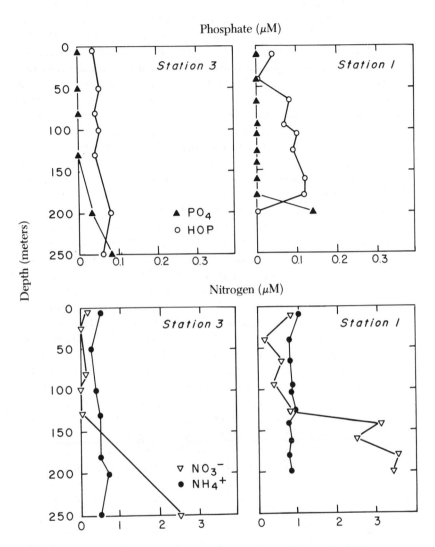

Figure 10. Hydrographic information for Stations 1 and 3.

"subtropical front" comes from the southeastern Sargasso Sea region as does the subtropical underwater. In this region, winter mixing of the 18 °C water with the surface water does not occur, nor does annual renewal of nutrients, as seen farther north. In our summer survey, flash frequency and total integrated bioluminescence was lower in the southern region. These observations reflected a lower standing stock of organisms and a change in the community of bioluminescent organisms (unpublished data).

The regional patterns that we suggest for stimulated bioluminescence are, in part, based on our preconceptions about the patterns of oceanographic conditions. However, some justification exists for believing that the typical stations selected here are representative of others in the same region. The expected difference in bioluminescence per square meter of sea surface at a single station is about 15%. The differences that we have suggested may be regional are larger, twofold or more. The agreement between stations in the "same" regions is almost the same as between the "same" stations in different regions, as described in Table III. For example, although hundreds of kilometers apart, Stations 4 and 1 and Stations 2 and 3 differ by less than 15% in total bioluminescence per square meter of sea surface (Figure 4). Not only the total stimulated bioluminescence per square meter of sea surface, but other aspects of the bioluminescence as well, seem to show "regional consistency," including flashing rate (Figure 5).

We have suggested that pronounced peaks in the subsurface-stimulated bioluminescence profiles are a feature of the northern Sargasso Sea, at least in the late summer. In support of that hypothesis, pronounced subsurface peaks were found in bioluminescent flashing rate at the three northernmost stations (out of five) in a section from south of Bermuda to the Gulf Stream in July 1975 on an R.V. *Trident* cruise (unpublished data).

In both sets of stations, organisms brighter than *P. noctiluca* produced the majority of the bioluminescence in the peaks. This bioluminescence suggests that many of the patterns may be caused by the zooplankton community. Seasonal changes in the abundance of bioluminescent zooplankton occur in the northern Sargasso Sea, including the Bermuda region (50–54). Plankton studies suggest that spatial variability in the northwest part of this region may be due to mesoscale patchiness in the plankton communities, which is introduced by the presence of cold core rings (54).

Less seasonal variation in bioluminescence patterns may occur in the southern Sargasso Sea. Profiles taken in the southern Sargasso Sea in February 1979 on Cruise 49 of R.V. *Endeavor* resemble those shown here in that they do not have pronounced subsurface peaks (28), and they produce about the same amount of stimulated bioluminescence per square meter of sea surface (Table III). The low amounts of bioluminescence may be a regional feature. In October 1976, profiles taken north of the Caribbean Sea (Stations Y and Z) on the R.V. *C.O. Iselin* cruise, including Station W (Table III), showed reduced bioluminescence activity and a lack of pronounced subsurface peaks. As more information on the seasonal and regional variability of bioluminescence becomes available in the Sargasso Sea, it will be interesting to discern to what extent they reflect the regional patterns described by biogeographers.

Acknowledgments

This work was supported by grants from the Oceanography Program of the National Science Foundation, OCE–8117744 and the Oceanic

Biology Program (Code 480) of the Office of Naval Research, N00014–81–C–0062.

Literature Cited

1. Haedrich, R. L.; Judkins, D. C. In "Zoogeography and Diversity of Plankton"; van der Spoel, S.; Pierrot-Bults, A. C., Eds.; Wiley: New York, 1979; chap. 2.
2. McGowan, J. A. In "Marine Science"; Anderson, N. R.; Zaharanec, B. J., Eds.; Plenum: New York, 1978; vol. 5, pp. 423–43.
3. Buck, J. B. In "Bioluminescence in Action"; Herring, P. J., Ed.; Academic: New York, 1978; chap. 13.
4. Boden, B. P. *J. Mar. Biol. Assoc. U.K.* **1969**, *49*, 669–82.
5. Tett, P. B. *J. Mar. Biol. Assoc. U.K.* **1971**, *51*, 183–206.
6. Rudyakov, Yu. A. *Okeanologia (Moscow)* **1968**, *7*, 710–14.
7. Seliger, H. H.; Fastie, E. G.; Taylor, W. R.; McElroy, W. D. *Limnol. Oceanogr.* **1969**, *14*, 806–14.
8. Backus, R. H.; Yentsch, C. S.; Wing, A. *Nature (London)* **1961**, *192*, 518–21.
9. Biggley, W. H.; Swift, E.; Buchanan, R. J.; Seliger, H. H. *J. Gen. Physiol.* **1969**, *54*, 95–122.
10. Sweeney, B. M. "Rhythmic Phenomena in Plants"; Academic: New York, 1969.
11. Tett, P. B.; Kelly, J. G. *Oceanogr. Mar. Biol. Annu. Rev.* **1978**, *11*, 89–173.
12. Kelly, M. G.; Tett, P. B. In "Bioluminescence in Action"; Herring, P. G., Ed.; Academic: New York, 1978; chap. 12.
13. Herring, P. J.; Morin, J. G. In "Bioluminescence in Action"; Herring, P. J., Ed.; Academic: New York 1978; chap. 7.
14. Herring, P. J. In "Bioluminescence in Action"; Herring, P. J., Ed.; Academic: New York, 1978; chap. 7.
15. Clarke, G. L.; Backus, R. H. *Deep-Sea Res.* **1956**, *4*, 1–14.
16. Clarke, G. L.; Wertheim, G. K. *Deep-Sea Res.* **1956**, *3*, 189–205.
17. Clarke, G. L; Hubbard, C. J. *Limnol. Oceanogr.* **1959**, *4*, 163–80.
18. Clarke, G. L.; Breslau, L. R. *Bull. Inst. Oceanogr.* **1960**, *1171*, 1–32.
19. Clarke, G. L.; Backus, R. H. *Bull. Inst. Oceanogr.* **1964**, *1318*, 1–36, 10 fig.
20. Clarke, G. L.; Kelly, M. G. *Bull. Inst. Oceanogr.* **1964**, *1319*, 1–20.
21. Kremer, P. M., Ph.D. thesis, Univ. of Rhode Island, Kingston, Rhode Island, 1975.
22. Losee, J. R.; LaPota, D. In "Bioluminescence: Current Perspectives"; Nealson, K. H., Ed.; Burgess: Minneapolis, 1981; pp. 143–52.
23. Gitel'zon, I. I.; Levin, L. A.; Shevyrnogov, A. P.; Utyushev, R. N.; Artemkin, A. S. In "Life Activity of Pelagic Communities in the Ocean Tropics"; Vinogradov, M. E., Ed.; Acad. Sciences USSR; 1971 (orig. in Russian, in English by: Israel Program for Scientific Translations, Jerusalem, 1973, pp. 51–66).
24. Seliger, H. H.; Fastie, W. G.; McElroy, W. D. *Science (Washington, D.C.)* **1961**, *133*, 699–700.
25. Seliger, H. H.; Fastie, W. G.; Taylor, W. R.; McElroy, W. D. *J. Gen. Physiol.* **1962**, *45*, 1003–17.
26. Clarke, G. L.; Kelly, M. G. *Limnol. Oceanogr.* **1965**, *10(Suppl.)*, R54–R66, 26.
27. Broenkow, W. W.; Lewitus, A. J.; Yarbrough, M. A.; Krenz, R. T. *Nature (London)* **1983**, *302*, 329–31.
28. Swift, E.; Biggley, W. H.; Verity, P. E.; Brown, D. T. *Bull. Mar. Sci.* **1983**, *22*, 855–63.
29. Young, R. E. In "Bioluminescence: Current Perspectives"; Nealson, K. H., Ed.; Burgess: Minneapolis, 1981; pp. 72–81.
30. Widder, E. A.; Latz, M. I.; Case, J. F. *Biol. Bull.* **1983**, *165*, 791–810.
31. Herring, P. J. *Proc. R. Soc., London, Ser. B* **1983**, *220*, 183–217.
32. Swift, E.; Biggley, W. H.; Seliger, H. H. *J. Phycol.* **1973**, *9*, 420–26.

33. Swift, E.; Meunier, V. A. *J. Phycol.* **1976**, *12*, 9–13.
34. Swift, E.; Meunier, V.; Biggley, W. H.; Hoarau, J.; Barras, H. In "Bioluminescence: Current Perspectives"; Nealson, K. H., Ed.; Burgess: Minneapolis, 1981; pp. 95–105.
35. Lee, J.; Seliger, H. H. *J. Chem. Phys.* **1964**, *40*, 519–23.
36. Seliger, H. H.; Biggley, W. H.; Swift, E. *Photochem. Photobiol.* **1969**, *10*, 227–32.
37. Solorzano, L. *Limnol. Oceanogr.* **1969**, *14*, 799–801.
38. Johnson, D. L. *Environ. Sci. Technol.* **1971**, *5*, 411–14.
39. Murphy, J.; Riley, J. P. *Anal. Chem. Acta* **1962**, *27*, 31–36.
40. Strickland, J. D. H.; Solorzano, L. In "Some Contemporary Studies in Marine Science"; Barnes, J., Ed.; Allen and Unwin: New York, 1966, pp. 665–74.
41. Wood, E. D.; Armstrong, F. A. J.; Richards, F. A. *J. Mar. Biol. Assoc. U.K.* **1967**, *47*, 25–31.
42. Steele, J. J.; Yentsch, C. S. *J. Mar. Biol. Assoc. U.K.* **1960**, *39*, 217–26.
43. Wüst, G. "Stratification and Circulation in the Antillean Caribbean Basins, I. Spreading and Mixing of the Water Types with an Oceanographic Atlas"; Columbia University Press: New York, 1964.
44. Carpenter, E. J.; Price, C. C. *Limnol. Oceanogr.* **1977**, *22*, 60–72.
45. Backus, R. H.; Craddock, J. E.; Haedrich, R. L.; Robinson, B. H. In "Fishes of the Western North Atlantic"; Gibbs, R. H. et al., Eds.; *Mem. Sears Found. Mar. Res.* **1977**, *7*, 266–86, 41, 46.; Backus, R. H.; Craddock, J. E.; Haedrich, R. L.; Shores, D. L. *Mar. Biol.* **1969**, *3*, 87–106.
46. Yentsch, C. S.; Menzel, D. W. *Deep-Sea Res.* **1963**, *10*, 221–31.
47. Ryther, J. H.; Menzel, D. W. *Bull. Mar. Sci. Gulf Caribb.* **1960**, *1*, 381–88.
48. Ryther, J. H.; Menzel, D. W. *Deep-Sea Res.* **1960**, *6*, 235–38.
49. Bottazzi, E. M.; Androeli, M. G. *J. Protozool.* **1982**, *29*, 162–69.
50. Napora, T. A., M. S. thesis, Univ. of Rhode Island, Kingston, Rhode Island, 1954.
51. Deevey, G. B. *Limnol. Oceanogr.* **1971**, *16*, 219–40.
52. Deevey, G. B.; Brooks, A. L. *Bull. Mar. Sci.* **1977**, *27*, 256–91.
53. Grice, G. D.; Hart, A. *Ecol. Monogr.* **1962**, *32*, 287–309.
54. Ortner, P. B.; Wiebe, P. H.; Cox, J. L. *J. Mar. Res.* **1980**, *38*, 507–31.

RECEIVED for review July 11, 1983. ACCEPTED February 28, 1984.

Fluorescence Spectral Signatures for Studies of Marine Phytoplankton

CHARLES S. YENTSCH and DAVID A. PHINNEY

Bigelow Laboratory for Ocean Sciences, McKown Point, West Boothbay Harbor, ME 04575

With fluorescence spectral signatures of phytoplankton, a distinction can be made between organisms in natural populations with differing accessory pigmentation. This technique is often referred to in remote-sensing circles as the color group model. This model is too crude to be of any great value in the characterization of marine phytoplankton. To refine the model further would require separation of fluorescence from accessory pigments that overlap in light absorption. A close relationship exists between features of the excitation spectra for chlorophyll a and phycoerythrin and mean cell size in natural populations. Therefore, cellular area–volume relationships in natural populations can be estimated with fluorescence signatures. Thus, although they do not satisfy characterization in the context of species diversity, fluorescence signatures are potentially useful in the study of the physiology of natural populations.

T he measurement of fluorescence from phytoplankton chlorophyll is one of a few biological oceanographic measurements that can be made in a continuous fashion. A better understanding of the time and space distributions of natural phytoplankton populations is the goal of this technique. But as important as those problems are, perhaps more important is the understanding of the distribution of phytoplankton assemblages. There are two questions: What causes the different phylogenetic assemblages to occur in different water masses, and why do different species appear and disappear in time and space as environmental conditions change? The fluorescent properties of algae, that is, their spectral signatures, might solve these problems by providing a means for continuous mapping and identification of species assemblages. Historically, early studies of plant physiology established the potential of utilizing fluorescence signatures. The pioneer-

0065-2393/85/0209-0259$06.00/0
© 1985 American Chemical Society

ing research of Engleman in the 1860s, demonstrated that, through their diverse pigmentation, plants and algae could absorb a range of colors of sunlight for photosynthesis; more physiological evidence that accumulated showed that different morphological groups of plants possess different suites of pigments. The pigmented composition of algae became an important criterion in distinguishing one phylogenetic group from another.

If a technique is to exploit differences in pigmentation, it needs prior measurements of the light-absorbing characteristics of the algae. Hence, the action spectrum for fluorescence of chlorophyll a is a tool for identifying the color types of algae (1).

This chapter will update the degree of success this fluorescent technique has had in aiding oceanographers to understand phylogenetic distributions of photosynthetic organisms in the oceans. For those who are knowledgeable of taxonomic problems, the apparent taxonomic simplification may seem very unsophisticated. Our goal is not to emphasize the degree of this oversimplification, nor to dwell on the degree to which the so-called color group hypothesis does not readily fit, but to illustrate how spectral measurements can aid in understanding how different wavelengths of sunlight are partitioned among a variety of photosynthetic algae in the sea. In the long run, this insight may go further in explaining why so many species exist, and how and why they have evolved in the oceans.

The Differentiation of Color Groups

The bulk of photosynthetic autotrophs were taken in the ocean; apparently, light is harvested in three basic ways. Phylogenetic color groups may be partially differentiated by light-absorbing pigments that are accessories to the major pigment chlorophyll a. In the simplest physiological sense, the accessory pigmentation can be thought of as additional light-harvesting units that allow the individual algae to utilize a wider range of wavelengths. Individual color groups can be indexed with specific types of pigmentation, namely, the accessory chlorophylls b and c, carotenoid proteins such as fucoxanthin, peridinin, and the biliproteins phycoerythrin and phycocyanin. This accessory pigment group classification is complete when the pigment groups are given a simplified color code. Hence *green* algae are those with only chlorophyll b as accessory pigments, *brown* algae have carotenoid proteins, and *blue-green* algae have biliproteins.

The measurements of fluorescence signatures of color groups show that differentiation involves varying the excitation wavelength E and observing the maximum emission of fluorescence F (Figure 1). This approach argues that in a mixture of algae such as found in natural populations, specific color groups can be sorted out by a combination of excitation and emission wavelengths. The magnitude of the fluorescence emission is assumed to be a function of fluorescent yield (Φ) and the absorption of light by the accessory pigments, such that $F = \Phi E$, where E is the light absorbed

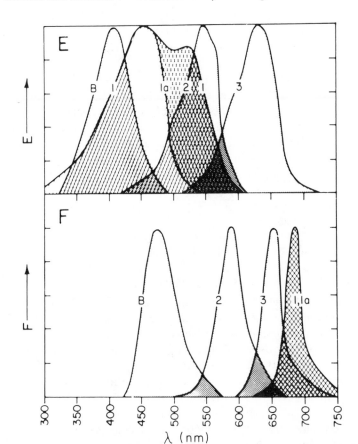

Figure 1. Fluorescence spectral signatures for different algal color groups (E = excitation, F = fluorescence). Key to schemes: 1, diatoms and dinoflagellates; 1a, green algae; 2, phycoerythrin of cyanobacteria; 3, phycocyanin. (Reproduced with permission from Ref. 5. Copyright 1981, Burgess Publishing Company.)

by the pigment ($A = E = \log I_o/I$) and is equal to A/LC, where A is absorbance, L is length of the light path, and C is concentration of the pigment.

To differentiate between brown and green algae, we assume that brown algae contain a sizable fraction of carotenoid protein and green algae do not. Thus, the relative abundance of brown algae (C_B) is

$$C_B = F_{680} = \Phi_B E_{530}$$

where 680 nm is the wavelength emission of chlorophyll a, and 530 nm is the wavelength of maximum absorption by the carotenoid protein. The relative abundance of green algae (C_G) is estimated by

$$C_G = F_{680} = (\Phi_G E_{450}) - (\Phi_B E_{530} 0.7)$$

The difference term is needed because chlorophyll a is excited at 450 nm in both color groups—brown and green; however, 530 nm excites chlorophyll a in only the brown algae. The factor 0.7 accounts for the contribution of absorption of 450 nm by brown algae in the mixture.

To measure the relative abundance of blue-green color groups (C_{BG})

$$C_{BG} = F_{580} = \Phi_{BG} E_{530}$$

The wavelength 530 nm is the maximum for light absorption, and 580 nm is the maximum fluorescence by phycoerythrin.

Potential Problems with Color Differentiation

Fluorescent Yield Variation and the Consistency of Pigment Ratios. Differentiation depends upon the fluorescent yield remaining relatively constant under various environmental conditions. A great amount of information exists on fluorescent yield changes with response to light and nutritional stress (2). In the most general sense, light fluoresced from the pigment system is lost from the photosynthetic carbon fixation pathway. To put it another way, any factor that influences the rate of carbon fixation will potentially influence the yield of fluorescence.

Nevertheless, the variability observed in chlorophyll fluorescent yield in natural populations has not seriously dissuaded workers from correlating these levels of fluorescence with total algae concentrations. Most researchers recognize the errors but would rather not lose the potential for continuous measurement.

The color differentiation method also depends upon the ratios of pigments accessory to chlorophyll a remaining constant, that is, constant with regard to changes in environmental extremes in light, temperature, and nutrients. In our laboratory, this constant nature is being tested in a variety of species of microalgae and macroalgae. To date, the results suggest that the fluorescence excitation signatures (i.e., ratios between accessory pigments) do not change markedly under environmental stress (3). Therefore, the major changes in the ratios of accessory pigmentation probably are due to phylogenetic and not environmental differences (3).

Usefulness of Color Group Differentiation. We could conclude that the color differentiation method has a reasonably good chance of working. But in light of what we now know about assemblages of planktonic organisms, is knowing the distribution of color groups useful?

Microfiltration techniques and epifluorescence have enhanced our knowledge of the distribution and diversity of species of marine phytoplankton. The diversity of species in the sea is much greater than heretofore expected. Reviewing these advancements, we can establish a baseline for what now is or could be useful in equating color strategies to species composition. First, for the pigment color differentiation method to be useful it should separate diatoms, dinoflagellates, and coccolithophores. These algae are at times extremely abundant in natural ocean waters. Second, the color techniques should also distinguish between cryptomonads and the cyanobacteria, both of which possess biliproteins as accessory pigments. With these two goals as the acceptable minimum, the color group technique described *cannot* adequately differentiate for the following reasons.

Diatoms and dinoflagellates are abundant at times in ocean waters and possess large quantities of carotenoid proteins. These organisms also have chlorophyll c, which is the accessory to chlorophyll a. Another group of abundant but quite different organisms, coccolithophores, contains chlorophylls a and c but a lower relative concentration of carotenoid protein (Table I). The problem of differentiation is illustrated by examining the range in the excitation ratio $(E_{530}:E_{450})$ for species of phytoplankton (Table I). The ratio represents the efficiency of light at 530 and 450 nm to excite chlorophyll fluorescence at 680 nm. The measurement is made from spectra not corrected for differences in Xe-source output at the two wavelengths. Phytoplankton with carotenoid proteins have a higher ratio than those without, and there is no basis to distinguish between these organisms. But phycoerythrin of cryptomonads absorbs in the same region, and therefore becomes equally important as carotenoids in the interpretation of this excitation ratio. Thus, using this single index (ratio of $E_{530}:E_{450}$) can cause considerable confusion.

In cryptomonads, phycobilins transfer the light they absorb to chlorophyll a. Phycoerythrin of cryptomonads also fluoresces (Figure 2). The pathway for these organisms is

$$E_{450} \rightarrow F_{680} \uparrow \text{chlorophyll a fluorescence}$$

$$E_{530} \rightarrow F_{680} \uparrow \text{chlorophyll a fluorescence}$$

$$E_{530} \rightarrow F_{580} \uparrow \text{phycoerythrin fluorescence}$$

These organisms may exhibit 680-nm fluorescence either by the absorption of the short blue band of chlorophyll and/or the transfer from the accessory pigment phycoerythrin.

In filamentous and coccoid forms of the marine cyanobacteria, another important group in the oceans, fluorescence is due primarily to E_{530} $\rightarrow F_{580} \uparrow$ (Figure 3). Therefore, for cyanobacteria no appreciable chloro-

Table I.   Chlorophyll–Accessory Pigment Relationships

Organism	Primary Accessory Pigments	Chlorophyll a Fluorescence $(E_{530}:E_{450})$
Dinoflagellates	Carotenoids, chlorophyll c	0.8–0.9
Diatoms	Carotenoids, chlorophyll c	0.7–0.8
Coccolithophores	Carotenoids, chlorophyll c	0.3–0.4
Green flagellates	Chlorophyll b	0.1–0.2
Cryptomonads	Phycobilins	0.7
Cyanobacteria	Phycobilins	—

phyll fluorescence occurs at 680 nm as the result of the light absorbed by phycoerythrin. These two groups of organisms cannot be differentiated on the basis of chlorophyll a fluorescence and/or independent phycoerythrin fluorescence.

However, fluorescent signatures can separate some groups as follows. With an excitation wavelength of 450 nm, chlorophyll fluorescence at 680 nm will arise from all groups of algae except cyanobacteria. With an excitation wavelength of 530 nm, the chlorophyll a fluorescence at 680 nm could be caused by either cryptomonads, dinoflagellates, diatoms, or prymnesiophytes. With excitation at 530 nm and emission at 580 nm, the phycoerythrin fluorescence could be caused by either cyanobacteria or cryptomonads.

It could appear that differentiation might not be such a problem if principal phylogenetic groups occupied similar ecological niches. The evidence, however, suggests the opposite; for example, the photosynthetic growth characteristics of the cyanobacteria appear to be very different from the eucaryote autotrophs (4). If fluorescent signatures are to be useful, it will have to be on some basis other than conventional taxonomy. A variable of natural populations worthy of investigation is cell size.

Fluorescence Signatures, Cell Size, and Growth Processes

For the past 4 years, in connection with multidisciplinary studies of remote sensing, we have measured the spectra of light attenuation and fluorescence of particulate matter collected from the world's oceans (5). The number of fluorescent cells for many of the open ocean areas is so low that technical problems quickly occurred in the attempt to obtain monochromatic spectra of light emission and absorption. Furthermore, when the cells are placed in cuvettes they tend to settle out; therefore, spurious results are obtained. To overcome these problems we elected to concentrate the cellular particles on membrane filters, which in turn were mounted in the light path of a conventional spectrofluorometer. (For details of this method see Reference 1.)

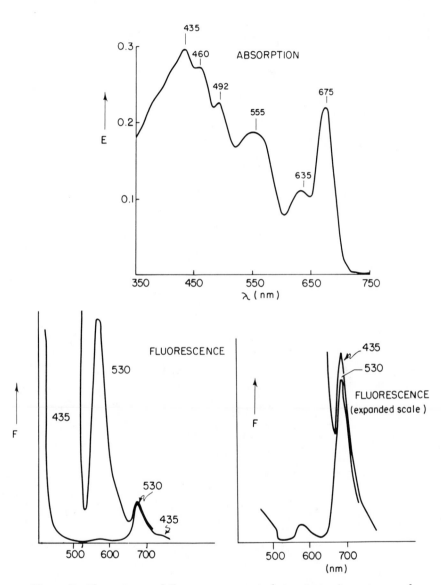

Figure 2. Absorption and fluorescence spectral signature of cryptomonad Chroomonas salina *(Bigelow Laboratory, Clone 3C).*

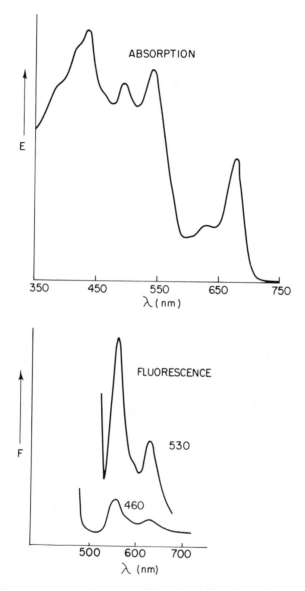

Figure 3. Absorption and fluorescence spectral signature of cyanobacteria Synechococcus sp. (Bigelow Laboratory, Clone L1601). Note the fluorescence emission of phycocyanin at longer wavelengths.

These studies have shown relationships between properties of fluorescence signatures and other properties of phytoplankton biology that would not have been anticipated from the color group hypothesis. These properties are summarized as follows:

1. The ratio of $E_{530}:E_{450}$ is strongly correlated with the chlorophyll biomass in the world's oceans (1).
2. Chlorophyll a fluorescence is excited much more efficiently at green wavelengths (e.g., 530 nm) in populations found in coastal waters than in populations found in the open ocean (1).
3. With the exception of filamentous cyanobacteria, fluorescence from biliproteins such as phycoerythrin and/or phycocyanin is not observed in samples taken by fine plankton nets.

To summarize our results, examine the extremes in spectral signature that we observed (Figure 4). The extremes are defined by the spectral differences observed for populations in eutrophic waters as opposed to oligotrophic waters. The examples of spectral signatures shown in Figure 4 were taken on either side of a color front that delineated Mississippi River water and the oligotrophic waters of the Gulf of Mexico. The front covered the distance of a few kilometers. These signatures are designated as green and blue waters. The concentration of chlorophyll in green water was 10 times greater than that in the blue water populations; the major difference in the spectral signature was the effectiveness at which light at 530 nm excited chlorophyll a fluorescence. At 530 nm, excited chlorophyll was approximately 10 times greater in green than in blue water populations. The excitation signature in green as opposed to blue water is distinguished by the strong shoulder at 530 nm.

When relying on color group hypothesis alone, one interpretation of these observations is that the green water populations contain organisms having carotenoid proteins as accessory pigments, namely, diatoms and dinoflagellates, whereas the blue water oligotrophic populations contain proportionally less of these species. Microscopic observations of water samples taken from these two water types confirmed that the green water population was indeed largely made up of diatoms and dinoflagellates; few of these species were found in the blue water samples. Both populations contained considerable quantities of phycoerythrin, but in terms of the total chlorophyll present in each population, the phycoerythrin content in blue water organisms was greater than that in green water organisms.

Other studies showed that the oligotrophic ocean areas are dominated by phytoplankton organisms of relatively small size, generally less than 5 μm. Therefore, the spectral signatures of the oligotrophic small cell size populations should resemble that of the blue water population, and the eutrophic populations typical of coastal and upwelling regions, which gen-

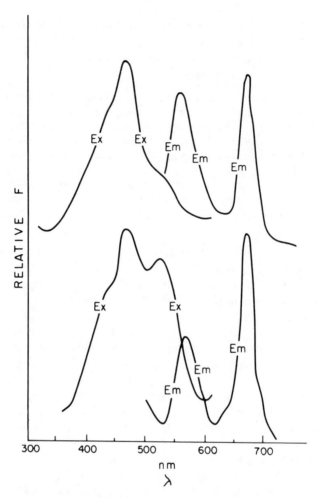

Figure 4. Fluorescent spectral extremes across a color front (Gulf of Mexico, October 25, 1977). Key: top, blue water, chlorophyll 0.2 μg/L; and bottom, green water, chlorophyll 2.0 μg/L.

erally have large cells, should have the spectral signature of the green water population.

To test this hypothesis we measured the relationship between fluorescent signatures and organism size. We selectively sieved and filtrated, with both plankton netting and membrane filters, phytoplankton populations of different cell size. An example of these results is shown in Figure 5. For organisms retained by a 10-μm sieve, a conspicuous shoulder is apparent in the excitation spectra in the 530-nm region. In this population, the wavelength ratio $E_{530}:E_{450}$ is greater than 0.5. Those organisms that pass through the 10-μm sieve and are retained by the 1-μm membrane have a

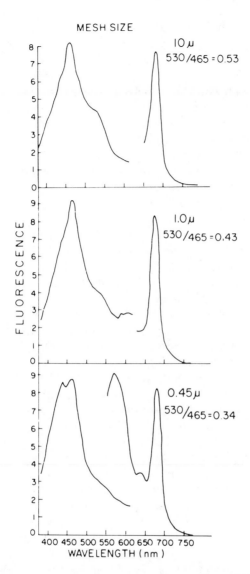

Figure 5. Fluorescence spectral signatures of natural populations as a function of cell size (Gulf of Maine, 1981).

ratio slightly greater than 0.4, but less than the larger organisms. In the larger organisms, fluorescence emission other than that by chlorophyll a was not indicated. In the organisms that pass through the 1-μm sieve and are retained on the 0.45-μm membrane, phycoerythrin is greatly increased and a small amount of fluorescence from phycocyanin occurs at 650 nm. The $E_{530}:E_{450}$ ratio is decreased to 0.3, and the excitation spectrum is characterized by the absence of any strong shoulder absorption at 530 nm. These data reveal that sizes greater than 1 μm contain accessory pigments to chlorophyll that strongly absorb at 530 nm; smaller sizes, that is those less than 1 μm, do not. Below the 1-μm size, the primary pigment accessory to chlorophyll is phycoerythrin. Referring back to the blue and green water extremes (Figure 4), we can explain these spectral signatures in terms of the sizes of cells that make up the population. The strong spectral shoulder at 530 nm in the green water population, as opposed to the blue water population, means that the former is made up of proportionally larger carotenoid-containing cells. The phycoerythrin concentration in both samples is representative of the fraction of small cells in each population.

By size fractionating a large number of samples of ocean populations we have extended the size-to-fluorescence relationship so that significant means and range of values can be established (Figure 6). If we consider only $E_{530}:E_{450}$ and a size range between 0.45 and 58 μm, this fluorescence ratio changes approximately 40%. At 58 and 10 μm the ratio is 0.6 and decreases to a value of 0.4 at smaller mesh sizes. In organisms retained by the 58-μm sieve the ratio range is 0.4–0.9. For those organisms that pass through and are retained by the 10-μm mesh, the spread in the ratio changes to 0.4–0.7. For those organisms that pass through this mesh and are retained by the 1.0-μm sieve, the ratio spread is 0.4–0.6. The extremely small organisms that pass through the 1.0-μm and are retained by a 0.45-μm membrane the spread in the ratio is 0.2–0.6. These data illustrate that the major change in the ratio occurs between about 0.5 and 10 μm.

The importance of demonstrating this relationship concerns the potential regulation of growth by the area-to-volume relationship. In an attempt to establish the mechanisms that control phytoplankton growth, considerable effort was spent to establish physiological differences (if any) between large and small photosynthetic autotrophs (6). Traditionally, these studies are referred to as net-to-nanoplankton studies. More vigorous analyses have featured fractionation throughout a wide range of cell sizes (7). These studies have shown a rather diverse set of physiological processes that can be related to cell size and are consistent with the area–volume hypothesis. For example, chlorophyll per unit cell volume decreases with increasing cell diameter (Figure 7). When compared to $E_{530}:E_{450}$ versus retention size, low ratios appear to be associated with high chlorophyll content per unit volume. In populations with high ratios, the chlorophyll content can be expected to occupy a smaller portion of the cell volume.

The significance of the relationships shown in Figure 7 will not be lost on those who have struggled with problems in marine phytoplankton growth processes (6). To date, experimental findings suggest that with increasing cell size, photosynthesis per unit chlorophyll decreases, the cells' affinity for nutrient assimilation decreases (high K_s, where K_s is the half-saturation constant), and respiration increases.

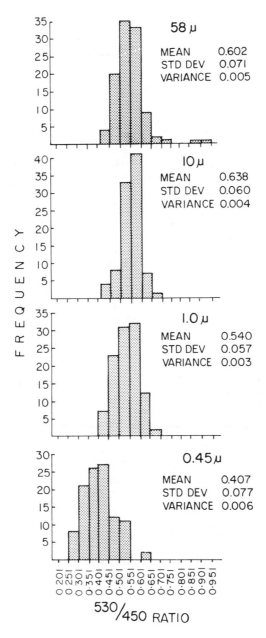

Figure 6. The ratio $E_{530}:E_{450}$ as related to cell size.

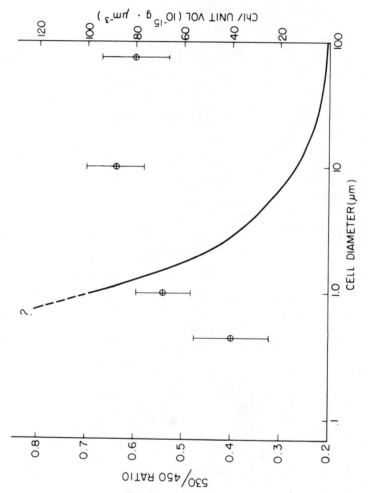

Figure 7. The relationships between E₅₃₀:E₄₅₀ and chlorophyll per cell volume. (Data taken from Ref. 6.)

According to the fluorescent extremes in Figure 4, cell size kinetics agree with diagnostic features in the spectra. Low $E_{530}:E_{450}$ ratios correlate with populations of small cell size and are characteristic of open ocean populations where nutrient impoverishment means cells should have a high affinity (low K_s) for the nutrients present. The capacity for photosynthesis per unit chlorophyll would be higher in warm waters provided a nutrient source was not totally limiting. Conversely, this capacity in larger cells should be lowest in cold, nutrient-rich water.

Over 500 observations of phycobilin distributions show that phycoerythrin is the dominant biliprotein in oceanic populations. How it relates to the environmental factors in the sea is not clear. Undoubtedly the difficulty in separating the source of the pigment, that is, cryptomonad versus cyanobacteria, is impeding understanding. The size-fractionation data indicate that most of this pigment is found in cells smaller than 10 μm. In the open ocean the small coccoid cyanobacteria are probably the major contributor; in coastal waters the larger cryptomonads may be a major contributor.

Although numerous claims exist as to the importance of cyanobacteria to total oceanic productivity (8), the light harvesting capacity and pigment morphology are different from conventional red algal pathways (9). Until the degree to which they are unconventional is known, establishing their role in marine populations is not possible. Evidence suggests that because of their shade characteristics, cyanobacteria augment primary production at depth (*10,11*), but the degree of this augmentation is uncertain.

Conclusions

Fluorescence spectral analysis is too crude to be of much use to those who wish to pinpoint large-scale distributions of important species assemblages as is done in principal component analyses. However, spectral changes are the results of changes in the species composition (i.e., certain species exploit certain environmental regimes).

Changes in these spectra are probably more directly in tune with factors that control and dictate environmental physiology. Ratios and the appearance of emission peaks can be used to physiologically characterize populations on the basis of cell features such as area–volume relations. Such a tool is a natural extension to the continuous mapping of chlorophyll and can play a role in understanding growth processes in the sea.

Acknowledgments

This work was supported by NASA Contract NAG 6–17, Wallops Flight Center and the state of Maine.

Literature Cited

1. Yentsch, C. S.; Yentsch, C. M. *J. Mar. Res.*, **1979**, 37, 471–83.
2. Yentsch, C. S. in "The Physiological Ecology of Phytoplankton"; Morris, I., Ed.; Blackwell: London, 1980; vol. 7, p. 95.

3. Topinka, J. A.; Korjeff, W. A. *J. Phycol.*, **1983**, in press.
4. Morris, I.; Glover, H. E. *Limnol. Oceanogr.*, **1981**, *26*, 957–61.
5. Yentsch, C. S.; Phinney, D. A. in "Bioluminescence: Current Perspectives"; Nealson, K. H., Ed.; Burgess: Minneapolis, 1980; p. 82.
6. Malone, T. C. in "The Physiological Ecology of Phytoplankton"; Morris, I., Ed.; Blackwell: London, 1980; vol. 7, p. 433.
7. Blasco, D.; Packard, T. T.; Garfield, P. C. *J. Phycol.*, **1982**, *18*, 58–63.
8. Li, W. K.; Rao, D. V. S.; Harrison, W. G.; Smith, J. C.; Cullen, J. J.; Irwin, B.; Platt, T. *Science*, **1983**, *219*, 292–95.
9. Kursar, T. A.; Swift, H.; Alberte, R. *Proc. Natl. Acad. Sci.*, **1981**, 6888.
10. Glover, H. E.; Phinney, D. A.; Yentsch, C. S. *Biol. Oceanogr.* **1984**, in press.
11. Platt, T.; Rao, D. V. S.; Irwin, B. *Nature* **1983**, *301*, 702–4.

RECEIVED for review July 5, 1983. ACCEPTED January 11, 1984.

14

Underway Analysis of Suspended Biological Particles with an Optical Fiber Cable

YOSHIMI KAKUI, AKIO NISHIMOTO, JUNZO HIRONO, and MOTOI NANJO[1]

Electrotechnical Laboratory, Osaka Branch, Nakoji, Amagasaki, Hyogo, 661, Japan

A probe has been developed to measure two important phytoplankton characteristics in situ and while underway in marine waters. The unit consists of a shipboard laser source and signal processor that are connected to a submerged sensory unit by an optical fiber cable. The submerged unit consists of two sensors: one measures particle size from the intensity of individual fluorescence pulses and the other obtains excitation spectra for chlorophyll a emission. The submerged unit is towed by a ship, and the measured data, such as the fluorescence pulse-height distribution and the excitation (emission) spectra, are obtained aboard ship.

IN SITU SENSING TECHNIQUES ARE HIGHLY DESIRED in oceanography. This chapter describes a method for measuring phytoplankton distribution by using optical fiber cables with laser techniques. The technique may be applicable for measuring the distribution, abundance, and size of phytoplankton for most areas of the oceans.

The blooming of red-tide phytoplankton causes problems in our local coastal environment. Bloom is believed to be a result of eutrophication or changes in temperature, salinity, vitamins, metals, or conditions that affect the amount of solar radiation reaching the organisms. The combination of these parameters makes prediction of these blooms extremely difficult. Monitoring such parameters as salinity, pH, dissolved oxygen, or light in a continuous fashion is already practical; however, we need to measure parameters of phytoplankton in the same real-time manner to predict accurately the cause of these blooms.

[1]To whom correspondence should be directed.

We have developed a new instrument that is applicable to in situ use and can analyze phytoplankton color groups. This device counts particles ranging from 5 to 100 μm and characterizes the particles in terms of their color groups.

Measuring System

Figure 1 shows a block diagram of the prototype measuring system, which is composed of two subsystems. The shipboard unit consists of a laser light source and signal processors. The underwater unit consists of a towed pressure vessel and the detecting optics. These two subsystems are connected by an optical fiber cable and have the following virtues: The submersible sensing part is compact and is ideally suited for towing, and the sensing part contains only optics and hence, no electronics (the electronic components are held above the surface in a shipboard laboratory).

Figure 2 shows the structure of the towed body, which is made of stainless steel (SUS 304) to withstand a pressure equivalent to a 30-m depth. When towing the vessel, the seawater flows into the two water channels mounted at the right and left side of the bottom. The length of the water channel makes the flow laminar at the measuring position where two optical windows are placed. The window material is Pyrex glass plate. The sizes of the windows are 56 mm in diameter and 10 mm thick, and 25 mm in diameter and 4 mm thick. The optical fiber cable is introduced through the head of the vessel and is fastened by a cylindrical buckle that maintains a watertight condition. A curved cable guide is mounted in front of the vessel head to prevent abnormal cable bending and excessive loading during towing. When the vessel is towed just beneath the sea surface, auxiliary planes are attached to the horizontal stabilizing wing to keep the towed body away from the ship-trail region where the water contains many air bubbles.

The optical fiber cable must transmit both the excitation and emission signals between the subsystems. Of the many fibers available, fused pure-silica core fiber was adopted because it has the lowest loss in the visible-wavelength region when compared with the transmission loss of multicomponent glass fibers or plastic fibers. The cable structure is also important because the cable must have enough mechanical strength to withstand towing stresses. A cross section of the cable is given in Figure 3. The cable has a concentric structure from the center to the outside layer; at the center is a tension member made of fiberglass-reinforced plastic (FRP) for mechanical reinforcement. Adjacent to the center are bundles of fine fibers, each containing 84 polymer-clad fibers 200 μm in diameter. Surrounding these fibers are eight silica clad, single-core fibers with a 400-μm diameter. Finally, a protecting sheath made of polyethylene is at the outermost side. The bundle fiber measures the excitation spectra. The single-core fibers are used as follows: one for transmitting the Ar-ion laser light, four for the dye

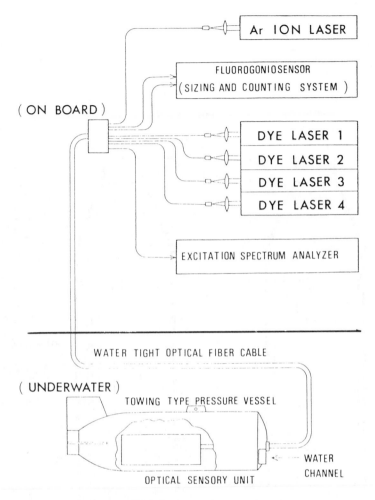

Figure 1. Block diagram of measuring system.

laser sources, one for the signal of the fluorescent emission, and one for the signal of the scattered light. The total length of the cable is 30 m.

The terminals of the cables are designed to connect and disconnect conveniently with other optical connectors. The power loss due to connection is approximately 1 dB and the total loss at the 488-nm excitation wavelength of the Ar-ion laser is estimated at about 30 dB/km.

Particle-Size Analysis Subsystem

Principles of the System.   The scattering of visible light has been used to extract some useful information on the nature of suspended particles in the sea (*1*). However, mixed suspensions of living and nonliving matter

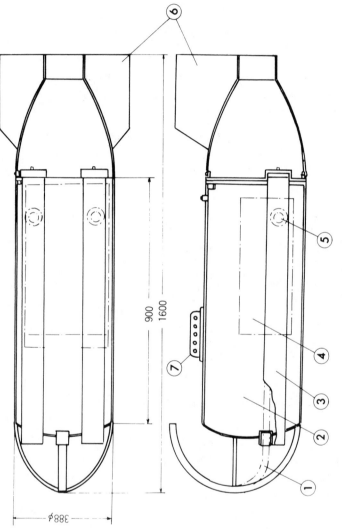

Figure 2. Structure of pressure vessel. Key: 1, watertight fiber cable; 2, compartment; 3, water channel; 4, sensing optics; 5, optical window; 6, horizontal wing; and 7, towing wire holder.

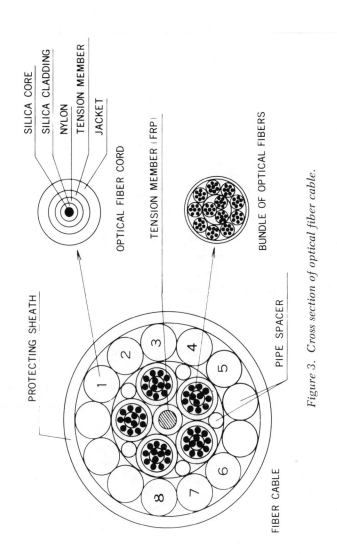

Figure 3. Cross section of optical fiber cable.

yield nonresolvable relationships for scattering at the same wavelength as the incident radiation. However, with phytoplanktonic particles, size information can be extracted from the fluorescence radiation emitted (2).

Our system works by focusing a laser beam into a stream of seawater and then measuring the fluorescent emission and the scattered laser radiation from the irradiated small-spot volume. As the irradiated spot moves through the water, the particles crossing the volume generate optical pulses whose peak height correlates with particle size. The concentration of the particles is also obtained from the counts, the measured volume, the flow speed, and the period of measurement.

Ar-Ion Laser Source and Electronics.   The main components of this subsystem are the Ar-ion laser source, the irradiating and collecting optics, and the associated electronics for signal processing. The Ar-ion emissions at 488 and 514 nm are used for the excitation. The laser tube is cooled by circulating water and is then put in an airtight box to reduce performance degradation from such causes as vibrations, salt air, and moisture.

Figure 4 (bottom) shows the configuration of the optics to irradiate the seawater and to detect the subsequent signal through the window. A set of lenses, L_1 and L_2, focuses the fiber image on the water channel. (The diameter of the laser beam at the measuring position was determined to be about 0.6 mm.) The fluorescence emitted and the light scattered in the focused volume are collected by the lens L_3 on the surfaces of the fibers F_1 and F_2 with a two-to-one reduction factor. A dichroic filter transmits the fluorescence radiation to the fiber F_1 and the scattered radiation to the fiber F_2.

Figure 5 shows a block diagram of the electronics for processing signals; the electronics are located in the ship's laboratory. The light transmitted through the optical fiber cable is converted to an electrical photocurrent pulse by the photomultipliers. A logarithmic amplifier in the processing system is used to obtain a larger dynamic range for the signals. The dynamic range of all the electronic components is three decades. Moreover, in the case of the logarithmic amplifier, the variation in the laser power or in the amplification factor does not affect the shape of the pulse-height distribution, but translates it along the horizontal axis. Because scattered light of the same wavelength as the incident light includes not only the light scattered by phytoplankters but also light scattered by very fine particles (even by water molecules), the average DC component is subtracted before the pulse signal is processed by the logarithmic amplifier.

Although the intensity of the scattered light was much stronger than that coming from the fluorescence emission, it was nevertheless too difficult to extract useful biological information from it in the presence of inorganic particles, even when the scattering signal was gated to the fluorescence component to exclude other causes of scattering. Thus, particle-size analysis by scattering was not pursued in the coastal waters described in this work.

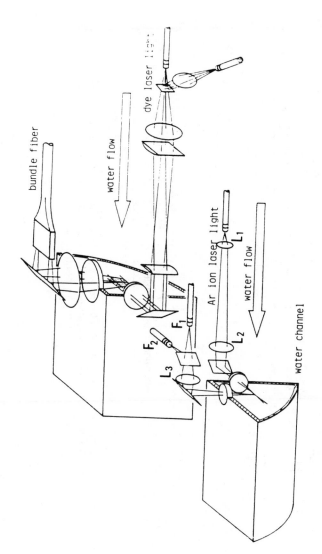

Figure 4. Optical sensor unit of irradiating and collecting optics. (Reproduced with permission from Ref. 2. Copyright 1983, Japan Association of Automatic Control Engineers.)

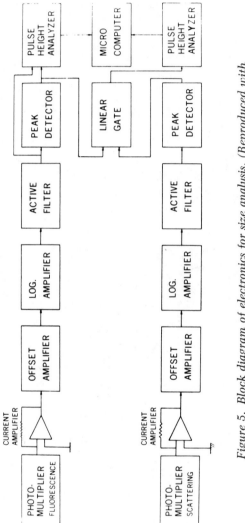

Figure 5. Block diagram of electronics for size analysis. (Reproduced with permission from Ref. 2. Copyright 1983, Japan Association of Automatic Control Engineers.)

Method of Data Processing. The amplitude of the signal pulses depends on the size of the phytoplankters, their trajectory, and their position in the measured volume because of the nonuniformity of the irradiated intensity and the characteristics of the collecting optics (Figure 6). This section presents our method to obtain size information of the particles from the raw data of the pulse-height distribution.

If the pulse-height distribution is denoted by $S(I_i)$, where the pulse height is I_i in the ith channel, and the size distribution of the phytoplankton is denoted by $N(r)$, then the relation between $S(I_i)$ and $N(r)$ is expressed by

$$S(I_i) = \int_0^{r_{max}} R(I_i, r)N(r)dr \tag{1}$$

where $R(I_i, r)$ is the pulse-height distribution of the monodisperse particle with the size r (hereafter denoted as monodisperse response function) and r_{max} is the maximum size of the particle.

Expressing Equation 1 in the matrix form, we obtain

$$y = Rx \tag{2}$$

where

$$y = \begin{pmatrix} S(I_1) \\ \vdots \\ S(I_m) \end{pmatrix} \quad \text{and} \quad x = \begin{pmatrix} N(r_1) \\ \vdots \\ N(r_n) \end{pmatrix} \tag{3}$$

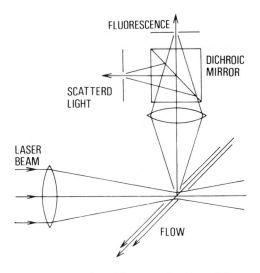

Figure 6. Irradiated beam and collected flux.

and

$$R = \begin{pmatrix} R(I_1, r_1)\Delta r_1 & \cdots & R(I_1, r_n)\Delta r_n \\ \vdots & & \vdots \\ R(I_m, r_1)\Delta r_1 & \cdots & R(I_m, r_n)\Delta r_n \end{pmatrix} \qquad (4)$$

Equations 2–4 assume that the pulse-height distribution has m divisions, the size distribution has n divisions, and the Δr is the bin width of the size classification. In our actual calculations, we had 20 divisions.

The size matrix is formally obtained by use of the inverse matrix and is given by

$$x = R^{-1}y \qquad (5)$$

Thus, the size information is obtained through the monodisperse response function. According to the analyses for the inverse problem of forward scattering (3), when a logarithmic amplifier is used the monodisperse function for a specified size differs from one for a different size only by the translation along the horizontal axis. Thus, assuming that the detected fluorescent intensity of the phytoplankters is proportional to the surface area or the laminar layer rather than the total volume of the particles [this assumption conforms to the results that the increase of the fluorescence excitation cross section with the phytoplankton particle size is not as much as that of the scattering cross section with the size (4)], and because 10 years of input to the logarithmic amplifier corresponds to 200 channels, we define the monodisperse response function as

$$R(I_m, r) = R\left(I_m + 400 \log \frac{r}{r_0}, r_0\right) \qquad (6)$$

where r_0 is the specified size, and I_m is the channel number of the maximum height for the r_0-size particle. The shape of the monodisperse response function is determined by calculating the dependence of the image size of the particle cross section that overlaps the input size of the collection optics on the trajectory. The calculated result shows that the shape of the monodisperse response function is almost independent of the specified particle when the largest size of the particle concerned is less than 20% of the size of the beam and the collecting optics (5). The shape of the monodisperse response function used in this study is shown in Figure 7.

Calibration Procedures.   Cultured phytoplankters were used to calibrate the size analysis subsystem. An example of the pulse-height distribution of the cultured phytoplanktons (green algae) is shown in Figure 8. The size ranged from about 10 to 15 μm as determined from microscopic observations. Using the function shown in Figure 7 and Equation 6, we obtained

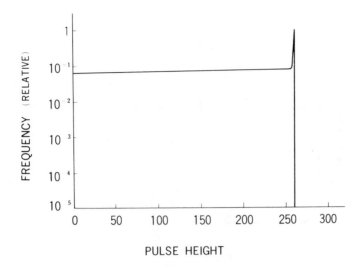

Figure 7. *Monodisperse response function.*

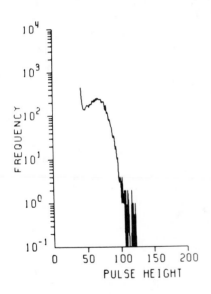

Figure 8. *Pulse-height distribution for system calibration (green algae). (Reproduced with permission from Ref. 2. Copyright 1983, Japan Association of Automatic Control Engineers.)*

the size distribution (Figure 9). The absolute value in the horizontal axis is scaled to fit the mean diameter obtained by microscopic observation. Another example of the pulse-height distribution is shown in Figure 10, where two kinds of phytoplankters of different size are mixed in almost equal concentration. The computed results of the size distribution are shown in Figure 11. The size distribution of the larger plankters (*Peridinium* sp.) calculated in this way appears to agree well with microscopic observations.

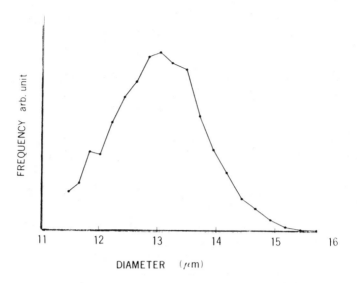

Figure 9. Particle-size distribution for pulse-height distribution shown in Figure 8. (Reproduced with permission from Ref. 2. Copyright 1983, Japan Association of Automatic Control Engineers.)

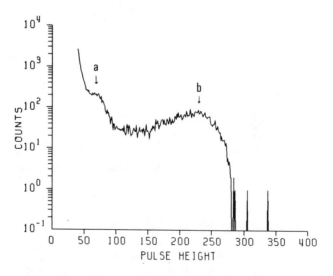

Figure 10. Pulse-height distribution, mixture of green algae (a) and Peridinium *sp. (b). (Reproduced with permission from Ref. 2. Copyright 1983, Japan Association of Automatic Control Engineers.)*

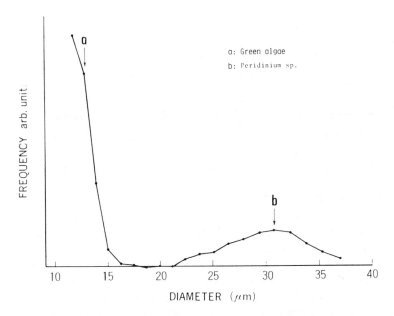

Figure 11. Particle-size distribution for pulse-height distribution shown in Figure 10. Mixtures are the same as in Figure 10. (Reproduced with permission from Ref. 2. Copyright 1983, Japan Association of Automatic Control Engineers.)

Excitation Spectra Measurement Subsystem

The excitation spectra of marine phytoplankton were found to be broad (6). Considering this feature, we emphasized continuous and rapid measurement at discrete excitation wavelengths. Flash-lamp pumped dye lasers were used because of their simplicity and their high efficiency. The excitation wavelengths were at 457, 499, 536, and 578 nm.

A part of the excitation laser beam transmitted through the fiber cable is returned by a beam splitter to monitor its fluctuations (*see* Figure 4). The laser beam passing through the beam splitter is formed into a rectangular shape by a set of cylindrical lenses and irradiates the seawater in the starboard side of the water sampler. The emitted fluorescence is then measured on board the ship. For this determination, the volume of water to be measured is made relatively large to give higher sensitivity and to reduce the statistical fluctuation of the number of plankters in the water.

The duration of the laser pulses is 0.3 μs; the time interval between each shot of different wavelength is 20 μs. Thus, a set of the laser shots finishes in about 60 μs. At a towing speed of 5 knots, the displacement during 60 μs is estimated to cause a change of 3% in the measurement volume; thus, the four laser shots of different wavelengths irradiate almost the same volume.

The photoelectric pulses corresponding to the fluorescence induced by the laser shots are integrated and synchronized. The integrated signal is converted to match the recording and the processing by the microcomputer. Interferences from external sunlight, the dark noise in the detector, and the offset in the amplifier are reduced or compensated. In addition to the laser beam reflected by the beam splitter, Raman-scattered radiation from the seawater was also useful for monitoring the excitation intensity. Figure 12 shows the excitation spectra for the different kinds of cultured phytoplanktons. The pattern of the excitation spectra shows good agreement with that obtained by a conventional spectrofluorometer.

Field Test in Osaka Bay

The performance of the prototype underwater remote-sensing system was tested in Osaka Bay (Figure 13). The output power of the Ar-ion laser was set at 500 mW. The data for the pulse-height analysis were accumulated for 100 s. In the excitation spectra, measurements were made at an interval of 6 s, and the data were recorded by a magnetic data recorder. Figure 14 shows an example of the pulse-height distribution of fluorescent emission. Figure 15 shows the size distribution obtained from the inversion technique and the data in Figure 14. The horizontal axis was scaled by the calibration described previously. The result shows that there were suspended phytoplankters with an average size of about 22 μm. According to the microscopic observation of the sampled seawater, the population was dominated

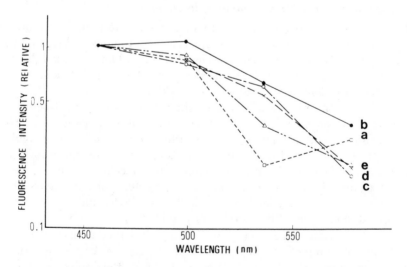

Figure 12. Fluorescence excitation spectra. Key: a, green algae; b, Peridinium sp.; c, Chattonella sp.; d, Chaetoceros sp.; and e, Gymnodinium sp. (Reproduced with permission from Ref. 2. Copyright 1983, Japan Association of Automatic Control Engineers.)

Figure 13. *Towed pressure vessel being lowered into Osaka Bay.*

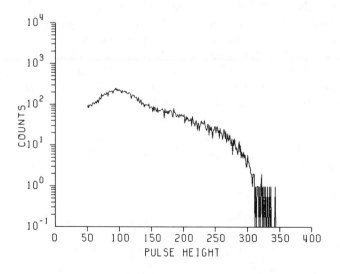

Figure 14. *Pulse-height distribution (field data).*

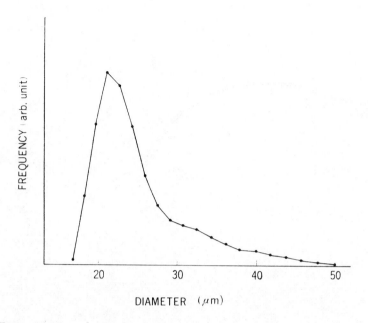

Figure 15. Particle-size distribution from pulse-height distribution shown in Figure 14.

by *Skeletonema* sp., a marine diatom that usually forms a filament composed of a chain of cells (7) and has a size that typically ranges from 18 to 35 μm. The size was also determined with the microscope on board, which confirmed the results of the instrumental analysis.

Figure 16 shows an example of the excitation spectra measurement in the same bay. The continuous dot plot shows the fluorescent emission intensity obtained every 6 s. The upper half shows the results of excitation at 499 nm, and the lower half shows the results at 578 nm. The horizontal axis shows the distance in meters. The three filled circles show the chlorophyll a concentration of the sampled water (chlorophyll a was obtained by extraction in acetone). The results show a decrease where the vessel crossed the estuarine region. Thus, continuous measurement by this technique can determine the distribution of phytoplankton, including the patch size, and the small-scale variability.

Conclusion

Underwater remote-sensing techniques using an optical fiber cable have several advantages. Useful data on the in situ size distribution, concentration, and algal color group of phytoplankton in the marine environment may be obtained. Measurements at high speed can be carried out because the towed pressure vessel can be streamlined by limiting it to the optical

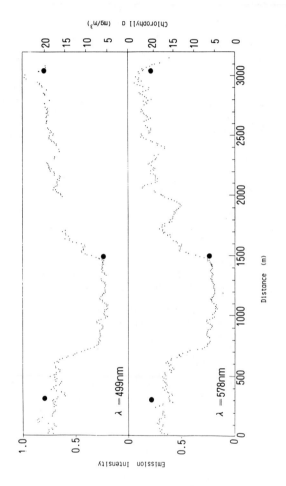

Figure 16. Continuous measurement of fluorescence intensity and extracted chlorophyll a concentration.

components. In addition, the measurement of biological particles is not affected by the existence of other kinds of suspended matters.

The function of the prototype system was proven in the field test in Osaka Bay. Although the watertight optical fiber cable in the system is 30 m long and the more flexible optical fiber to connect the watertight cable with the electronic component is about 10 m long, the signal pulse height was still large enough to detect even only one particle. Considering the low optical attenuation in the fibers, we expect that this type of remote-sensing technique will be promising even at a depth of 150 m where the photosynthesis rate is still active.

Acknowledgments

The authors thank R. Okaichi in Kagawa University, A. Murakami (now at Sanyo Suiro Sokuryo Co. Ltd.), S. Uno in Nansei Fisheries Research Laboratory, and M. Furuki in the Environmental Science Institute of Hyogo Prefecture for teaching us the techniques to keep cultured phytoplanktons.

Literature Cited

1. Shifrin, K. S.; Kopelevich, O. V.; Burenkov, V. I.; Mashtakov, Yu. L. "Using the Light Scattering Function for Studying Suspended Matter in the Sea," NASA Technical Translation, NASA TT F-14, NASA: Washington, D.C., 1973; p. 781.
2. Kakui, Y.; Nanjo, M. *Systems and Control*, **1983**, *27*, 96 (in Japanese).
3. Holve, D.; Self, S. A. *Appl. Opt.*, **1979**, *18*, 1632; and **1979**, *18*, 1646.
4. Kakui, Y.; Nanjo, M.; Minato, H.; Majima, T. *Acta IMEKO*, **1979**, 585.
5. Kakui, Y.; Nishimoto, A.; Hirono, J.; Minato, H.; Nanjo, M. "High Speed Measuring Technique of Suspended Particles in Sea Water," Research Report in 1980, Comprehensive Research for the Prevention of Pollution in the Seto Inland Sea and Coastal Waters, Environmental Agency, Japan.
6. Yentsch, C. S.; Yentsch, C. M. *J. Mar. Res.*, **1979**, *37*, 371–403.
7. Bold, H. C.; Wynee, M. J. "Introduction to Algae," Prentice Hall: Englewood, N.J., 1978; chapter 7.

RECEIVED for review July 5, 1983. ACCEPTED March 28, 1984.

Biological Profiling in the Upper Oceanic Layers with a Batfish Vehicle: A Review of Applications

ALEX W. HERMAN

Department of Fisheries and Oceans, Atlantic Oceanographic Laboratory, Bedford Institute of Oceanography, Dartmouth, Nova Scotia, Canada B2Y 4A2

Biological and physical parameters consisting of copepods, chlorophyll, temperature, salinity, and depth are profiled continuously with a Batfish vehicle towed behind a ship in a sawtooth undulating pattern. A description of the vehicle and mounted sensors (conductivity, temperature, and depth unit, Variosens fluorometer, and in situ electronic zooplankton counter) is presented. Methods of copepod and chlorophyll calibration and verification of sampled data are described. Several applications of the biological Batfish sampler to ecological studies are presented. Intensive Batfish sampling of a frontal boundary zone south of Nova Scotia indicates that vertical-mixing events driven by the M_2 internal tides give efficient vertical mixing of nutrients (e.g., nitrates and inorganic phosphates) and subsequent high primary productivity along the boundary. Simultaneous copepod and chlorophyll profiles measured with a vertical resolution of ~ 1 m indicate that copepods aggregate at depths shallower (~ 3–8 m) than the depth of the chlorophyll maximum.

Interactions of phytoplankton and zooplankton with each other and their physical environment are studied continuously by measuring their biomass, counts, and sizes over a wide range of spatial and temporal scales. Operational criteria require a horizontal spacing of successive vertical profiles of ≤1 km, which is sufficient to resolve dominant phytoplankton patch scales in the range of 5–10 km (1). The sampler system should also provide spatial resolution of ~1 m while rapidly profiling in the vertical plane. The problem of sampling wide ranges of spatial scales prompted the

0065-2393/85/0209-0293$06.50/0
© 1985 American Chemical Society

development of the Batfish vehicle at the Bedford Institute of Oceanography. This vehicle is capable of continuously profiling in a sawtooth undulating pattern while towed behind a ship.

This chapter describes the Batfish vehicle with its complement of sensors measuring zooplankton abundance, phytoplankton biomass, and their physical environment. The Batfish sampler system is described as well as methods of measurement, deployment, sampling, and calibration. The sizing and copepod species identification potential of the electronic zooplankton counter is demonstrated by comparing the electronic measurements with microscopic measurements of the same sample. Application of the Batfish sampler to various ecological studies on the Scotian and Peruvian shelves and in the eastern Canadian Arctic is described. Continuous Batfish tows for ~ 30 h through a shelf-break front south of Nova Scotia have shown, for example, movements of intrusions at the base of the front coherent with the M_2 internal tides. These intrusions cause continuous nutrient replenishment at the subsurface front by efficient vertical mixing, and they account for the high biological productivity there.

Instruments and Methods

Batfish Vehicle.   Figure 1 shows the Batfish vehicle with its full complement of sensors—a Variosens fluorometer, a digital conductivity, temperature, and depth (CTD) unit, and a zooplankton sampler and electronic sensor. The vehicle is towed on an armored seven-conductor cable with flexible plastic fairing and is capable of cycling between two preset depths to a maximum of 400 m. Diving is accomplished by rotation of the hydroplanes; mechanical power is derived from hydraulic pressure supplied by an impeller-driven pump situated at the tail of the vehicle. Further details of the Batfish design and operation can be found elsewhere (2).

Variosens Fluorometer.   The Variosens fluorometer (Impulsphysik GmbH, Hamburg, FRG) is used to measure in situ chlorophyll a, the pigment most commonly used to estimate the standing crop of phytoplankton. The transmitter emits a filtered blue light in an optical band from 350–550 nm, and the receiver component of the fluorometer utilizes an optical bandpass filter that peaks at 680 nm, with a bandwidth of 20 nm (full-width at half-maximum). The transmitter and receiver employ a lens system that optically subtends a measured sampling volume of ~ 1 mL. The Variosens fluorometer mounted on Batfish is used in conjunction with a Turner 111 fluorometer (flow-through system); the latter provides a reference calibration (3) for the Variosens. Both the Variosens and Turner fluorometers are calibrated simultaneously prior to a cruise by using laboratory-grown cultures of phytoplankton. During a cruise, replicate 1-L seawater samples from various depths (0–50 m) are filtered from the effluent of the Turner fluorometer and chlorophyll a is measured by the acetone

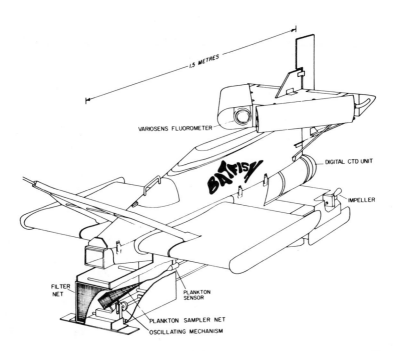

Figure 1. The Batfish vehicle with its full complement of sensors: a Variosens fluorometer, a digital CTD unit, and the prototype electronic zooplankton counter with its sampler net (7).

extraction method (*4*). A sea calibration for the Turner fluorometer is obtained from the linear relationship between fluorescence and chlorophyll. A comparison of the slopes of the fitted calibration lines for both the laboratory and sea calibration gives the change in fluorescence intensity per unit mass of chlorophyll from laboratory to sea conditions. Thus, a scaling factor for the laboratory-to-sea calibration of the Variosens fluorometer is determined. A laboratory calibration for the Variosens fluorometer is presented in Figure 2, which shows the detection range of chlorophyll from ~0.3 to ~30 mg/m^3. Further details are discussed elsewhere (*5*).

A dynamic calibration check was obtained at sea for both fluorometers from a chlorophyll transect obtained by towing a Batfish and submers-

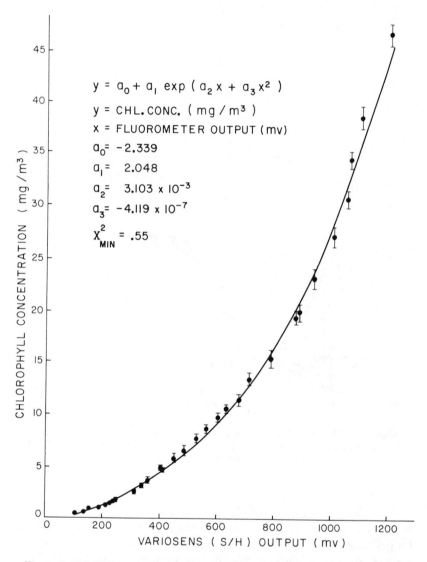

Figure 2. The Variosens fluorometer calibration curve consisting of chlorophyll a concentration vs. the sample/hold output voltage. The general equation for an exponential function was chosen to describe the data, and the coefficients were obtained from a multiparameter fit that minimized χ^2 (5).

ible pump (the latter continuously supplying seawater to the Turner fluorometer on deck) at a constant depth of 11 m in the coastal waters of Nova Scotia. Curves a–c in Figure 3 show the chlorophyll transect sampled with both instruments for a 30-min tow. Curve 3b represents the raw Variosens data filtered to match the frequency response of the Turner data in Curve

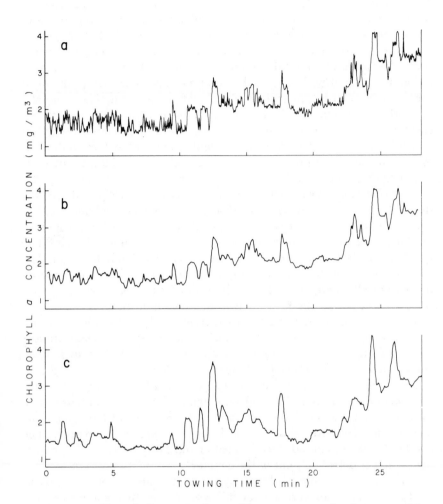

Figure 3. Chlorophyll transect obtained from a constant depth tow at ~11 m. Key: a, raw Variosens data; b, the same data filtered with a binomial smoothing function having the best-fit value of σ = 3.4 s; and c, data from the Turner 111 fluorometer where samples are transmitted continuously (5).

3c. Not only is there good statistical agreement (5) between the two data sets, but the raw Variosens data of Curve 3a indicates a higher frequency response (cut-off frequency f = 0.025 Hz) and, therefore, higher data resolution.

Electronic Zooplankton Counter. The electronic zooplankton counter (prototype) mounted on Batfish operates on a principle similar to the Coulter counter where an insulating particle, such as a zooplankter, displaces seawater in a tubular cell and causes a change of electrical conductance throughout the cell. The resultant change can be electronically translated into a proportional measurement of volume of the zooplankton. The plankton sensor is mounted on the undercarriage of the Batfish (shown in Figure 1) and is fed by a sampler net that filters seawater and concentrates the animals. Behind the sensor is attached a 0.5-m length of plastic tubing terminated with sampler bag that entraps the electronically measured plankton for future analysis. Zooplankton sampler nets are generally susceptible to clogging by algal material. The sampler net used on Batfish is cleaned and kept free from clogging by a side-to-side motion (~ 10 Hz) which maintains open meshes. This cleaning method enables continuous towing for periods up to 3–4 h. The electronic zooplankton counter, its components, and calibration have been described in detail elsewhere (6).

The electronic zooplankton counter counts and sizes copepods ranging from 0.8 to 4 mm (equivalent spherical diameter) over a volume range of 0.2–100 mm^3. The counter can discriminate between the dominant copepod taxa (four or five) in samples from various geographical areas. The species identification procedure was described in detail in Reference 7. The volumes of the copepods taken from the Batfish sampler bag are measured microscopically, and the resulting volume distributions are compared with those obtained from the in situ electronic measurements. An example of volume distributions derived from electronic and microscopic measurements for copepods sampled with Batfish in the eastern Canadian Arctic is shown in Figure 4. The dominant species consisted of *Calanus finmarchicus* Stage V, *Calanus glacialis* Stage V, and *Calanus hyperboreus* Stages V and VI; lesser concentrations were found among *Pseudocalanus minutus* and *Calanus hyperboreus* III. The most dominant peaks (3–6, Figure 4a) in the microscopic measurements indicate good agreement with the electronic measurements (Figure 4b) both in peak position and relative intensity. The less dominant animals indicated by Peaks 1 and 2 (Figure 4a) are approaching the lower detection limit of the counter.

Two Batfish profiles of chlorophyll and copepods (sampled in the eastern Canadian Arctic) were separated by volume into their component species and are shown in Figure 5. The separation shows that both *Calanus finmarchicus* Stage V and *Calanus glacialis* Stage V are situated ~ 5 m above the chlorophyll maximum, whereas *Calanus hyperboreus* Stages V and VI were situated primarily at or below the chlorophyll maximum. Such examples illustrate the species identification potential of the electronic counter and the resolution capabilities of the system for isolating species in the water column.

CTD and Data Logging.   A digital CTD unit (Model 8705, Guildline Instruments Ltd., Smith Falls, Ontario, Canada) mounted on the Batfish (*see* Figure 1) measures in situ conductivity, temperature, and depth (CTD) with accuracies of ± 0.01 ppt*, ± 0.01 °C, and ± 1 dbar (full scale

* parts per thousand.

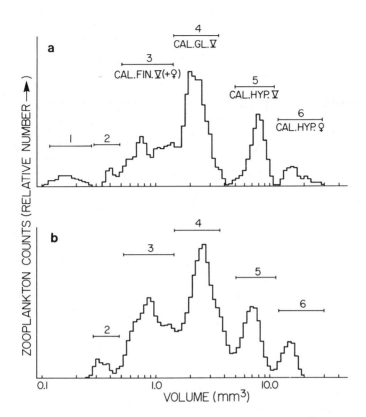

Figure 4. (a) Microscopic measurements of volume distribution of copepods sampled by Batfish in the eastern Canadian Arctic. The dominant copepods identified were: 3, C. finmarchicus V; 4, C. glacialis; 5, C. hyperboreus V; and 6, C. hyperboreus VI. Low abundances were found among 1-Pseudocalanus minutus and C. hyperboreus III. (b) Volume distribution of copepods from the same samples as in a as measured by the electronic zooplankton counter (15).

0–400 dbar), respectively, and with resolutions of ± 0.005 ppt, 0.003 °C, and ± 0.2 dbar, respectively. The CTD is also used to digitize the analog signal from the fluorometer and has seven available channels for additional inputs. During towing, digital data from the zooplankton counter and CTD deck units are transferred to a Hewlett–Packard 21MX minicomputer for on-line plotting and data storage on magnetic tape. Fluorometry and CTD data are digitized at a rate of 10 Hz. Zooplankton data occur in random sequence and are therefore tagged with their arrival time.

Applications to Ecological Studies

Scotian Shelf. PRIMARY PRODUCTION.   Figure 6 shows the sharp boundary zone separating coastal water and warmer slope water charac-

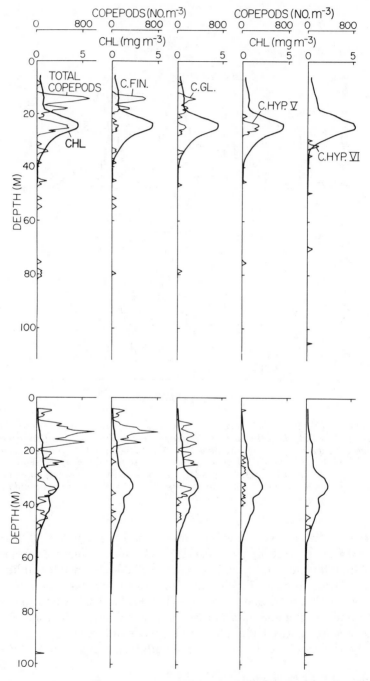

Figure 5. Two Batfish profiles of copepods, chlorophyll, temperature, and salinity. The chlorophyll maximum is located at the base of the thermocline (15).

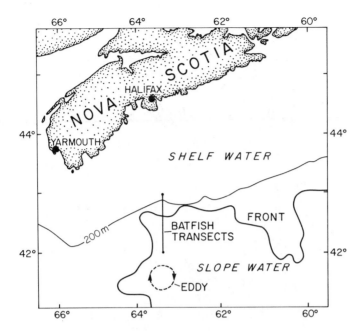

Figure 6. Scotian Shelf and slope-water region. The mean position of a thermal front with surface temperatures of ~ 9–11 °C was contoured from satellite imagery and partly from Batfish profiling. The work area for Batfish transects is shown along 63°25′ W (10).

teristic of the shelf-break region south of Nova Scotia. During the spring-bloom period, these fronts become a region of highly localized aggregations of phytoplankton with concentrations higher than in nearby coastal or shelf waters. The source of productivity enhancement is evidently the nutrient-rich slope waters (8), which also appears to be the important source of enrichment for the entire Scotian Shelf (9). High productivity at the frontal boundary requires vertical transport of nutrients; however, the underlying mechanisms are not well understood.

The shelf-break front was intensively profiled with a Batfish vehicle during May 1977; the operational area is shown in Figure 6. Contoured profiles of temperature, density, salinity, and chlorophyll a on depth (10–100 m) traced by the Batfish are shown in Figure 7. Slope water can be identified by its temperature–salinity characteristics and is located south of the frontal zone with a temperature range of ~ 10–13 °C and salinity values > 34.5 ppt (*see* Figure 7). The frontal zone was characterized in the surface layer by a horizontal temperature contrast ranging from ~ 7 to 11 °C and a salinity contrast ranging from ~ 32.5 to 34.5 ppt. In coastal waters, the subsurface chlorophyll maximum had concentrations of ~ 1–2 mg/m^3 and was situated at ~ 30 m. The highest chlorophyll concentrations

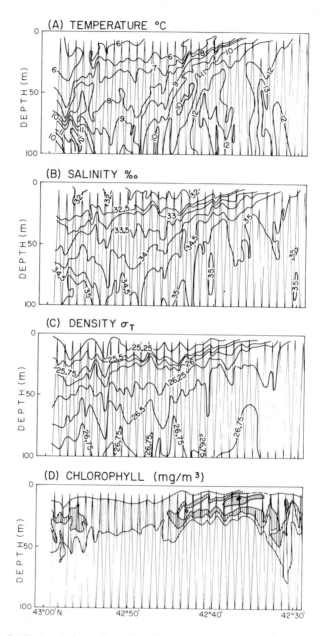

Figure 7. Contoured profiles of temperature (A), salinity (B), density (C), and chlorophyll a (D) on depth (10–100 m) between 42°30' N and 43°00' N on the Scotian Shelf. Increased shading indicates concentrations ranging from 1 to 5 mg/m³ (10).

were measured in the front itself (at 42°40′ N in Figure 7) and ranged from 4 to 5 mg/m³ with localized maxima sometimes as high as 7 mg/m³. The subsurface chlorophyll maxima were distributed obliquely along the front extending ~ 10 km in the horizontal and over a depth range from surface to 40 m. The chlorophyll maxima were consistent with temperature and depth.

Nutrient concentrations were mapped simultaneously with CTD/bottle sampling along the identical Batfish transect. Deeper slope water situated below the front generally contained high nutrient concentrations: nitrates, 12–16 mg/m³; nitrites, i.e., 1 mg/m³; silicates, 6–14 mg/m³; and inorganic phosphates, 0.8–1.4 mg/m³. Surface waters (0–20 m depth) to the north of the front (left in Figure 7) were depleted in nutrients such as nitrates (< 0.5 mg/m³) that are typical of the post-spring-bloom period and the onset of stratification. Nutrients within the pycnocline were also high, for example, nitrates (~ 4–10 mg/m³), and were supporting high chlorophyll concentrations within the front.

Following the initial transect shown in Figure 7, Batfish sampling was intensified in the boundary zone along a 20-nautical mile transect bounded by 42°40′ N to 43°00′ N. Seven subsequent tows were made along the identical transect of Figure 7 in 24 h. The data revealed (10) cyclic variations in the horizontal and vertical scales for both chlorophyll and temperature. Further analysis indicated that the cyclic variations had a periodicity of ~ 12 h and had resolved the constituent of M_2 internal tide. The eight successive Batfish transects of the thermal front depict the movements coherent with the semidiurnal tidal cycles (Figure 8). In Panels 1–3 of Figure 8, the front appears relatively passive; in Panel 4 there appears a bimodal intrusion consisting of a 9 °C plume of water apparently descending from surface water, and a 10 °C plume of water ascending from slope water. The front also appears in its highest vertical position in Panel 4, which corresponds with a measured tidal mode (low). The remaining panels (5–8) also indicate coherence with the tidal cycles; in Panel 7 the intrusions appear totally receded.

Additional calculations (10) indicated that the movements of the intrusions produced localized turbulent mixing ascribed to shear instabilities. These mixing events, driven by the M_2 internal tides, caused a potentially efficient mechanism for vertical transport of heat and salt as input to the Scotian Shelf. These vertical-mixing events, which continually replenish nutrients (e.g., nitrates, nitrites, silicates, and inorganic phosphates) from slope waters below, can explain the anomalously high productivity occurring at the front.

ZOOPLANKTON.   Studies of the outer edge and shelf-break region of the Scotian Shelf were also extended to the vertical and horizontal distributions of zooplankton. The most intriguing feature of these vertical profiles was the consistent location of copepods above the subsurface chlorophyll

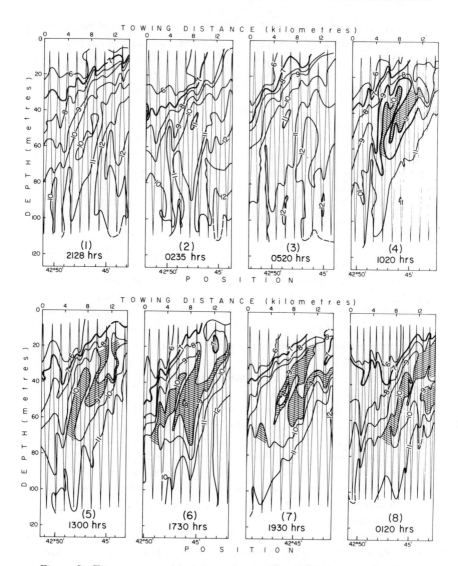

Figure 8. Time sequence of temperature profiles of the front south of Nova Scotia taken by Batfish over ~ 30 h. A bimodal intrusion occurs in Panel 4 at 1020 h (10).

maximum. Figure 9 shows a sample of five profiles taken along the same transect in Figure 7. The profiles indicate that the main copepod layers (indicated on the left) were situated on the upper chlorophyll gradient (indicated on the right) and not at the depth of maximum concentration. Therefore, the copepods are not keying on maximum biomass for grazing purposes but on some other cue, perhaps biological or chemical, at shal-

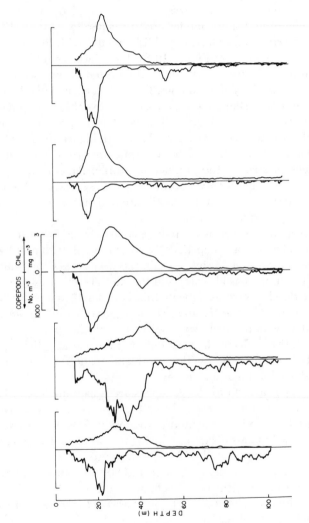

Figure 9. *Vertical Batfish profiles taken from the transect of Figure 7 on the Scotian Shelf. Chlorophyll is represented on the right scale and copepods on the left (13).*

lower depths. Similar observations were made (11) in the eastern tropical Pacific where the depth of zooplankton maximum lies ~20 m above the subsurface chlorophyll maximum, which was coincident with the depth of maximum ^{14}C fixation and supported an earlier observation (12).

Subsequent calculations (13) and modeling (14) enable the production profile to be computed from a corresponding chlorophyll and irradiance profile. The chlorophyll light model used to generate these profiles has been verified (Herman and Platt, unpublished data) with in situ observations. An example is given in Figure 10 for four Batfish profiles sampled along the transect of Figure 7. Two methods (13) have been used to generate the production profiles (dashed lines, Figure 10); each method indicates that the maximum in carbon production occurs above the chlorophyll maximum at the upper gradient. Figure 10 shows that the production profiles are correlated with the copepod profiles. Panel c in Figure 10 indicates that copepods are not distinctly layered but decrease uniformly with depth following the general form of the production profile. Panel d in Figure 10 represents high chlorophyll concentrations near the surface in the frontal zone; however, the copepod layers are still situated on the chlorophyll gradient above the subsurface maximum.

A statistical comparison was made between the depth of maximum chlorophyll, production, and copepods for continuous Batfish profiles sampled along the main transect. Figure 11 shows two passes of the transect in which a tow began in coastal water and the ship's course was reversed at the front. The depth centroids indicate that both small copepods (*Pseudocalanus minutus* and *Clausocalanus arcuicornis*) and *Calanus finmarchicus* Stages IV and V were more closely associated with the production maximum than with the chlorophyll maximum. On the average, the copepods were within 1–2 m of main production depth and about 8 m shallower than the depth of the chlorophyll maximum.

It is difficult to ascertain why copepods are responding to production rates directly; they probably respond to other biological or chemical cues associated with the depth of maximum production. However, these copepods do bypass the biomass maximum during migration and aggregate in a lower biomass region associated with maximum production.

Eastern Canadian Arctic. In August 1980, ecological studies were conducted in Baffin Bay near the Greenland coast bounded by 75–76° N and 68–72° W. Figure 12 shows two typical profiles of temperature, salinity, chlorophyll, and total copepods from near surface to 100-m depth. The surface-mixed-layer depth extended to only ~15 m with salinity values from 31.5 to 33.0 ppt, which indicate the influence of glacial runoff. The subsurface chlorophyll maximum was situated below a sharp thermocline that ranged from 1.0 to −1.5 °C. However, copepods were mainly situated above the chlorophyll maximum within the thermocline itself.

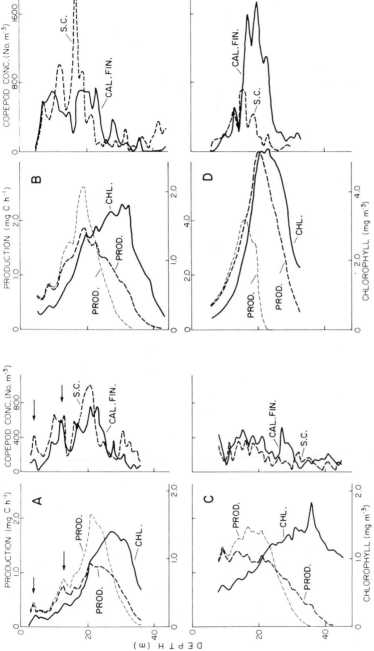

Figure 10. Vertical Batfish profiles taken from the transect of Figure 6 on the Scotian Shelf. The profiles are separated into the following components: chlorophyll, production estimated in two ways from a chlorophyll/light model, small copepods (S.C.), and all stages of C. finmarchicus (13).

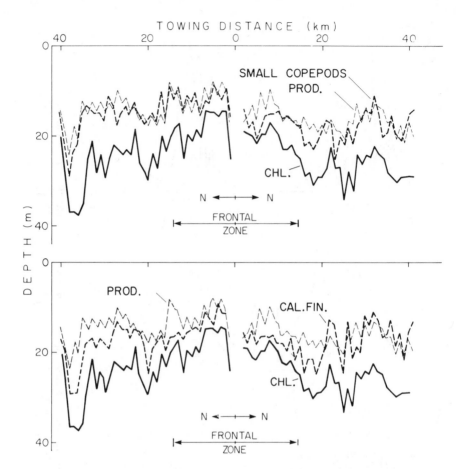

Figure 11. The depth of maximum chlorophyll, production relative to small copepods, and C. finmarchicus. The tow began in coastal waters of the Scotian Shelf (left side of plot) and the course was reversed at the front. There are ~ 90 vertical Batfish profiles in the transect (13).

A statistical analysis was made (15) similar to that of the Scotian Shelf comparing the depth of maximum chlorophyll, production, and copepods for two Batfish transects consisting of ~ 90 profiles near the Greenland coast. Figure 13 shows the plotted depth centroids for the two transects. Again, the depth centroids of *Calanus finmarchicus* and *Calanus glacialis* are more closely associated with the depth of maximum production than with the depth of the subsurface chlorophyll maximum. The major difference in the Arctic data (as compared to Scotian Shelf data) was that the mean depth differential between copepods and chlorophyll was only 3–4 m and required considerably higher measurement resolution.

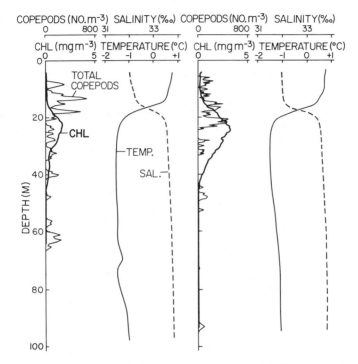

Figure 12.   Two Batfish profiles of copepods, chlorophyll, temperature, and salinity sampled in the eastern Canadian Arctic. The chlorophyll maximum is located at the base of the thermocline (15).

A similar pattern emerged once more, however, when copepods were above the biomass maximum. The situation did not apply to all species measured. However, a similar analysis was made for *Calanus hyperboreus* Stages V and VI; these larger animals were located at, or several meters below, the subsurface chlorophyll maximum which placed them approximately in a high biomass region but below the depth of maximum production.

Peruvian Shelf.   The coastal waters of Peru between 6 and 11° S were the subject of an ecological study called Investigation Cooperativa de la Anchoveta y su Ecosistoma (ICANE) during November 1977. This study was designed to study anchovy larvae and their ecosystem. Geostrophic estimates of the alongshore velocity field on the shelf at 9° S were made *(16)* with high-resolution Batfish data. Geostrophic estimates obtained by conventional sampling (i.e., CTD stations ~ 10 km apart) were highly variable and grossly misleading. The high variability in measured chlorophyll distributions were correlated to the variability in geostrophic estimates.

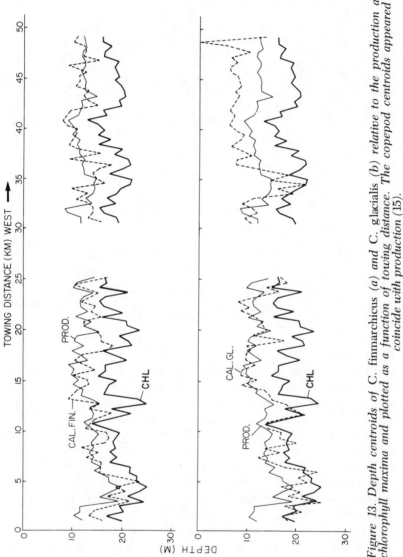

Figure 13. Depth centroids of C. finmarchicus (a) and C. glacialis (b) relative to the production and chlorophyll maxima and plotted as a function of towing distance. The copepod centroids appeared to coincide with production (15).

A geostrophic velocity section of the Peruvian Shelf at 9° S is shown in Figure 14. By using ~150 continuous Batfish profiles separated by ~0.5 km in the horizontal scale, a high-resolution geostrophic field could be contoured in Figure 14. The velocity field in Figure 14 is highly complex and indicates poleward flow with some near-surface equatorward flow (shaded areas). High variability was found by moving only several kilometers in the horizontal scale, which resulted in large changes in velocity structure and indicated the grossly misleading estimates that can be made by sampling across the shelf with five or six CTD profiles separated by ~10 km.

Temporal variability was investigated by successive sampling of a short transect bounded by 78°52′ W and 79°05′ W in Figure 14. Successive Batfish sampling indicated that the shaded lenses of equatorward flow were highly variable in both horizontal position and in physical size. Surface lenses of chlorophyll at the same position were highly coherent with these horizontal movements observed in the geostrophic field. This observation explained, in part, the vertical variability observed in other station work performed during the same cruise.

An example of eight sequential Batfish profiles sampled along 9° S on the Peruvian shelf is shown in Figure 15a and consists of chlorophyll and total copepods. The chlorophyll structure indicated a surface layer from surface to 15-m depth with concentrations of ~5–10 mg/m^3 and a subsurface layer at ~30 m with similar concentrations. A chlorophyll minimum at ~20 m was occupied by a distinct layer of copepods, a feature that persisted for several days. The dominant copepods in our samples consisted of *Centropages brachiatus*, *Calanus chiliensis* Stages V and VI, and *Eucalanus inermis* Stages V and VI. Estimates of the production profiles shown in Figure 15b indicated that the copepods were not located at the production maximum but at the base of the production layer that occurred in the surface layer only (0–10 m). Several red-tide events consisting of *Mesodinium rubrum* were observed in the near surface region during the cruise; these copepods may have avoided the surface layer for that reason.

Although the vertical features of Figure 15 were common during our sampling period, they rarely persisted for more than 1 or 2 days. They were then replaced by totally different vertical structures and concentrations of both copepods and chlorophyll by as much as an order of magnitude. Such variability was consistent with the observed variability of the geostrophic velocity estimates.

Discussion

Separation of the vertical interactions of copepods and chlorophyll requires a measurement resolution of ~1 m. Such a requirement was apparent in the Arctic data, which indicated small vertical separations of ~3–4 m between copepods and chlorophyll. Continuous profiling with the Batfish provides adequate statistical sampling of a large number of profiles

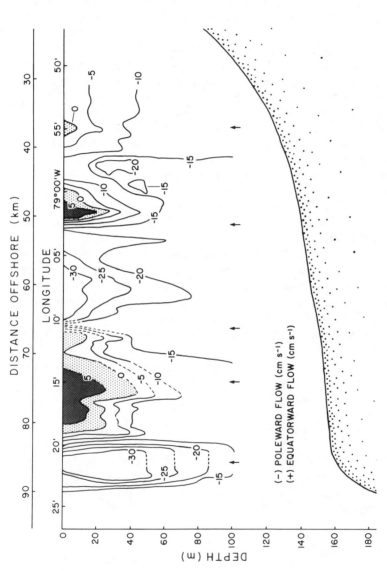

Figure 14. *The geostrophic velocity field generated for an across-shelf Batfish transect on the Peruvian shelf at 9° S. Equatorward flow is indicated by the shaded area (16).*

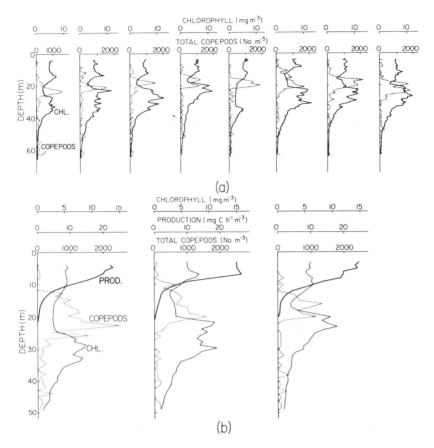

Figure 15. *Copepod and chlorophyll profiles sampled at 1900 h (a) and the corresponding estimated production profiles (b) from the Peruvian shelf at 9° S. The vertical chlorophyll structure shows both a surface and subsurface layer (15).*

separated by 0.5 km; our hypotheses can then be tested over large horizontal scales such as the comparison of depth centroids in Figures 11 and 13.

The species identification potential of the zooplankton counter was demonstrated by the clear separation of copepod peaks in Figure 4. These Arctic communities, however, have a very low species diversity and therefore the few peaks representing copepod species were easily separated. However, oligotrophic waters possess a high diversity of copepod species, and thus a higher degree of smearing of the copepod peaks that often contain many species of similar size. As a result, the volume distribution can only be separated into two or three dominant groups containing large and small copepods where each group represents several species. Therefore, the species identification capability of the counter is limited and varies with the geographical area sampled.

The Variosens fluorometer mounted on Batfish and the deck-mounted Turner 111 fluorometer require continuous sea calibration to monitor changes in fluorescence efficiency per milligram of chlorophyll. Day–night fluorescence changes as large as twofold have occasionally been observed on the Scotian Shelf, and continuous monitoring is required. We routinely filter 1 L of seawater samples from the effluent of the continuous flow Turner fluorometer (see "Instruments and Methods") over a range of depths to ~50 m. A comparison of the slopes of the day–night linear calibration curves indicates any fluorescent changes present.

The in situ Batfish sampler does not provide seawater samples for on-deck analysis. Therefore, we have also developed a continuous profiling pumping system (17) that measures biological profiles of copepods and chlorophyll and temperature while on station. Copepods and chlorophyll are measured on deck from the hose effluent with a deck-mounted electronic zooplankton counter and Turner 111 fluorometer. Measured profiles of copepods and chlorophyll indicate a vertical resolution of ~3–4 m while the pumping system delivers ~50–60 L/min of seawater on deck.

Acknowledgments

I wish to thank A. S. Bennett, E. Horne, and D. D. Sameoto for their comments and review of the manuscript.

Literature Cited

1. Denman, K. L.; Okubo, A.; Platt, T. *Limnol. and Oceanogr.* 1977, 22, 1033–38.
2. Dessureault, J.-G. *Ocean Eng.* 1976, 3, 99–111.
3. Herman, A. W. "*In situ* chlorophyll and plankton measurements with Batfish vehicle," MTS-IEEE Oceans '77 Conference Record, Session 39B, 1977, 1–5.
4. Yentsch, C. S.; Menzel, D. W. *Deep-Sea Res.* 1963, 10, 221–31.
5. Herman, A. W.; Denman, K. L. *Deep-Sea Res.* 1977, 24, 385–97.
6. Herman, A. W.; Dauphinee, T. M. *Deep-Sea Res.* 1980, 27, 79–96.
7. Herman, A. W.; Mitchell, M. R. *Deep-Sea Res.* 1981, 28, 737–55.
8. Fournier, R. O.; Van Det, M.; Wilson, J. S.; Hargreaves, N. B. *J. Fish. Res. Board of Can.* 1979, 36, 1228–37.
9. Smith, P. C.; Petrie, B. "A proposal to study the dynamics at the edge of the Scotian Shelf," Bedford Institute Oceanographic Report Series, BI-R-76-4, 1976.
10. Herman, A. W.; Demnam, K. L. *J. Fish. Res. Board of Can.* 1979, 36, 1445–53.
11. Longhurst, A. R. *Deep-Sea Res.* 1976, 23, 729–54.
12. Hobson, L. A.; Lorenzen, C. J. *Deep-Sea Res.* 1972, 19, 277–306.
13. Herman, A. W.; Sameoto, D. D.; Longhurst, A. R. *Can. J. of Fish. and Aquatic Sci.* 1981, 38, 1065–76.
14. Herman, A. W.; Platt, T. *Ecol. Modell.* 1983, 18, 55–72.
15. Herman, A. W. *Limnol. and Oceanogr.* 1983, 3, 131–46.
16. Herman, A. W. *J.M.R.* 1982, 4, 185–207.
17. Herman, A. W.; Mitchell, M. R.; Young, S, W. *Deep-Sea Res.* 31, 439–50.

RECEIVED for review July 5, 1983. ACCEPTED January 10, 1984.

The Undulating Oceanographic Recorder Mark 2

A Multirole Oceanographic Sampler for Mapping and Modeling the Biophysical Marine Environment

J. AIKEN

Natural Environment Research Council, Institute for Marine Environmental Research, Prospect Place, Plymouth, United Kingdom

The undulating oceanographic recorder (UOR) Mark 2 is a self-contained, multirole, oceanographic sampler that can be towed by merchant ships or research ships at speeds from 4 to 13.5 m/s. The UOR is independent of the vessel for any service, can undulate between preset depth limits from 0 to 200 m with an undulation length between 800 and 4000 m, and can record (internally on a miniature digital tape recorder, resolution 0.1 % full scale) measurements by sensors for temperature, salinity, depth, chlorophyll concentration, radiant energy, or other variables. The application of the UOR to specific problems of biological oceanography, such as studies of ichthyoplankton research and ocean color (compatible with satellite remote sensing) and the modeling of primary production, is discussed.

Tʜᴇ ᴘʟᴀɴᴋᴛᴏɴ ᴏꜰ ᴛʜᴇ sᴇᴀs ᴀɴᴅ ᴏᴄᴇᴀɴs adjacent to the British Isles have been studied for over 50 years with the continuous plankton recorder (CPR) (*1*). Continuous plankton recorders are towed at a fixed depth of 10 m at monthly intervals by ships-of-opportunity (merchant ships and weather ships) on standard routes. They retain the plankton on a mechanically propelled, continuously advancing band of silk for subsequent laboratory analysis (*2*). A few CPRs have carried temperature recorders and sensors for other environmental parameters (*3, 4*).

The undulating oceanographic recorder (UOR) (*5–7*) was conceived as a variable-depth sampler, with an internal data logger and sensor system for a range of environmental parameters, to complement and enhance the fixed depth CPR (*8*). Undulating oceanographic recorder Mark 1 (*5*) used

0065-2393/85/0209-0315$06.00/0
© 1985 American Chemical Society

variably inclined diving planes to generate the hydrodynamic dive and lift forces to achieve depth undulation (9), but its large size and weight, its limited range of towing speeds (4–7 m/s, 8–14 knots), and its need of trained personnel have limited its use to research vessels (10). Undulating oceanographic recorder Mark 2 (7) was designed to overcome these limitatings so that it could be towed by fast merchant vessels. Power supplies for the servomechanism are generated internally, and sensor measurements are recorded in situ. Since its initial development, UOR Mark 2 has been deployed from fast merchant vessels as originally intended and from a number of slower research vessels on intensive cruises to study areas of special ecological interest.

This chapter describes the design and performance of the UOR Mark 2, the data logger, some specially developed sensors, and some of the different system configurations and sensor suites employed to meet various scientific objectives. Results are presented from some of the different modes of deployment, and enhanced interpretations of the data are discussed with simple physical models.

System Design and Performance

The general design considerations for a ship-of-opportunity towed vehicle, such as UOR Mark 2, require that it be small, lightweight, robust, and stable at speeds up to 10 m/s (20 knots). It must be easily and safely launched and recovered by unsupervised, nonscientist, ship's crew while the vessel is underway at full speed. The data logger and sensors should be small and reliable, have low electrical power consumption, and be capable of unattended operation for periods up to 1 day.

The component parts of the UOR Mark 2 are an underwater-towed vehicle, a plankton sampling mechanism (PSM), an undulation servocontrol system (USS), and a marine environmental recorder (MER) (see Figure 1). With a total system approach, the data replay and data processing system are considered a component part of the system.

The design of the UOR vehicle was influenced by three hydrodynamic factors. First, the lateral equations of motion for underwater towed bodies show that the wingspan of any planes used to produce hydrodynamic dive or lift forces is a major parameter responsible for lateral instabilities (11). In UOR Mark 2, the wingspan has been minimized by eliminating any diving planes, the body itself acting as principal generator of hydrodynamic forces. Second, tail-fin surface area is a major parameter in restoring roll and yaw perturbations; UOR Mark 2 has twin tail fins to maximize this parameter. Third, the depth performance available from any underwater-towed body is dominated by the drag of the towing cable so that the diving forces that the body can generate (limited for safety considerations to one-fifth of the breaking strength of the cable) are of lesser importance, provided the overall mass and drag of the vehicle can be kept to a minimum.

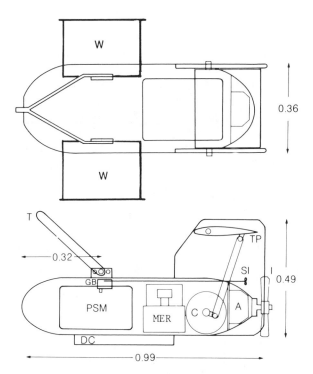

Figure 1. Schematic plan (top) and long section (bottom) of the UOR Mark 2 (dimensions in meters) showing the positions of the towing arm T, the diving chute DC, the tailplane TP, the plankton sampling mechanism PSM, the PSM gearbox GB, the subsidiary impeller SI, the alternator housing A, the impeller I, the control unit C, and the marine environmental recorder MER; the optional fixed wings W (span 0.75 m) for slow speed towing are shown in plan.

Further, for ship-of-opportunity acceptability, the size and weight should be comparable to the CPR (length, ~ 1 m; weight, 90–100 kg).

To meet these design criteria, the body was constructed in glass-reinforced plastic with a double-fin box tail, a stainless steel diving chute (DC) to give the vehicle stability when launched at high speed, and a symmetrical towing arm (*see* Figure 1). Normally the plankton sampler from the CPR is used (*1*) and is mechanically propelled by a subsidiary impeller at the rear. Spatial resolution of the plankton samples can be adjusted from 18 to 1.8 km by varying the silk transport rate. A flowmeter is included to determine accurately the volume of water filtered per sample. Plankton samples are analyzed by the methods described by Rae (*12*) and Colebrook (*13*).

The undulation servocontrol system comprises a motorcycle alternator (A) housed in an oil-filled pressure vessel at the rear, an electronic control system, and a servomechanism (C) fitted transversely in the rear section of the instrumentation hold. The alternator is impelled by a proprietary outboard motor propeller, which provides electrical power (threshold speed ~ 3.5 m/s) and the distance reference to the control system. The target undulation profile is a linear up–down sawtooth between a preset minimum depth (range 0–90 m) and a preset maximum depth (10–200 m); the length of the undulation can be preset to vary from 800 to 4000 m.

The actual depth achieved depends on the towing speed, cable diameter (cable drag if no fairing is used), and length of cable in the water. Table I compares the depths achieved by Batfish (9), UOR Mark 1 (5), and UOR Mark 2 for towing speeds at the lower end of their operational range (~ 4 m/s for all three instruments) and at the upper end of the range (7 m/s for

Table I. Depths Achieved by Batfish, UOR Mark 1, and UOR Mark 2

Sampler	Speed (m/s)	Cable Length[a] (m)	Depth Achieved (m)
Batfish	3.6	65	35
		150	—
		300	100
		330	—
	7	65	—
		150	—
		300	60
		330	200
UOR Mark 1	4	65	35
		150	55
		300	80
		330	—
	7	65	32
		150	46
		300	65
		330	—
UOR Mark 2	4	65	32 (50)
		150	55 (70)
		300	74 (90)
		330	240[b]
	7	65	26
		150	39
		300	54
		330	205[b]
	10	65	22
		150	36
		300	48
		330	185[b]

Note: The cable was of 8-mm diameter; drag coefficient was 0.1; numbers in parentheses indicate with wings on 6.5-mm diameter cable.
[a]All cable lengths were unfaired except at 330 m (faired, drag coefficient was 0.1).
[b]Computed by extrapolation of data using computer simulation of cable profiles.

Batfish and UOR Mark 1, and 10 m/s for UOR Mark 2). Below 5 m/s, fixed-attitude wings (span, 0.75 m) are fitted to provide increased hydrodynamic lift for long cable lengths and heavy instrumentation loads up to 150 kg. Fitting produces an undulation amplitude of 80 m at 4 m/s with as little as 180 m of unfaired 6.5-mm diameter cable in water.

In other respects UOR Mark 2 meets its design targets with towing stability at speeds up to 13.5 m/s; it can be launched and recovered safely at speeds up to 10 m/s. Figure 2a shows the microcomputer-generated data replay (discussed later) for part of a UOR Mark 2 tow from the ferry vessel M.V. *Cornouailles* at 7 m/s in the western English Channel in July 1982; Figure 2b shows the computer plot (before contouring) of the digital recording of depth (2–68 m) for a tow at 5.5 m/s (FRV *Cirolana*, May 1981).

Sensors and Instrumentation

Sensors and instrumentation carried by the UOR are located in the foremost half of the instrument hold, on each side of the PSM in the forward hold, or the whole of the PSM hold is available if the PSM is omitted or attached to the outside of the vehicle.

Sensor measurements are recorded in situ by a battery-powered, eight-channel, miniature digital tape recorder (MDTR) (Oxford Medical Systems Ltd.) with range, 0–1.023 V; resolution, 0.1%; scan period, 15 s (alternatively 5 s); duration (C120 cassette), 25 h (alternatively 8 h); and data capacity, 6000 scans. Direct (bandwidth 100 Hz) or pulse-width modulated analog signals (range, 120 mV; bandwidth, 0–8 Hz; and resolution, 1–2%) can be recorded on the additional three tracks of the same cassette.

The recorded cassettes from the MDTR are replayed at 25 times the recording speed on a special replay unit linked to a digital-to-analog converter to provide "quick-look" records on a multipen chart recorder, and a microcomputer translation and data processing system (14). The system is fully portable and can be used at sea.

Computer programs have been developed that provide quick-look tabulations of the data in raw numbers or in calibrated physical units, produce vertical plots of the data (on the printing terminal), draw histograms of individual vertical profiles for each undulation or group of undulations as required, and calculate derived parameters (e.g., salinity from conductivity and temperature measurements, temperature gradients and eddy diffusion coefficients, attenuation coefficients from light sensor measurements, or chlorophyll concentration from the measurements of the blue- and green-light sensors). The programs can calculate statistics on the data, for example, the number of measurements (or average time spent) in each depth interval or average or total flow measured.

In the laboratory, the microcomputer communicates the data to a mainframe computer to produce contoured vertical sections of the measured parameters within 1 h. Examples of data processed by each of the methods described will be presented in the following sections.

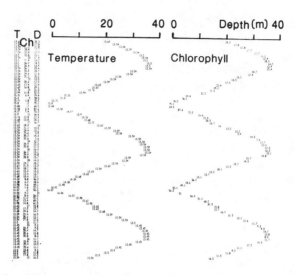

Figure 2a. Microcomputer-generated undulation profile (0.2–38 m) for part of a UOR Mark 2 tow from the ferry vessel M.V. Cornouailles at 7 m/s in the western English Channel, July 1982. On the left are tabulated the temperature (degrees Celsius), depth (meters), and chlorophyll (arbitrary); plotted in the center is the temperature (degrees Celsius); plotted on the right is chlorophyll (arbitrary; 1 unit ≃ 0.1 mg/m³).

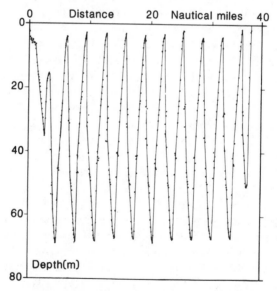

Figure 2b. Computer-plotted undulation profile (2.5–68 m) for a UOR Mark 2 tow at 5.5 m/s with 200-m, 6-mm outside diameter (OD) cable outboard of 200-m, 8-mm OD cable (about 300 m of cable in water). Taken from F.R.V. Cirolana, Celtic Sea, May 1981.

Standard sensor suites include salinity, temperature, and depth probes (proprietary sensors are used), one or two chlorophyll sensors (range, 0–20 mg/m^3; resolution, 0.1 mg/m^3) (4), and light sensors for solar radiant energy (range, 0–1000 μE/m^2/s; resolution, 0.1%).

The chlorophyll sensor measures the red fluorescence radiation emitted by chlorophyll molecules (at 680 ± 10 nm) when excited by blue radiation (at 435 ± 50 nm). The fluorescence signal is optimized by using a single very intense pulse of blue radiation (~ 1–2 J per flash) repeated every 5–15 s, and two, specially selected filters in both blue and red channels to eliminate the detection of scattered radiation from any particulate material in the sample. The chlorophyll sensor optical system and filter combination are shown in Figure 3.

Both broad-band light sensors, measuring total quanta 400–700 nm, and narrow-band light sensors, at 460 ± 10 nm and 550 ± 10 nm, are used. From the measurement at 460 nm, the total quanta 350–700 nm can be calculated (15). From the measurements at 460 and 550 nm, the attenua-

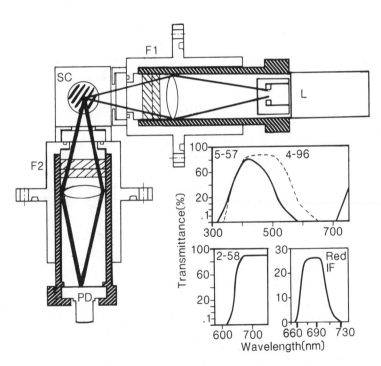

Figure 3. Cross-sectional diagram of the optical system of the chlorophyll sensor showing the flash lamp and Litepac L, the blue filters F1, the sample cell SC, the red filters F2, and the photodetector PD. The transmission spectra of the blue filters (Corning 5-57, solid line; and Corning 4-96, dashed line), the red interference filters (IF), and the blocking filter (Corning 2-58) are inset.

tion coefficients at these wavelengths and the total chlorophyll in the water column above the sensors can be calculated from (16)

$$C_{tot} = \frac{\ln(I_G/I_B) - (K_{WB} - K_{WG})Z + \ln(A)}{Z(K_{CB} - K_{CG})}$$

where I_G is the measurement by green (550 nm) sensor at depth Z (m); I_B is the measurement by blue (460 nm) sensor at depth Z (m); K_{WB} and K_{WG} are the attenuation coefficients of 1 m of water for blue and green radiation, respectively; K_{CB} and K_{CG} are the attenuation coefficients for 1 mg/m^3 of chlorophyll for blue and green light, respectively (17); and A is the ratio of the measurements of surface sunlight at 460 and 550 nm, which is a constant for most conditions except for very low solar elevations when the spectral composition of sunlight changes (16). Narrowband light sensors, for both upwelling and downwelling light and other wavelengths, can be added to meet specific objectives such as ocean color measurements with the coastal zone color scanner (discussed later).

Optional, Additional Sensors

Dissolved oxygen sensors, pH electrodes, or specific ion electrodes can be fitted, and a fluorometer for petroleum oils was developed from the chlorophyll sensor (sensitivity, 1 μg of naphthalene/L). A solid-state bioluminescence sensor was developed (18) with a minimum detection of 5×10^{-10} W equivalent to about 10^9 quanta/s at 480 nm, sufficiently low for the detection of a single dinoflagellate flash (19).

For quantifying zooplankton abundance and depth distribution, particle counters have been developed on the basis of commercially available five-electrode conductivity cells, which are similar in concept to those used on Batfish (20–22). Particle counters have particular application for counting fish eggs or larvae that fall in narrow bands (mackerel eggs are 1.3 ± 0.1 mm in diameter), or the dominant species of copepod in any plankton assemblage.

The Ship-of-Opportunity UOR

Undulating oceanographic recorder Mark 2 was designed for ship-of-opportuity use, and since April 1979 it has been deployed 45 times (including some day/night pairs) from the ferry vessel M.V. *Cornouailles* (Brittany Ferries Ltd) operating from Plymouth to Roscoff in the western English Channel (*see* inset Figure 4), launched and recovered by the crew of the vessel at full speed (up to 9 m/s, 18 knots).

The northern part of the area with low tidal flows ($\bar{u} < 0.5$ m/s) stratifies in the summer, but the area adjacent to the French coast remains mixed throughout the year due to the high tidal stream velocities (>2 m/s) (23).

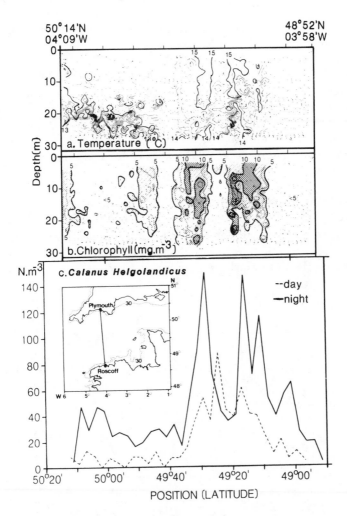

*Figure 4.   Computer-processed and contoured vertical sections of tempera-
ture (0.5 °C intervals) (a) and chlorophyll concentration (2.5-mg/m³ intervals) (b) for a day tow by the UOR Mark 2 from Plymouth to Roscoff in the
western English Channel in August 1979 with the abundance of* Calanus
helgolandicus *Claus for the corresponding day and night tows (c). The inset
shows the track of the ferry M.V.* Cornouailles *from Plymouth to Roscoff.*

The UOR transect often makes an oblique section through the frontal
boundary between the stratified and mixed areas. Simultaneous vertical
sections of temperature and chlorophyll can be an aid to the interpretation
of the complex structure at the boundary (7, 24). The spring bloom is dominated by diatom species, but in most summers a bloom of the dinoflagellate

Gyrodinium aureolum (Hulbert) saturates the surface waters in the northern stratified area.

The UOR tows of August 1979 (Figure 4), July 1981, and July 1982
(Figures 5 and 6) showed chlorophyll concentrations of 10–20 mg/m³ in the
stratified waters (usually highest adjacent to the tidal front), and in June
1980 chlorophyll concentrations close to 10 mg/m³ were measured at the
front. The increased abundance of zooplankton, coincident with the high
concentrations of chlorophyll measured at the front, is shown in Figure 4c
(abundance of *Calanus helgolandicus* Claus, the major herbivorous copepod in the plankton samples).

Applications to Remote Sensing

The UOR Mark 2 measurements complement remotely sensed, satellite
measurements of sea surface temperature and ocean color. Although satellites provide synoptic, large area, limited-depth measurements, for a few
parameters, the UOR Mark 2 provides a depth-resolved (below the eupho-

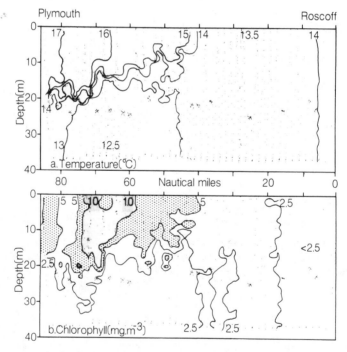

*Figure 5. Computer-processed and contoured vertical sections of tempera-
ture (0.5 °C intervals) (a) and chlorophyll concentration (2.5-mg/m³ intervals shaded > 5 mg/m³, cross-hatched > 10 mg/m³) (b) for a UOR Mark 2
tow from Roscoff to Plymouth in the western English Channel July 7–8, 1982
(M.V. Cornouailles). (See Figure 4 for vessel track inset.)*

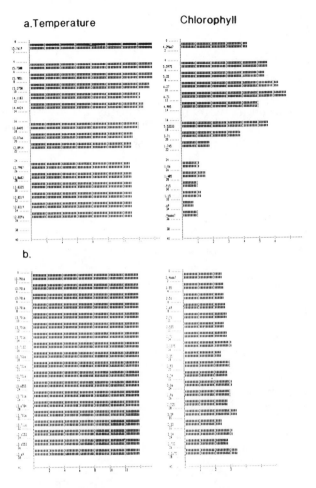

Figure 6. Microcomputer-generated vertical profiles of temperature (°C) (left) and chlorophyll (mg/m^3) (right) on July 7–8, 1982 (as in Figure 5) for a single undulation, in the stratified water in the center of the bloom (a), and in the mixed water at the southern end of the course (b).

tic zone to 40 m or more depending on towing speed) quasisynoptic (top speed ~ 36 km/h) vertical slice, of a similar or enhanced set of parameters. From in situ measurements of downwelling and upwelling solar radiant energy in the narrow wavebands of the coastal zone color scanner (CZCS) at 443 ± 10, 520 ± 10, 550 ± 10, and 670 ± 10 nm, true water-leaving radiances can be calculated to provide a more accurate basis for the evaluation of CZCS images (24). A UOR has been fitted with similar narrowband light sensors to fulfill this role.

The Ichthyoplankton UOR

The UOR Mark 2 was used in a survey of the mackerel spawning grounds in the Celtic Sea in May 1981 (*see* inset Figure 7) (*25*), in which simultaneous measurements of temperature, chlorophyll concentration, and plankton samples (including mackerel eggs and larvae) were recorded (Figure 7). The significance of the UOR Mark 2 for ichthyoplankton research is two-fold. First, conventional sampling for stock assessment by oblique net hauls for plankton and stationary vertical profiles and water samples for physi-

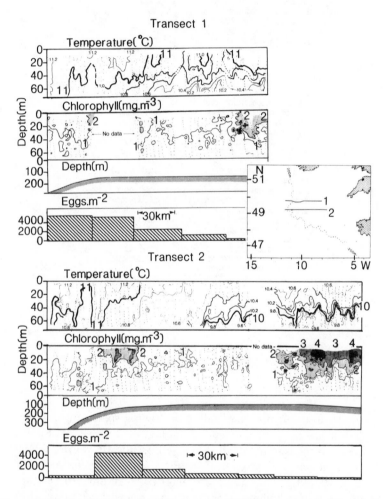

Figure 7. Vertical sections of temperature (0.2 °C intervals) and chlorophyll (1.0-mg/m³ intervals) and abundance of Stage 1 mackerel eggs (number per square meter) from UOR tows along two transects in the Celtic Sea in May 1981 (F.R.V. Cirolana). The inset shows the cruise track.

cal, chemical, and biological parameters is expensive and time consuming for the research vessel and scientific analysis effort. In contrast, the UOR Mark 2 is a high-speed survey tool with considerable automation of the data analysis. Second, in vertically stratified seas the gradients of variability of fish egg and larval abundances and their potential food (phytoplankton and zooplankton) are generally greater vertically than horizontally (26). Thus, to match the searching potential of a fish larva, vertical resolution of measurements needs to be on a scale of meters, which conventional sampling fails to provide. (Conventional sampling provides depth-integrated data, spatially resolved on length scales of several kilometers.) The UOR Mark 2, with its standard sensors and an electronic counter for vertically resolved measurements of fish eggs, larvae, and zooplankton, provides a nearly complete set of measurements with the necessary spatial resolution to support the standard theories of the availability of food for first-feeding larvae in the critical period.

Discussion

The UOR Mark 2 can be applied simultaneously to the specific problems in biological oceanography just presented. The UOR can carry up to 16 sensors, although with more than 8 sensors a second data logger or sensor multiplexing is required. Computer modeling, even as part of the routine data processing, is an important requirement for the interpretation of the data, for example, to compute primary production from theoretical, light, chlorophyll, and temperature relationships (27–30), or from empirical relationships of carbon, chlorophyll, and light determined in local sea truth measurements programs (31).

An alternative approach to modeling primary production of phytoplankton was suggested by the data acquired in the summer of 1982 in an area of the Celtic Sea in the vicinity of 50°30′ N, 7°00′ W (*see* inset Figure 8). The area was chosen for its predicted stability; it is thermally stratified throughout the season, lying within the >3 contour of the $h/\bar{u}^3$ models (32–34), well away from the transitional or frontal regions in the environs of the 1.5 contour (h = water depth, $\bar{u}$ = average tidal stream velocity). The UOR measurements of temperature and chlorophyll concentration for all three visits to the area (June 12, July 7, and September 22) confirmed the predicted homogeneity of the area (*see*, for example, the July sections, Figure 8) although the vertical structures were different for the different seasons: in June there was a broad thermocline (15 °C at ~12 m to 9.5 °C at ~35 m) with a broad subsurface chlorophyll maximum ~2.5–3 mg/m^3 at 30–35 m, just above the base of the thermocline (*see* Figure 9a); in July the thermocline had sharpened (15 °C at ~25 m to 9.5 °C at ~32 m) with a sharp chlorophyll peak, 2–4 m thick of concentration of 5 mg/m^3 in the thermocline (Figure 9b); in late September, the thermocline was deeper and broader than the July structure (15 °C at 15 m to 9.5 °C at 35 m) with

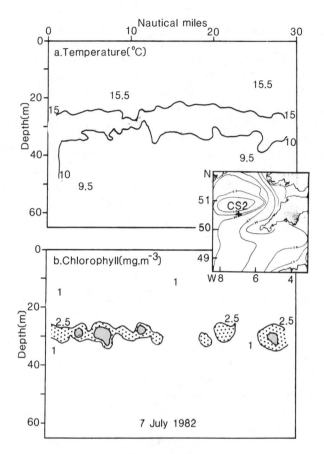

*Figure 8. Vertical sections of temperature (0.5 °C intervals) (a) and chloro-
phyll concentration (1-, 2.5-, and 5-mg/m³ contours, shaded > 2.5 mg/m³)
(b) for a UOR Mark 2 tow in the Celtic Sea, July 7, 1982, RRS* Frederick
Russell. *The inset shows the cruise area.*

chlorophyll surface biased (concentration up to 3 mg/m³) in the mixed water above the thermocline, going to a minimum (near zero at times) in the
thermocline (Figure 9c). The data shown in Figure 9 are examples of nearly
identical profiles for every undulation (30–50 in all) from each tow, and
show a homogeneity of vertical structure even when a small variation of
peak height may indicate a patchy horizontal structure in the vertical
sections.

From this series of measurements, two conclusions can be drawn.
First, the vertical distribution of temperature and chlorophyll is homogeneous spatially throughout this area of the Celtic Sea; second, a close relationship exists between the vertical structure of temperature and chlorophyll for all the seasons, although the exact relationship may differ for the

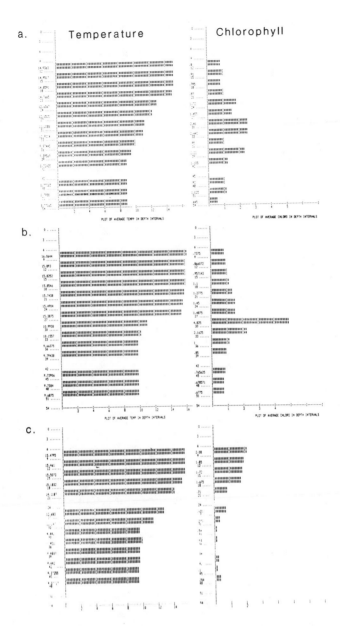

Figure 9a. Microcomputer-generated vertical profiles from near surface to 54 m (in 3-m depth averages) of temperature (°C) (left) and chlorophyll concentration (mg/m³) (right) from a single undulation of a UOR tow in the Celtic Sea. Key: a, June 12, 1982; b, July 7, 1982; and c, September 22, 1982. Cruise area is shown in the inset of Figure 8.

different seasons. These conclusions imply that for this area a spatial coherence in the physical and environmental processes regulates the horizontal and vertical structures of temperature and chlorophyll and that these processes must persist with time to produce the uniform changes observed throughout the seasons. Because physical processes regulate the vertical distribution of temperature (35), phytoplankton (chlorophyll concentration/primary biomass) probably is distributed by similar processes.

Qualitatively, in an isothermal and totally mixed layer (as is the case for the surface mixed layer), or throughout the water column of a shelf sea in winter, or all season if tidal mixing is high, the vertical distribution of chlorophyll is uniform, independent of the vertical distribution of light, photosynthetic production of phytoplankton, phytoplankton sinking (or swimming), or grazing by zooplankton (*see* Figure 6b). Thus, mixing forces appear to control the distribution, and even photoinhibition and nutrient limitation are unimportant.

As stability and a shallow thermocline develop, lower mixing forces in the stable region (of the thermocline) cause an accumulation of cells at these depths. The effects of photoinhibition or surface nutrient limitation in this case will accentuate the generation of a subsurface maximum, and once started, the processes will sharpen the peak further; less chlorophyll at the surface will mean less production of phytoplankton, even if productivity is highest.

A simple, quantitative, steady-state diffusion model (36) demonstrates the importance of physical processes in shaping the vertical distribution of phytoplankton. This model uses values of the eddy diffusion coefficient K from the theoretical model of James (35), which reproduces accurately the annual cycle of vertical temperature structure for this area of the Celtic Sea. The submodels for photosynthetic production, light, and grazing can be varied to any of the established models; nutrient luxury or nutrient limitation of growth can be included. The model reproduces the main features of the UOR observations in the Celtic Sea and English Channel.

1. In a mixed water column (or mixed layer) with relatively high values of K, the model generates a uniform phytoplankton distribution throughout the column/layer, independent of production, light attenuation, or distribution of grazing organisms.

2. With a shallow, surface-mixed layer and broad thermocline (early season case), a broad, deep phytoplankton maximum in the thermocline is produced.

3. With a deeper surface-mixed layer and a sharp thermocline (July case), a narrow phytoplankton peak is generated in the thermocline with relatively low, uniform concentrations above and below the peak.

4. With a deep, surface-mixed layer, deeper thermocline and

smaller productivity (September case), phytoplankton are concentrated in the surface-mixed layer with a minimum near the base of the thermocline.

These models are continuing to be developed with the introduction of a time-dependent element and with a closer simulation of the actual conditions.

Acknowledgments

I acknowledge the collaboration of many colleagues in various aspects of this program: Gerry Robinson, Bob Bruce, Steve Coombs, Ian Joint, Arnold Taylor, Bob Clarke, Charlie Fay, Bob Williams, Mike Jordan, Patrick Holligan, and Jeff Kelly. I thank the officers and crew of the many vessels that have participated in the data acquisition, in particular M.V. *Cornouailles* (Brittany Ferries Ltd), F.R.V. *Cirolana* (Ministry of Agriculture, Fisheries, and Food), R.R.S. *Frederick Russell* (Natural Environment Research Council), and R.M.A.S. *Auricula* (Ministry of Defence, Navy). This work forms part of the program of the Institute for Marine Environmental Research, which is a component body of the Natural Environment Research Council.

Literature Cited

1. Hardy, A. C. *Hull Bull. Mar. Ecol.* 1939, *1*, 1–57.
2. Edinburgh Oceanographic Laboratory *Bull. Mar. Ecol.* 1973, 7, 1–174.
3. Aiken, J. *Mar. Biol.* 1980, 57, 231–40.
4. Aiken, J. *Mar. Ecol. Prog. Ser.* 1981, 3, 235–39.
5. Bruce, R. H.; Aiken, J. *Mar. Biol.*, 1975, 32, 85–89.
6. Aiken, J.; Wood, G.; Jossi, J. *IEEE Oceans' 80*, 1980, 837–43.
7. Aiken, J. *J. Plankton Res.* 1981, 4, 551–60.
8. Glover, R. S. *Symp. Zool. Soc. Lond.* 1967, *19*, 189–210.
9. Dessureault, J.-G. *Ocean Eng.* 1976, 3, 99–111.
10. Aiken, J.; Bruce, R. H.; Lindley, J. A. *Mar. Biol.* 1977, *39*, 77–91.
11. Laitinen, P. O. "Cable-towed underwater body design" Rept 1452, U.S. Naval Electronics Lab., San Diego, 1967.
12. Rae, K. M. *Hull Bull. Mar. Ecol.* 1952, 3, 135–55.
13. Colebrook, J. M. *Bull. Mar. Ecol.* 1960, 5, 51–64.
14. Aiken, J.; Fay, C. W. *IERE Proc. Electron. for Ocean Technol.* 1981, 159–66.
15. Jerlov, N. G. *Inst. for Fysisk Oceanografi Rep. No. 10.* 1974.
16. Aiken, J. "A feasibility study for the design and development of a total chlorophyll sensor" Inst. for Mar. Environ. Res., unpublished report, 1980.
17. Yentsch, C. S. *Deep-Sea Res.* 1960, 7, 1–9.
18. Aiken, J.; Kelly, J., *Cont. Shelf Res.* 1984, 3, in press.
19. Lynch, R. V. "The occurrence and distribution of surface bioluminescence in the oceans during 1966 through 1977," NRL Report 8210, 1978.
20. Dauphinee, T. M. *MTS-IEE Oceans '77 Conference Record* 1977, 1–5.
21. Herman, A. W.; Dauphinee, T. M. *Deep-Sea Res.* 1980, 27, 79–96.
22. Herman, A. W.; Mitchell, M. R. *Deep-Sea Res.* 1981, 28, 739–55.
23. Pingree, R. D. *J. Mar. Biol. Ass. U.K.* 1975, 55, 965–74.
24. Holligan, P. M.; Viollier, M.; Dupouy, C.; Aiken, J. *Cont. Shelf Res.* 1983, 2, 81–96.
25. Coombs, S. H.; Aiken, J.; Lockwood, S. J. *ICES C.M. 1981/H: 32 (mimeo)* 1981.

26. Coombs, S. H.; Lindley, J. A.; Fosh, G. A. "Proceedings of F.A.O. Expert Consultation to examine changes in abundance and species composition of neritic fish stocks," 1983.
27. Bannister, T. T. *Limnol. Oceanogr.* **1974**, *19*, 1–12.
28. Bannister, T. T. *Limnol. Oceanogr.* **1974**, *19*, 13–30.
29. Bannister, T. T. *Limnol. Oceanogr.* **1979**, *24*, 76–96.
30. Smith, R. A. *Ecol. Modell.* **1980**, *10*, 243–64.
31. Jordan, M.; Joint, I. *Cont. Shelf Res.*, **1984**, *3*, 25–34.
32. Simpson, J. H.; Hunter, J. R. *Nature (London)* **1974**, *250*, 404–6.
33. Fearnhead, P. G. *Deep-Sea Res.* **1975**, *22*, 311–21.
34. Pingree, R. D.; Griffiths, D. K. *J. Geophys. Res.* **1978**, *83*, 4615–22.
35. James, I. *Est. and Coast. Mar. Sci.* **1977**, *5*, 339–53.
36. Aiken, J.; Taylor, A., in press.

RECEIVED for review September 16, 1983. ACCEPTED January 18, 1984.

Sampling the Upper 100 m of a Warm Core Ring with a Towed Pumping System

D. R. SCHINK, P. J. SETSER, S. T. SWEET, and N. L. GUINASSO, JR.

Department of Oceanography, Texas A&M University, College Station, TX 77843

A towed pumping system can provide water from throughout the upper 100 m of the water column for shipboard analysis. These analyses, combined with data from in situ sensors, provide information on small-scale and mesoscale features not otherwise available. Signals in the sample stream are modified by passage through the hose. This modification and time delays introduced by analysis must be considered in sampling strategy and data management approaches. This system has been used to determine nutrients, in vivo fluorescence, and temperature in a warm core ring. Examples of the results are provided; the fluorescence, temperature, and nitrate distributions show considerable independence.

OCEANOGRAPHERS HAVE DETERMINED vertical and horizontal variability in the ocean with discrete samples obtained from sampling bottles during an exercise called the "hydrographic cast." Measurements thus obtained were expected to follow a fairly regular progression and if not, they often were "smoothed" on the assumption that an error had occurred. The use of the bathythermograph (BT), the expendable bathythermograph (XBT), and later the conductivity, temperature, and depth (CTD) sensor revealed a striking degree of small-scale variability in the vertical structure of salinity and temperature that could never have been seen with sampling bottles and reversing thermometers. When an oxygen probe was added to the CTD, similar variability was observed in the dissolved oxygen (*1*).

Such measurements tended to focus attention on the vertical dimension. After so many hours staring down a wire, oceanographers were naturally inclined to feel that the "answers" lay directly below. But recently emphasis has shifted. Physical studies have demonstrated the strong impedance to

0065-2393/85/0209-0333$06.00/0

vertical mixing encountered in densely stratified waters. Concurrently, remote-sensing images of surface temperature and color have revealed the rich detail of horizontal variations. As a result, the attention has shifted to horizontal variability.

These studies, although refocusing attention, do not mean earlier workers completely neglected the problem. For example, Hardy (2) deployed a continuous plankton sampler; Lorenzen (3) spurred many underway fluorescence measurements; Weichart (4–6) conducted a number of continuous underway phosphate surveys; and Kelley (7) mapped biologically significant variables in ocean surface waters. Progress was most apparent with towed in situ sensors (8–11). A publication on water sampling while underway (12) reviewed some existing capabilities for horizontal sampling. Undulating bodies with in situ sensors have provided an impressive look at fine-scale and mesoscale oceanic features. But most continuous measuring systems have used only in situ sensors. Parameters that can be measured this way are few. To overcome this limitation, water must be delivered to shipboard analyzers. Surface water is continuously available, but sampling becomes more difficult in the deeper layers. Only a few systems are capable of sampling from depths of 100 m or more while underway. One such system (13) has been modified somewhat; the newer configuration is described here, along with some of the deployment experiences encountered during exploration of a warm core ring off the New England coast.

Equipment

The basic equipment (originally constructed by Fathom Oceanology and later donated to Texas A&M by Texaco) has been substantially modified for easier handling of the underwater vehicle and for easier access to the instruments (13). The deep-towed pumping system can operate safely and effectively to depths of 110 m while underway at 5.1 m/s (10 knots) or to 140 m at 2.1 m/s (4 knots). The system includes an underwater vehicle or "fish," a faired hose–cable combination, a shipboard winch, and associated analytical and signal processing equipment feeding into a data-logging system.

Winch and Framework. The winch and framework are shown in Figure 1. The upper end of the towing cable leads through the winch drum and then divides. The electrical connections pass through one end of the drum axle to a slip-ring assembly, and the hose connects to a watertight, rotary coupling on the other drum hub. Only one layer of hose–cable can be wound onto the winch drum because the rigid cable fairing stands straight out from the drum. The drum is 2.1 m in diameter and holds 27 turns of faired hose–cable.

An articulated U-frame provides flexibililty in handling the fish. The upper frame holds a 0.9-m diameter sheave on a screw-geared shaft that serves as a level wind for the cable. Hydraulic rams adjust the position of the upper frame and connect to a pneumatic accumulator that serves as a

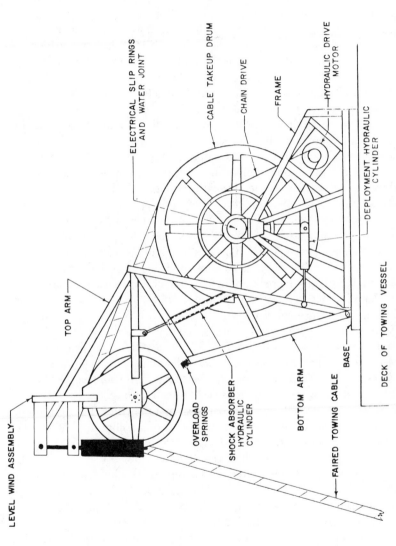

Figure 1. *Towed pumping system winch, framework, and cable. The system is ~5.5 m long, 3 m high, and 2 m wide, and weighs ~2800 kg. The winch is bolted to the deck of the ship.*

shock absorber for transient stresses on the cable. The lower part of the frame can be pivoted out (hydraulically) to extend substantially beyond the stern of the towing vessel.

Cable and Fairing. The hose–cable described by Setser et al. (*13*) consists of 180 m of stainless steel armored jacket enclosing 20 electrical conductors (No. 20 copper wire), imbedded in plastic and wrapped around a nylon hose. The two-layered armor braiding has a 4500-kg minimum breaking strength, far exceeding loads imposed by the underwater vehicle (fish). Power to, or signals from, the pump motor and instruments mounted in the fish are provided through the conductors. A nylon hose [0.95-cm inside diameter (ID)] transports water from the fish to the winch. The cable is enclosed in a segmented fairing (Fathom Oceanology's Flexnose fairing) that reduces the cable drag coefficient from 1.2 to 0.13 and approximately doubles the depth achieved by the fish.

A newer hose, also used with this sampling system, has 16 signal conductors and two shielded conductors. The stainless steel armor has been replaced by Kevlar with a specified breaking strength of 5500 kg.

We attempted to deploy the fish at reduced ship speeds without fairing on the hose–cable. Even at 3 knots (1.5 m/s) the need for fairing was dramatically apparent. Intense strumming occurred with the unfaired cable at towing speeds above 1.5 m/s. The amplitude of the oscillations did not appear to increase with ship's speed, but the frequency of the oscillations did. Strumming greatly increased when towing at depths less than 30 m. At 1.5 m/s the fish reached a depth of 90 m with 150 m of cable payed out; by comparison, the same length of faired cable had no strumming problem and would reach a depth of 110 m at a speed of 5 m/s.

Underwater Vehicle (Fish). The fish described by Setser et al. (*13*) has been replaced by a new, larger underwater vehicle (*see* Figure 2). This fish is constructed of a rolled stainless steel body with stainless steel wings and tail section. As with the previous design, the hydrodynamic stability of the fish is attained through dynamic depression of the wing. The position of the fish in the water column can be controlled by raising or lowering the hose–cable or by changing ship speed. Access to the array of in situ sensors inside the fish is obtained through a large hatch on top.

Instruments in the underwater vehicle include conductivity, temperature, dissolved oxygen, depth, and irradiance sensors (Table I). In addition to these sensors, two pendula inside the electronics package provide continuous readings as to bank and climb altitudes while towing.

The pumping system (Figure 3) has been modified from the one described by Setser et al. (*13*) to avoid contact between the sample stream and metallic surfaces. The 12-stage centrifugal pump is now used to power the hydraulically driven "clean-pump." Seawater from the centrifugal pump is exhausted in situ after driving a reciprocating piston in the sample pump. This drive piston shares a common shaft with two acetal resin (Delrin) pis-

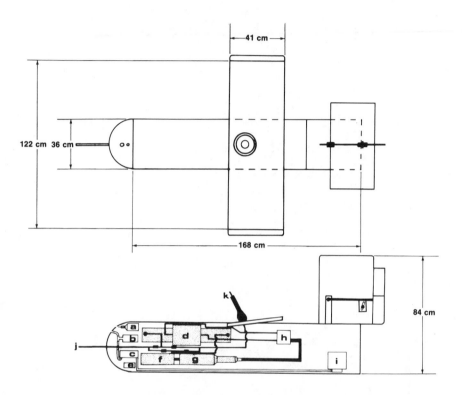

Figure 2. Towed pumping system underwater vehicle.

Instrumentation shown in side view includes: temperature sensor (a), dissolved oxygen sensor (b), conductivity sensor (c), clean pump (d), depth sensor (e), submersible motor (f), drive pump (g), solenoid switching valve (h), pump (Little Giant) providing constant flow regime across O₂ and conductivity sensor (intake is through nose near b) (i), "stinger" water intake for clean pump (water travels from j through a check valve to clean pump on intake stroke and through another check valve to the hose in the towing cable on the exhaust stroke) (j), and towing cable and termination (end cut away to show water hose) (k). For clarity, the underwater data-management package and electrical connections are omitted from the drawing.

Table I. Towed Pumping System In Situ and Associated Analytical Instrumentation

Measurement	Manufacturer/Model	Type	Accuracy	Response Time	Comments/Methods
IN SITU					
Conductivity	Seabird/SBE-3-01	Induction sensor	±0.01	170 ms when towed at 4 knots	Resolution: 5×10^4 mmoh/cm at 12 samples/s
Temperature	Seabird/SBE-4-01	Platinum resistance thermometer	±0.01 °C	70 ms at 1-m/s water velocity	Resolution: 0.0005 °C at 12 samples/s
Depth	Sensotec TJE	Pressure transducer	±0.1% FS[a]	10 ms	—
Dissolved O_2	Grundy Environmental/5175	Polarigraphic	2% FS[a]	6 s for 95% / 60 s for 98%	—
Irradiance	Biospherical Instruments, Inc./QSP-170	Teflon sphere collector–silicon photovoltaic detector	±6%	—	Measures only photosynthetically active light
ANALYTICAL					
In vivo fluorescence	Turner Designs/Model 10	Double-beam filter fluorometer	±1.0%	1 ± 0.02 s to 63% / 4 ± 1.0 s to 98%	Better than 5 ppt of detection Method: Ref. 14
Chlorophyll a	Turner Designs/Model 10	Fluorescence of acetone extracts	±8%	1 ± 0.02 s to 63% / 4 ± 1.0 s to 98%	Limit of detection with a 2-L sample is 0.01 mg/m^3 chlorophyll a Method: Ref. 14
Dissolved inorganic nutrients	Technicon Auto-Analyzer II	Photometric	Variable	—	Methods: $Si(OH)_4$ - Ref. 15 NO_3^- - Ref. 16 PO_4^- - Ref. 17
$CO-H_2$	In-house H_2–CO analyzer	HgO reduction–atomic absorption	±5%	5 min	Method: Ref. 18

[a]FS indicates of full scale.

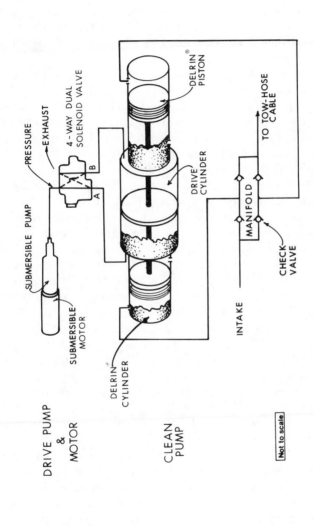

Figure 3. Towed pumping system clean pump. Clean-water flow lines represent PVC pipe (nominal 0.5 in. = 1.5-cm ID).

tons at either end; they are all working in Delrin cylinders. Piston motion is reversed by switching the drive fluid through a four-way dual solenoid valve (ASCO, Model 8344 366). A deck unit controls switching frequency. Four poly(vinyl chloride) (PVC) check valves rectify the clean-water flow.

The sample stream enters through a PVC pipe [1.2-cm outside diameter (OD)] extending 0.5 m in front of the fish and passes through the pump and up the hose, thus, contacting only PVC, Tygon, and Delrin. The hose and rotary joint at the winch are nylon.

Water pressure at the pump is approximately 830 kPa (120 psi). After passing through the 210 m of hose–cable, the rotary joint, and 30 m of 1.0-cm ID tubing on deck, the water arrives at a manifold in the laboratory (3 m above the waterline) with pressure of approximately 210 kPa (30 psi). Flow time from pump to manifold is 3 min. From the manifold, water is distributed to the various analytical instruments.

Down in the fish a second stream of clean water is drawn through the nose, across the oxygen electrode, and through the conductivity cell to discharge through a submersible pump (Little Giant 3E-12 NOVR).

Electrical System.   The conductors in the hose–cable carry power and control signals down to the underwater vehicle and return information. The drive-pump motor is powered by three-phase, 220 V carried down three paired conductors. The solenoid valve, which controls the clean pump, is energized by 24 V direct current (DC), which is controlled on deck. The submersible pump that circulates water through in situ sensors uses 110 V alternating current (AC) power from the hose–cable.

A watertight data management unit [underwater computer (UWC)] in the underwater vehicle takes DC power in the range from 12 to 32 V from the ship and converts this to 5 V DC with a DC–DC converter. This 5-V supply powers the UWC internal circuitry and two ± 15-V DC–DC converters. One converter is used for the requirements of the UWC; the other is used to supply + 30 V DC and + 15 V DC to the external sensors.

The UWC is designed around a 8085 microprocessor that controls an analog-to-digital converter and frequency-counting circuitry. The UWC accepts up to eight frequency signals and up to eight voltage signals from the underwater sensors. The frequency signals are fed through a multiplexer and then converted to five-character hexidecimal code. A separate multiplexer routes the analog signals to an analog-to-digital converter where they are converted to a four-character hexidecimal code. The frequency signals are coded to 15-bit precision; analog full-scale precision is 12 bits.

The 8085 microprocessor sends the coded frequencies and voltages serially up a shielded conductor in the hose–cable to the ship; a 1488 integrated circuit is used to drive the cable to standard RS232C signal levels. A similar conductor feeds ASCII control signals from the ship to the UWC where they are received by an RS232C standard 1489A receiver. The control signals consist of two-character commands that start and stop data transmission and select the frequency and voltage channels to be read. The system can scan and transmit the data from all 16 channels to the ship about three times per second at 2400 baud. Only data from activated channels are transmitted; therefore, higher data rates are possible if fewer channels are used. Because the UWC receives and transmits at RS232C levels, a computer terminal can be connected directly to it for testing purposes.

Data Management.   Management of data is more complex with a towed pumping system than with systems consisting entirely of in situ sensors. Depth, conductivity, and temperature and other in situ sensors all report in a coordinated fashion, and their signals are easily assembled in proper correlations. So, too, are the electronic navigation data. For samples sent up the hose, proper data correlation requires extra care. Failure to do so properly can generate gross distortions when the underwater vehicle is changing depth by 1 m every 6 s. Moreover, the system volume must be known and the flow rate must be monitored to calculate proper time delays. Each instrument has its own delay based on distance from the manifold and analysis time.

A coordinated time base is essential. This simple requirement is not necessarily trivial on-board ship. Many analytical instruments and recording devices keep time by counting electrical cycles. This timekeeping method uses the excellent frequency control commonly found in U.S. electric power systems. At sea, however, frequency drifts, thus, clocks drift too. If all laboratory data were logged immediately into a computer data base, it would be relatively easy to coordinate computer time and real time. However, the situation can be confused when signals first go to strip-chart recorders for subsequent evaluation and then transfer to the general files. Provision then must be made for regular time marks on the chart.

Effects of Pumping Systems on Sample Stream

If data from fast response, in situ sensors are to be combined with slower analyses on a pumped sample stream, the effects introduced during sampling must be considered; both delays and distortions occur in a pumped sample stream. Distortions result from smearing in the hose and from mixing in any dead volumes. These distortions tend to obliterate high-frequency signals. In effect the sampling system behaves as a linear filter by operating on the input signal (e.g., concentration variations) and by delivering the filtered output to an analyzer. Guinasso (*19*) used the mixing theory of Taylor (*20*) to derive the frequency response function for turbulent flow through a long hose in series with a dead volume. When dead volume is only 1 % of hose volume, significant attenuation occurs in signals with frequency of 20 cycles per hose travel time.

We conducted experiments involving passage of an impulse of dye through the hose–winch system. These experiments showed no perceptible change in dispersion whether the hose was entirely spooled on the drum or only 10% spooled. The observed dispersion—through the hose and out the rotary water joint—was approximately threefold greater than predicted by Taylor's smooth-hose theory (20). Guinasso (19) showed that the data are in close agreement to the model proposed by Yen (25).

Typical hose-flow characteristics of this system are summarized as follows: hose length, 210 m; inside radius, 0.48 cm; hose volume, 16 L; flow rate, 6 L/min; flow velocity, 130 cm/s; Reynold's number, 11,700 (at 20 °C); time in hose, 160 s; predicted and observed hose smear, 1.2 and 2.8 s, respectively; and estimated system dead volume (includes manifold in laboratory), 400 cm^3. After an impulse enters the hose, the predicted hose smear represents the exit interval between signal maximum and one standard deviation (for this experiment a 4.5-L/min flow rate was used). Flow becomes turbulent at about 150 L/min. The clean pump does not behave as a pure dead volume, so the effective volume is only an estimate; however, by using the estimates just discussed and considering a gain factor of 0.35 as a threshold, the spatial resolution can be determined for the system of 50 m in the horizontal scale at a 5-m/s (10 knot) ship speed or of 25 m at 2.5 m/s (5 knots).

Operation

The system was deployed during *Gyre* Cruise 82-G-8B (July 1982) while examining structures across the boundary of warm core Ring 81-G. Figure 4 (inset) shows the position of the ring as determined by remote sensing (21) before and after the cruise. During the cruise itself, clouds obscured the ring; location by satellite was, therefore, not possible.

Over the 14 days of the cruise, the ring moved approximately 148 km to the west. To interpret our observations with respect to the ring during this period, we assumed the ring advanced at a constant rate (~ 10.5 km/day), and we used this rate to relocate observations around the ring, that is, all positions of observations were shifted westward (by dead reckoning). The amount of the shift was determined by the time difference between an observation and the reference time, 1200Z, July 26; this value was multiplied by the ring translation rate of 10.5 km/day. Figure 4 shows the depth of the 10° isotherm as established from adjusted positions of expendable bathythermographs (XBTs) and by CTD lowerings. Contours of the 10° isotherm indicate the ring center location at 67.3° N, 39.5° W, slightly east of the ring center as inferred from remote sensing but perhaps consistent with the asymmetry evident in 10° isotherm contours.

The ship's track during the deployment (Figure 4) indicates some of the troublesome features encountered during a survey of this sort. The ring rotates at an appreciable velocity that varies almost linearly with the radius out to the ring edge. In principle, the ship's track might be adjusted to compensate for this set; however, the effort does not appear justified. Over 19 h,

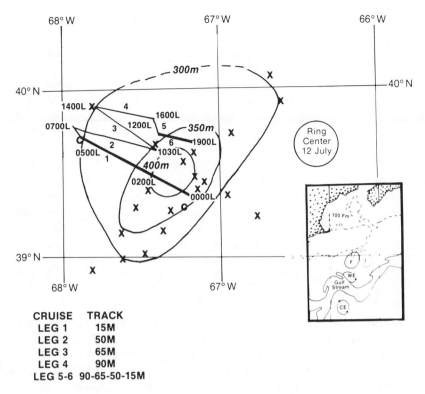

CRUISE   TRACK
LEG 1   15M
LEG 2   50M
LEG 3   65M
LEG 4   90M
LEG 5-6 90-65-50-15M

Figure 4. Depth of 10° isotherm and track of R.V. Gyre during part of
Cruise 82-G-8B, July 26, 1982.

The Xs represent a shifted XBT used in contouring (see text), and the Cs represent
CTD casts. Bold lines on cruise track denote night sampling. The inset shows ring posi-
tion determined by remote sensing (21). The ring labelled I (inset) represents the initial
ring position (July 12, 1982), and F represents the final ring position (July 26, 1982).

the uppermost ring waters move almost halfway around the circuit. Deeper
waters move a lesser distance. As a token accommodation to the moving
waters, reciprocal courses were sailed, and the resultant set was ignored.
Thus, the ship was set around the ring, but to a lesser degree than waters of
the ring travelled. Our sampling strategy included running back and forth
across the ring boundary and holding constant depth on each leg. Four
legs were run at 15-, 50-, 65-, and 90-m depth; then a final dogleg was
made while the underwater vehicle was raised in steps to 20 m and finally
retrieved.

To construct a cross-sectional representation of these observations, the
data were considered to lie along chords across a circle. The chords were assembled as if they lay in a single vertical plane; to do this, they were translated so that the closest point of approach (CPA) to the ring center for each

lay directly above or below the CPA for the others. Because water initially surveyed at the ring edge would be nearly halfway around the ring at the end of this survey, the sections presented represent true pictures of ring structure only if the ring displayed perfect radial symmetry; however, the ring does not display perfect radial symmetry. Rather, these figures offer guidance to the kinds of features that must occur at ring boundaries. The features shown probably never existed as detailed here. However, each parameter was assembled by similar methods; the correspondence, or lack thereof, offers some insights.

Data from the various parameters were contoured, and a final contour pattern was adopted after differences were resolved. The first parameter to be contoured was temperature, by taking XBTs into account. Subsequent contours were biased because the structure of the temperature field was known; wherever subjective decisions on contouring were required, the patterns were biased toward similarity of features from one parameter to the next. The resulting contours for temperature, nitrate, and in vivo fluorescence are shown in Figure 5; electronic problems rendered salinity data invalid over this portion of the cruise. The control data fall along the dotted line in each figure.

The most striking feature of these sections and the silicate and phosphate (not shown) is the considerable degree of independence in spite of our bias toward similarity as just noted. A strong correlation was previously shown (22) between temperature, nitrate, and phosphate in surveys across the California upwelling system. Correlations as strong as those did not occur here.

Neither did we find such correlation in waters west of Baja California during September 1979 (23). In the different water types identified in that survey of surface waters, the correlations between temperature, nutrients, and fluorescence were rather poor and varied from one water type to another (Table II). The close correlation between temperature, nitrate, and phosphate just discussed (22) probably exists in freshly upwelled water; then, as phytoplankton consume increasing amounts of the initial nutrients, and as solar heating modifies the initial temperature, these close relationships break down.

Optimizing Deployment

The reconstructed sections just described provide valuable insights into the variability existing across a ring, but the synthetic nature of the sections is not entirely satisfying. A more effective survey would involve a scan of the entire mixed layer as the ship proceeds on a single pass. Such a scan is possible, in principle, without much sacrifice in detail, because when towing on a horizontal line most of the data acquired strongly resemble the previous interrogation. Information yields would be optimized if the variables were continually changing. This variability would be achieved if the underwa-

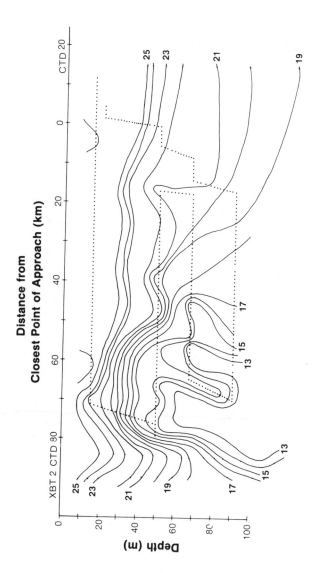

Figure 5a. Temperature data (°C) compiled from transects across warm core Ring 81-G. The upper scale represents kilometers from closest point of approach to ring center. The data were collected using an in situ temperature sensor, XBTs, and CTDs. Note the intrusion of cold water at depth in the ring boundary zone.

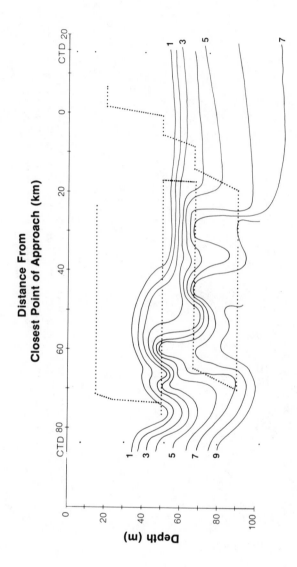

Figure 5b. Nitrate concentration (μM) data compiled from transects across warm core Ring 81-G. The data are from continuous analysis of pumped water with an autoanalyzer and from discrete measurements of bottle samples during CTD casts. Note the different shape of nitrate-rich intrusions.

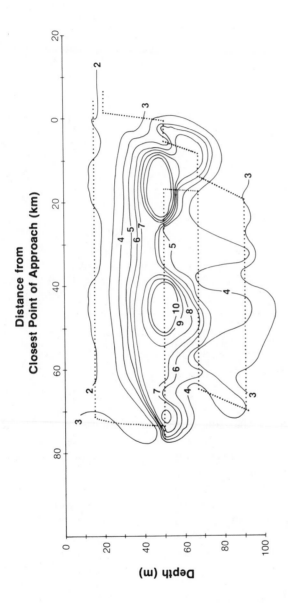

Figure 5c. In vivo fluorescence of the pumped water stream as determined by laboratory fluorometer from transects across warm core Ring 81-G. Fluorescence per unit of chlorophyll a varied with depth and thus resulted in a weak correlation between chlorophyll a and fluorescence. Fluorescence is reported in relative units, where one unit is equivalent to fluorescence of 4 μg/m³ of chlorophyll a from spinach extract. The three mid-depth maxima appear unrelated to temperature or nutrient contours.

Table II. Correlation Coefficients and Number of Observations
Between Parameters Measured in Various Surface Waters

Cold

	T	IVF	PO$_4$	SiO$_2$
			Correlation Coefficient (r)	
		IVF	PO$_4$	SiO$_2$
T		-0.11^a	-0.68	0.68^a
IVF	19		-0.78^a	0.00
PO$_4$	8	3		-0.08
SiO$_2$	8	3	8	
Number of Observations				

Cool

	T	IVF	PO$_4$	SiO$_2$
			Correlation Coefficient (r)	
		IVF	PO$_4$	SiO$_2$
T		0.22	-0.32	0.42
IVF	80		-0.04^a	0.29
PO$_4$	56	39		0.41
SiO$_2$	54	37	53	
Number of Observations				

Warm

	T	IVF	PO$_4$	SiO$_2$
			Correlation Coefficient (r)	
		IVF	PO$_4$	SiO$_2$
T		0.52	-0.38	-0.08^a
IVF	23		-0.39^a	-0.77
PO$_4$	26	14		0.62
SiO$_2$	27	15	26	
Number of Observations				

NOTE: Ship's track, course 025° from 630 km out, into San Diego. (*See* Reference 23 for water-type descriptions.) T is water temperature, IVF is continuous flow in vivo fluorescence, and PO$_4$ and SiO$_2$ are concentrations in solution.
[a]Not significant at the 0.10 probability level.

ter vehicle were constantly changing depth—undulating from surface to maximum depth. The underway systems mentioned earlier routinely operate in this fashion; however, a towed pumping system introduces new sampling problems that seem worthy of analysis.

The existing configuration is not capable of routine undulation during sampling. The fairing does not always spool smoothly onto the drum as the underwater vehicle is raised and requires special attention during this maneuver. Such mechanical problems might be readily eliminated.

The underwater vehicle can easily be lowered to 100 m in 5 min; a complete cycle should require 10 min with a minimum practical depth of 3–5 m. A more conservative sampling program might cycle at half this rate. Figure 6 illustrates such a path superimposed on a distribution pattern resembling one found in Ring 81-G. Many warnings have been issued on the pitfalls of undersampling the fine-scale features in the ocean. Denman and Mackas (24) showed how an artificial distribution may arise from undulat-

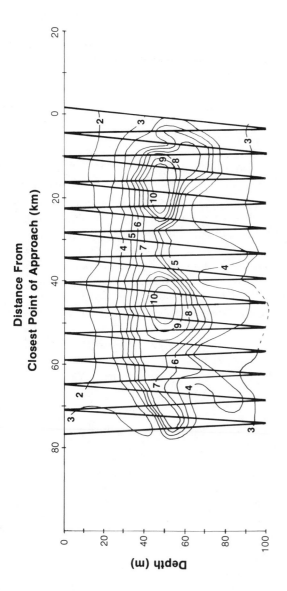

Figure 6. Computed section showing in vivo fluorescence as observed through hose dead volume filter with degraded resolution. The data points for calculation were taken by interpolation from values in Figure 5c at 2-m intervals along the track shown. This hypothetical data set was passed through a hose dead volume filter (computationally) at a flow rate of 1.8 L/min to depict poorer than expected resolution. The resulting data when compared to the initial data show the type of small-scale distortions that might appear under these conditions.

ing sampling through an internal wave regime. They also suggested an essential counter to such distortion—that distributions should be assembled according to density levels rather than depth. Moreover, they noted that most spectra are "red," that is, dominated by the longer wavelengths, and this condition tends to reduce the problem.

The effects of signal smearing in the hose would distort an observed distribution. Because vertical gradients are much greater than horizontal gradients, the sampling path in Figure 6 appears much more susceptible to such effects than a simple horizontal line would be. However, this concern arises principally from the vertical exaggeration. A ship advances 3 km during a lowering of 100 m to produce a 2° descent angle. The system can resolve features of ~50 m, so the fish would descend 1.7 m while sampling through this distance. To examine smearing effects, we have taken the interpolated distribution shown in Figure 5c as if literally true and have run it (computationally) through the hose dead-volume filter, described by Guinasso et al. (19), by using the characteristics listed on page 342. The signal that emerged was indistinguishable from Figure 5c. As a further test, smearing in the hose was increased threefold by decreasing the flow rate to near the turbulent–laminar flow crossover. The signal again was passed through the filter, and the results were plotted and contoured in Figure 6 as if no distortion had occurred. The new pattern still corresponds closely to the initial distribution, but alternate up-smearing on ascent and down-smearing on descent produce a perceptible sawtooth pattern to formerly flat contours.

The results are instructive and encouraging. However, the initial distribution, assumed to represent real values, was actually obtained through the hose filter so that the high-frequency components were absent from the start.

Conclusions

Continuous underway sampling using a deep-towed pump and a hose–cable can examine chemical patterns in the ocean that are nearly impossible to describe by other techniques. Growing realization of the importance of mesoscale features and oceanic fronts makes use of such sampling tools an essential part of oceanography. Effective data management must be an indispensable part of any deep-towed sampling system.

Acknowledgments

Recent modifications of the towed pumping system were performed with considerable assistance from Clarence Propp and Toby Selcer of Brazos Technology. Nonita Yap and Sue Schauffler also provided valuable assistance. John Stockwell demonstrated again that he is a great asset to any seagoing operation. Ernest Frank did the statistical analysis presented in Table II. Chester Kronke and David Carey designed, constructed, and programmed the circuit board for the underwater computer while working for Innovative Development Engineering Associates (College Station, Texas). Frank O'Hara designed and built the underwater computer housing.

Literature Cited

1. Kester, D. R.; Crocker, K. T.; Miller, G. R., Jr. *Deep-Sea Res.* 1973, *20*, 409–12.
2. Hardy, A. C. *Discovery Rep.* 1936, *11*, 457–509.
3. Lorenzen, C. J. *Deep-Sea Res.* 1966, *13*, 223–27.
4. Weichart, G. *Dtsch. Hydrogr. Z.* 1963, *16*, 272–81.
5. Weichart, G. *Dtsch. Hydrogr. Z.* 1965, *18*, 210–18.
6. Weichart, G. *Dtsch. Hydrogr. Z.* 1970, *23*, 49–60.
7. Kelley, J. C. *Deep-Sea Res.* 1975, *22*, 679–88.
8. Joseph, J. *Dtsch. Hydrogr. Z.* 1962, *15*, 15–23.
9. Bennett, A. S. In "Oceanology International 72"; London BPS Exhibitions Ltd., 1972; pp. 353–56.
10. Bruce, R. H. *Proc. Challenger Soc.* 1973, *4*, 153.
11. Dessureault, J. G. *Ocean Eng.* 1976, *3*, 99–111.
12. Steering Committee, Underway Water Sampling Technology, "Water Sampling While Underway"; proceedings of a symposium and workshops, Washington, D.C., 1980; National Academy Press: Washington, D.C., 1981.
13. Setser, P. J.; Guinasso, N. L., Jr.; Condra, N. L.; Wiesenburg, D. A.; Schink, D. R. *Environ. Sci. Technol.* 1983, *17*, 47–49.
14. Strickland, J. D. H.; Parsons, T. "A Practical Handbook of Seawater Analysis," 2nd Ed.; Bull. Fish Res. Board Canada, 1972, pp. 185–206.
15. Brewer, P. G.; Riley, J. P. *Anal. Chim. Acta* 1966, *35*, 514–19.
16. Armstrong, F. A. J.; Stearns, C. R.; Strickland, J. D. H. *Deep-Sea Res.* 1976, *14*, 381–89.
17. Murphy, J.; Riley, J. P. *Anal. Chim. Acta* 1962, *27*, 31–36.
18. Bullister, J. L.; Guinasso, N. L., Jr.; Schink, D. R. *J. Geophys. Res.* 1982, *87*, 2022–34.
19. Guinasso, N. L., Jr., Ph. D. Dissertation, Texas A & M University, College Station, Tex., 1984, 48 pp.
20. Taylor, G. I. *Proc. R. Soc. London, Ser. A* 1953, *219*, 186–203.
21. National Environmental Satellite Service, *Oceanographic Analysis* July 28, 1980, D127–N98–No. 387, National Weather Service.
22. Traganza, E. D.; Nestor, D. A.; McDonald, A. K. *J. Geophys. Res.* 1980, *85*, 4104–6.
23. Setser, P. J.; Bullister, J. L.; Frank, E. C.; Guinasso, N. L., Jr.; Schink, D. R. *Deep-Sea Res.* 1982, *29*, 1203–15.
24. Denman, K. L.; Mackas, D. L. In "Spatial Pattern in Plankton Communities"; Steele, J. H., Ed.; New York, 1978; pp. 85–109.
25. Yen, L. C. *J. Sanit. Eng. Div., Am. Soc. Civ. Eng.* 1969, *69*, 1105–15.

RECEIVED for review November 29, 1983. ACCEPTED March 5, 1984.

Airborne Mapping of Laser-Induced Fluorescence of Chlorophyll a and Phycoerythrin in a Gulf Stream Warm Core Ring

F. E. HOGE

NASA Goddard Space Flight Center, Wallops Flight Facility, Wallops Island, VA 23337

R. N. SWIFT

EG&G Washington Analytical Services Center, Inc., Pocomoke City, MD 21851

The National Aeronautics and Space Administration's airborne oceanographic lidar has been used to produce a wide-area map of the distribution of phytoplankton fluorescence from chlorophyll a and phycoerythrin photopigments. Following each 532-nm laser pulse, the chlorophyll a and phycoerythrin fluorescence and water Raman backscatter were transferred by telescope to a diffraction-grating spectroradiometer subsystem having 32 photomultiplier tube detectors covering 380–740 nm in contiguous 11.25-nm bands. Although a high degree of correlation or coherence existed between the phycoerythrin and chlorophyll a fluorescence over wide areas of the warm core ring (~ 100 km), variations in the chlorophyll a–phycoerythrin fluorescence ratio are sometimes observed on smaller spatial scales over sections of the warm core ring and adjacent water masses. The biological significance of these findings is still under investigation.

AIRBORNE LASER-INDUCED FLUORESCENCE (LIF) has been used to measure chlorophyll a concentration in water bodies. Results from airborne single-channel detection of laser-induced chlorophyll a fluorescence at 685 nm were first published in 1973 (1). A flashlamp pumped dye laser at 590 nm was used as the excitation source; a narrow-band interference filter and photomultiplier formed the basic single-channel detector. Descriptions of simultaneous dual-channel lidar detection of chlorophyll a and water Raman backscatter have been published (2, 3). The spectral separations were accomplished by using a 50:50, neutral beam splitter and by isolating the Raman and chlorophyll a lines with a 10-nm interference filter centered at 560 nm and a 23-nm filter centered at 685 nm, respectively. Limited field investigations (4, 5) were performed by using two laser excitation wavelengths (from a single laser) while flying over riverine and estuarine water bodies.

During 1982, the National Aeronautics and Space Administration's (NASA) Wallops Flight Facility (WFF) airborne oceanographic lidar (AOL) was flown on six missions over Gulf Stream warm core rings (WCRs). These missions were part of a multi-institutional investigation of these features known as "the Warm Core Ring Experiment" (6). The primary objective of the AOL participation was to provide relatively synoptic maps of surface layer chlorophyll a concentration and water surface temperature in the ring itself, as well as in adjacent shelf- and slope-water masses.

Results of AOL experiments designed to measure concentrations of chlorophyll a and other photopigments (related to phytoplankton) with a multichannel-receiver capability have been published for missions flown over both estuarine (7, 8) and open ocean water masses (9). The estuarine field work, as well as the work presented in this chapter, was conducted with a single laser excitation wavelength (532 nm). Open ocean investigations have been performed with the AOL configured with an additional XeCl-excimer pumped dye laser operated at 427 nm and fired on alternating pulses with the 532-nm laser (9). Other marine applications of the AOL include oil spill (10, 11), fluorescent tracer dye (12), and dissolved organic matter measurements (13).

Considerable attention has focused on WCR features related to the Gulf Stream. Warm core rings play an important role in the distribution of productivity and biota. Mechanisms for the formation of WCRs are similar to those mechanisms giving rise to cold core rings (14, 15). Briefly, a meander of the Gulf Stream to the north or to the northwest creates a large loop current that can close on itself, detach from the Gulf Stream, and thus trap relatively warm Sargasso Sea water within a swiftly flowing ring of Gulf Stream water. Warm core rings can reach sizes of 100–200 km in surface diameter and can extend to depths of more than 1000 m. These rings can persist for as long as 18–24 months until they eventually migrate to the southwest, recoalesce with the Gulf Stream (in the vicinity of Cape Hatteras), and lose their identity. During its lifetime, a WCR can become temporarily and loosely reattached to the Gulf Stream; thus, an infusion of low-nutrient stream water may occur in the ring boundary region. Likewise, as a WCR approaches the continental shelf, as it inevitably does, colder and more nutrient-rich shelf and/or slope water may be pulled around and possibly entrained within its extremity.

The large-scale contrast in nutrient levels between the WCR and surrounding slope- and shelf-water masses provides a potential for increased phytoplankton development, especially within the ring boundary region where the water masses are in contact. During the late winter and early spring, a high-temperature contrast within the near-surface water also exists between the WCR and cooler surrounding water masses. This contrast contributes further to conditions for increased productivity. The size of WCRs and dynamic productivity conditions that take place in and around them make WCRs difficult to study. Sampling these mesoscale features with conventional shipborne equipment yields results that are spatially undersampled and temporally smeared, and thus interpretation is difficult. The high data acquisition rate of an airborne laser-fluorosensing system and the relatively high speed of aircraft platforms provide a tool for assisting the oceanographer in resolving these sampling problems, especially when used with other airborne and spaceborne instrumentation, as was done during the WCR experiment. The combined remote-sensing and conventionally acquired data sets are currently being analyzed. This chapter discusses only the AOL results from the WCR mission flown on April 20, 1982.

Instrumentation

The airborne oceanographic lidar (AOL) is a state-of-the-art scanning laser radar system that has a multispectral time-gated receiving capability. The system is designed to allow adjustment in most transmitter and receiver settings. This built-in flexibility gives the AOL system potential application in many oceanographic areas. The AOL can be operated in two basic modes. In the bathymetry mode (*16*), a single backscattered wavelength (usually the on-frequency laser wavelength) is temporally resolved over a 90-ns interval. In the fluorosensing mode, the backscattered radiation from the sea surface and water column are spectrally resolved between 380 and 740 nm. The AOL was operated in the fluorosensing mode during the WCR investigations reported in this chapter. Portions of the hardware and software components of the AOL have been discussed elsewhere (*7–9*), but will be summarized and expanded as needed to illustrate important aspects of the fluorosensing mode of the instrument as used during these WCR experiments. Figures 1 and 2 illustrate this hardware. Figure 2 is a detailed schematic of the AOL, the general location of which is given within Figure 1.

A frequency-doubled Nd:YAG (neodymium: yittrium aluminum garnek) laser was used to provide excitation at 532 nm. A high-speed silicon photodiode views the radiation scattered from the first folding mirror to

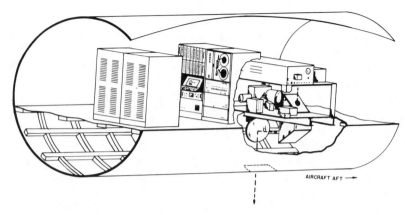

Figure 1. Cut-away illustration of the AOL system as configured on the Wallops Flight Facility P-3A aircraft.

provide the start pulse timing and monitoring of the laser-output pulse power. Digitization and recording of this signal allow the data to be corrected for laser-output power variations. The pulsed laser output is folded twice through 90° in the horizontal plane of the upper tier, into the vertical folding mirror, and then downward through the adjustable beam divergence–collimating lens. The laser-output beam divergence of the frequency-doubled YAG laser is controllable with the collimating lens only between 0.3 and 5 mrad. Minimum divergence was used during all of these field experiments. The beam then exits directly downward through the 9-cm opening in the main receiver folding flat, onto the scanner folding mirror, finally striking the angle-adjustable, nutating scan mirror. A setting of 15° off nadir was used for all WCR missions, and the data were obtained in a nonscanning mode. The scan mirror finally directs the beam to the ocean surface and water column. The total surface and volume backscattered signals return through the same path; but because of their uncollimated spatial extent, the signals are principally directed into the 30.5-cm Cassegrainian receiving telescope. The horizontal and vertical fields of view of the receiving telescope are each separately controlled by a pair of operator-adjustable focal plane knife edges.

The beam-splitting mirror (used only in the fluorosensing mode) directs a major portion of the excitation wavelength and the fluorescent return signal into the fluorosensing detector assembly. The laser excitation wavelength (532 nm) component of the return signal was rejected from the spectrometer by a high-pass (wavelength) filter (Kodak 21). This filter rejects radiation below ~540 nm. A small amount of the surface return signal is allowed to pass through a small 1-cm opening in the beam splitter where it is sensed by the bathymetry photomultiplier tube and is subse-

SECTION THROUGH A–A'

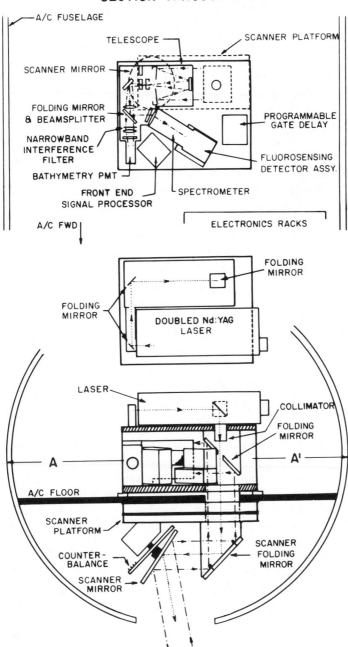

Figure 2. Detailed schematic of the transmitter and receiver components.

quently used to measure slant range and to generate the gate pulses for the fluorosensor analog-to-digital charge digitizers (CDs).

The fluorosensing detection assembly contains an 11-cm diameter transmission diffraction grating that is blazed for 480 nm and has 600 grooves/mm. An 11-cm diameter simple lens brings the dispersed radiation to the entrance surface of 40 quartz light guides. These guides are optically coupled to two separate banks of 20 RCA 8644 photomultiplier tubes (PMTs) of which only 32 were used.

A 32-channel LeCroy Model HV4032 high-voltage power supply was added to the AOL fluorosensor. The multichannel power supply allows the gain of 32 of the PMTs to be individually adjusted. This capability facilitates better real-time monitoring of the instrument during the field missions and significantly reduces the complexity of postmission calibration and data-processing corrections. The 32 channels, each having a spectral bandwidth of 11.25 nm, were arranged contiguously between 380 and 740 nm. The tubes are not shuttered or gated but remain active at all times. Ambient background radiation rejection is provided by the 0–20-mrad adjustable field of view (FOV) knife-edge pairs located at the focal point of the receiving telescope. The optimum operational FOV for our field tests was experimentally determined to be 4 mrad by observing and maximizing the water Raman signal-to-noise ratio (SNR).

The pulsed analog outputs of the entire bank of PMTs are routed to alternating current (AC)-coupled buffer amplifiers that drive each of the CD input channels. The amplifiers respond only to wide-bandwidth fluorescent pulses; thus, response to background noise is minimal, and permits full daylight operation. All CDs are simultaneously gated "on" to obtain the entire spectral waveform at a temporal or depth position determined by the surface return signal from the bathymetry PMT. With proper adjustments in delay relative to the bathymetry PMT-derived surface return, the spectral waveforms may be taken at any position above or below the ocean surface. In this experiment the spectral-waveform data acquisition was started 3 ns prior to encountering the surface and terminated 30 ns later. The CD output is directed through computer-aided measurement and control (CAMAC) standard instrumentation to a Hewlett-Packard (HP) 21MX computer and recorded.

A real-time spectral data monitor was placed in operation. The monitor consists of an independent microprocessor-based system that acquires real-time spectral data and additional housekeeping information flowing through the CAMAC/HP computer-interface lines. The complete laser-induced fluorescence spectrum produced by each laser shot is displayed in real time. The system also provides a two-channel output to an analog chart recorder that produces two profile traces (usually the laser-induced phycoerythrin and chlorophyll a spectral response peaks).

Although not directly related to the AOL system, an airborne expendable bathythermograph (AXBT) launch tube, receiver, and microprocessor-based data acquisition system were also built to provide valuable subsurface temperature data in the WCR boundaries. The AXBT data system multiplexes the temperature data into the AOL system HP computer for simultaneous recording with the laser data. The specially adapted radio receiver/data acquisition system is configured to handle standard AN/SSQ-36 (U.S. Navy) AXBTs. Surface temperature data, acquired by a Barnes platinum resistance thermometer (PRT)-5 IR radiometer, are also recorded by the AOL computer system.

Aircraft Field Experiment Results

Figure 3 shows the general location, size, and shape of the warm core ring designated by the National Oceanic and Atmospheric Administration (NOAA) as 82B. It evolved from the Gulf Stream in February 1982 at 39° N latitude and 69° W longitude and began its characteristic southwesterly drift toward Cape Hatteras. Our flight was conducted on April 20, 1982 during a 3-week WCR cruise conducted by investigators from a number of oceanographic institutions. Participating research vessels included the *Oceanus* and *Knorr* from Woods Hole Oceanographic Institution, and the *Endeavor* from the University of Rhode Island.

The flight lines occupied during the mission are shown in Figure 4. These flight lines were arranged to coincide as closely as possible with the chords of a "star"-sampling pattern occupied by the R.V. *Endeavor*. The symbols at the five points of the star indicate the locations of AXBT deployments. The flight lines referred to later are numbered within the figure. The airborne experiment was designed to provide a synoptic assessment of surface layer chlorophyll a, phycoerythrin, and water surface temperature. During the 36-h period surrounding the airborne mission, the R.V. *Endeavor* was engaged in deploying XBTs as well as gathering underway surface current, salinity, and phytoplankton fluorescence measurements. The results from shipborne instrumentation have not yet become available, thus, no comparison between the airborne and surface observations was possible.

The 532-nm excitation wavelength provided by the frequency-doubled YAG laser yields spectra similar to those obtained from Atlantic Ocean water in work performed at the NASA Wallops Laser Laboratory. Two LIF spectra, typical of results obtained throughout the mission, are shown in Figure 5. These spectra were collected at the locations shown in Figure 4 on Flight Line 7 at positions labeled "a" and "b." Each spectrum is a simple average of data gathered over 5 s. At a laser repetition rate of 6.25 pulses per second and a ground speed of 100 m/s, about 31 individual spectra were averaged over a 0.5-km distance. The spectrum in Figure 5a was obtained

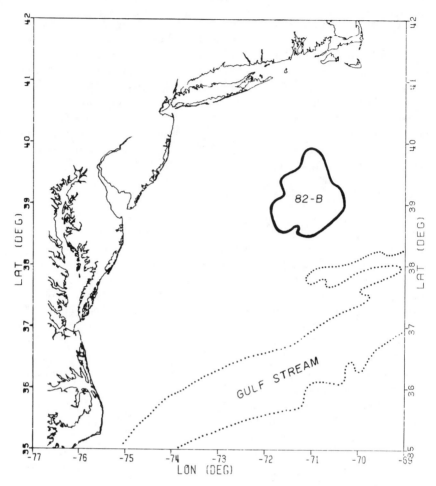

*Figure 3. Map showing location of warm core Ring 82B on April 28, 1982
according to the National Weather Service.*

in the ring boundary region where elevated chlorophyll a concentrations
were generally encountered. The spectrum in Figure 5b was acquired well
inside the WCR where lower, monotonous chlorophyll a signals were generally observed. The three spectral lines labeled R, C, and P in both Figures
5a and 5b denote the Raman backscatter (650 nm), chlorophyll a (685 nm),
and phycoerythrin (580 nm) bands, respectively. These spectra have not
been corrected for a slight distortion from the long-pass filter (Kodak 21)
used to partially reject the laser wavelength from the spectra. Also, cross-
channel interference between the Raman peak return and the chlorophyll a
return has not been removed. This interference may produce some minor
error in very clear, offshore waters where our relative chlorophyll values

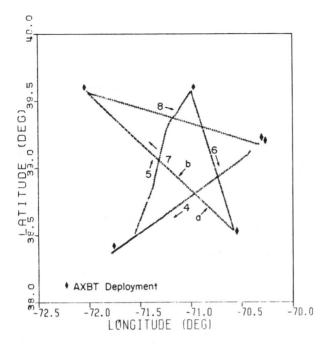

Figure 4. Aircraft flight lines executed over Ring 82B during the field mission of April 20, 1982. The symbols located at the end of the flight lines indicate where the AXBTs were launched. The numbers denote the flight lines as well as the sequence in which they were flown. The positions on Flight Line 7 marked "a" and "b" indicate where spectral waveforms were selected for Figure 5.

may be slightly elevated. A comparison of the spectra given in Figure 5 indicates that an increase in the amplitude of the phycoerythrin signal accompanies the higher chlorophyll a signal in the spectra obtained near the ring boundary region (Figure 5a). However, the ratio of the chlorophyll a to phycoerythrin signals may vary considerably within the ring boundary region and surrounding water masses (discussed later).

The techniques used in the processing and analysis of AOL data collected for mapping the distribution of chlorophyll a and phycoerythrin have been previously documented (7, 8). A review of these techniques is given to provide ample explanation of recent modifications made as a result of hardware changes presented in the preceding section. The implementation of the 32-channel, software-controlled voltage supply allowed the development of improved techniques for calibrating the spectral channels. During preflight procedures conducted on the ground, a fluorescent target of known fluorescence spectral characteristics was excited by the 532-nm

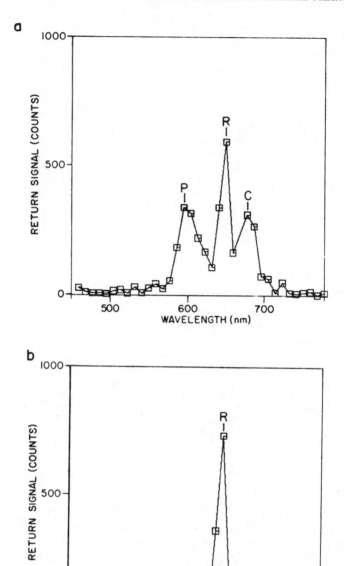

Figure 5. Top: Laser-induced spectrum obtained at position "a" on Flight Line 7 shown in Figure 4; at this ring boundary location the chlorophyll fluorescence was elevated. Bottom: Waveform obtained in a low chlorophyll a concentration region within the warmer core of the ring at position "b" in Figure 4.

laser. During this calibration procedure, the expected voltages required for the flight experiment were loaded separately into the 32-channel power supply. A short segment of data was acquired at each voltage level. Voltage levels required to equalize the sensitivity of the individual spectral channels were then selected from the recorded data. The voltage matrix resulting from the calibration was then inserted into the real-time AOL software package as a lookup table.

During transit to the study area, an optimal array of voltages was selected from the lookup table. These voltages allowed the backscattered water Raman spectral signal to vary comfortably, and without saturation, within the 1024-count digitization limitation of the individual fluorosensor channels. The development of the real-time display facilitated both the selection of the operating voltage array and the monitoring of the return signal during the experiment.

During the April 20th WCR flight experiments, a high-pass filter was placed in front of the AOL fluorosensor. As mentioned earlier, the filter was used to remove the on-frequency 532-nm backscatter return from the YAG lasers. The effect of the filter on the spectral response signal from the 532-nm laser was minimal except in the 585-nm spectral region where some minor roll-off distortion occurs in the phycoerythrin fluorescence band. The water Raman line (650 nm) and chlorophyll a band (685 nm) are equally suppressed by a small constant amount.

The backscatter data were corrected for changes in altitude and variations in laser-output power. Effects on the chlorophyll a and phycoerythrin laser-induced fluorescence (LIF) responses due to spatial variations in optical attenuation properties of the water column were removed by normalizing these fluorescence signals with the 650-nm water Raman signal. Bristow et al. (3) used this normalization technique to correct chlorophyll a fluorescence obtained with a two-channel laser fluorosensor in an experiment conducted over Lake Mead. They further reported a high correlation between their airborne water Raman backscatter signal and the reciprocal of the beam attenuation coefficient measurements acquired from a surface-truthing vessel. Computations supporting the application of the water Raman normalization technique for correcting fluorescence measurements of chlorophyll a were presented by Poole and Esaias (17). Their modeling results show, in part, that chlorophyll a LIF measurements from an aircraft fluorosensor are essentially confined to the upper several meters of the water column, even under open ocean conditions. This lack of penetration is due to the high absorption of seawater in the 670–690-nm spectral region. Poole and Esaias, therefore, concluded that improved results can be attained if the chlorophyll a LIF response measurements are normalized with water Raman backscatter signals obtained near the 680-nm spectral region; 532-nm excitation allows this response.

Figure 6 shows airborne data acquired on Flight Line 7 (identified in Figure 4). Flight Line 7 was flown from southeast to northwest and passed near the center of the WCR. The locations of the spectra shown in Figure 5 are identified in Figure 6a. Profiles of peak fluorescence response channels from the chlorophyll a and phycoerythrin bands are plotted in Figures 6a

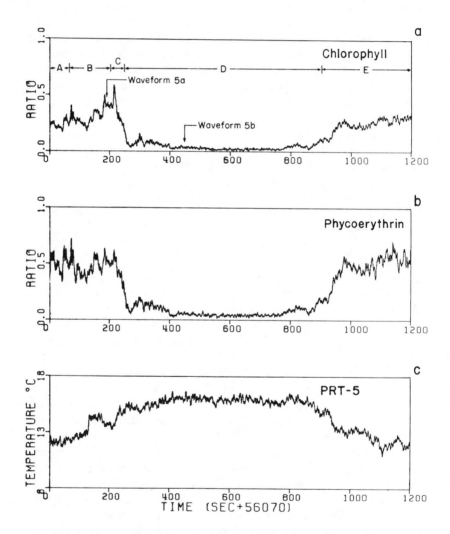

Figure 6.   Chlorophyll a fluorescence profile obtained along Flight Line 7 of Figure 4 (top). The positions labeled "a" and "b" show where the spectra of Figure 5a and 5b were obtained. Phycoerythrin fluorescence (middle) and temperature (bottom) profiles obtained along Flight Line 7 of Figure 4. The temperature profile was obtained with a PRT-5 radiometer.

and 6b, respectively, as a function of time. A profile of the surface temperature obtained with the PRT-5 IR thermal sensor is given in Figure 6c. All of the data in Figure 6 are 21-point moving averages. These averages correspond temporally to a sample every 3.4 s or spatially to a sample about every 0.34 km. This averaging reduces the higher frequency variations observed in the raw data, and thus provides a clearer picture of the features on the scale of these plots. The sea surface temperature values are approximately correct to within ±1 °C. The chlorophyll a and phycoerythrin fluorescence are expressed as normalized ratios of the raw fluorescence signal divided by the Raman backscatter signal (7–9). Supporting surface-truth chlorophyll a data from points with concurrent airborne and surface observations have not yet become available. The forthcoming surface-truth data will allow refinement of sea surface temperature measurements, as well as conversion of the airborne chlorophyll a ratio into absolute units of concentration for further analysis and scientific interpretation. Previous results from field experiments have shown the AOL chlorophyll a to be linear over a range of 0.1–6.0 μg/L and to have linear correlation coefficients ranging from 0.81 to 0.97 when compared to available surface-truth measurements (8). Preliminary results from discrete chlorophyll determinations furnished from vessels operating within the WCR during April 1982 indicate that the concentrations in and around the ring are within the lower portion of this range. Because no generally accepted method for the extraction and quantification of phycoerythrin has been adopted, the profile of phycoerythrin LIF will likely remain as a normalized ratio of fluorescence signal.

Additional profiles for the remaining chords of the star-sampling pattern are condensed in Figure 7. The general shape of the star-sampling pattern is somewhat distorted, and the center of the star is offset from the center (surface layer) of the WCR; thus these remaining flight lines contain varying amounts of ring and boundary segments. The PRT-5 surface temperature profiles provide good definition of the position of the WCR surface layer within each of the flight lines. The levels of chlorophyll a and phycoerythrin concentrations within the higher temperature core of the ring appear to be relatively low and rather monotonous. Higher concentrations and rather pronounced inflections in the profile records are shown in the ring boundary region. These inflections in the concentration of the photopigments are abrupt and appear to be nearly coincident with the strong thermal gradients associated with the ring boundary. Moreover, on flight lines (Figures 6 and 7a, c, and d) that extend beyond the WCR into surrounding shelf- and/or slope-water masses, a decrease in the level of both chlorophyll a and phycoerythrin photopigment concentrations is evident outside of the boundary region. A similar distribution in the photopigment concentration was observed in the results obtained on a mission flown on April 29, 1982 (9); however, on the third mission flown over warm core Ring 82B on June 24, 1982, the chlorophyll a concentration was higher and

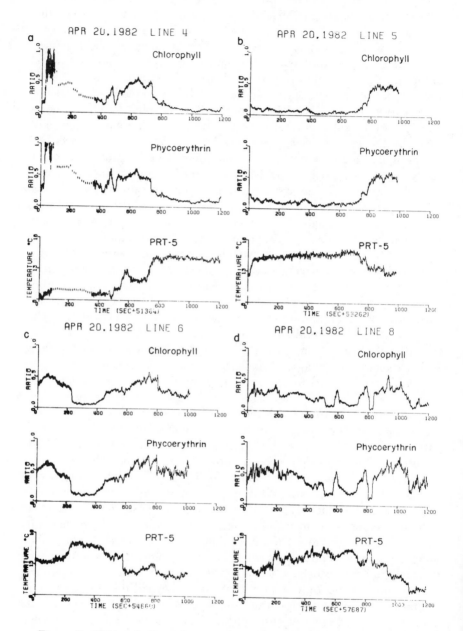

Figure 7. Chlorophyll a, phycoerythrin, and surface temperature along Flight Lines 4, 5, 6, and 8 shown in Figure 4.

more variable within the ring itself than in the boundary region or surrounding water masses.

The general distribution of photopigment concentration can be assessed more clearly in Figures 8 and 9, where contoured projections (made from the five profile cross sections shown in Figures 6 and 7) are provided for chlorophyll a and phycoerythrin, respectively. In the construction of the contours, excellent agreement between the profiles was observed at all

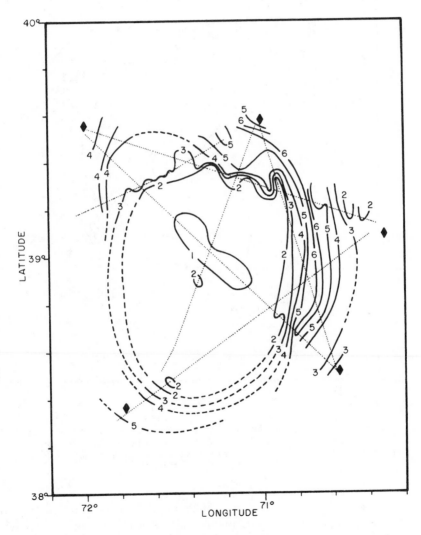

Figure 8. Contour plot of the chlorophyll a fluorescence of warm core Ring 82B. The fluorescence levels are higher in the boundary region, particularly along the eastern wall of the ring.

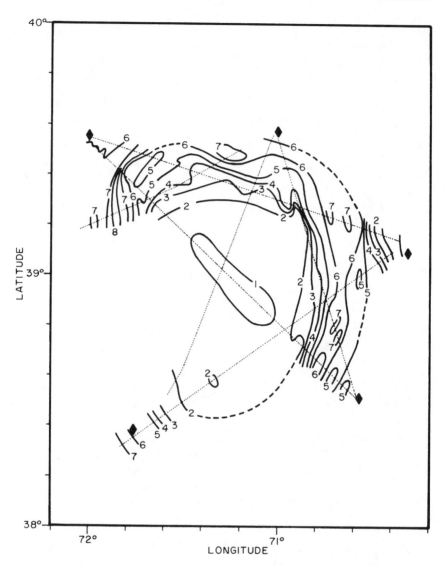

Figure 9. Contour plot of the phycoerythrin fluorescence in warm core Ring 82B on April 20, 1982. The phycoerythrin fluorescence is more uniform along the circumference of the boundary than was the chlorophyll a fluorescence shown in Figure 8.

flight-line crossing points. In Figure 8, chlorophyll a concentrations appear higher in the boundary regions on the eastern side of the ring. No such distinction is evident in the phycoerythrin contour in Figure 9. The inclusion of "streamers" of shelf water into the WCR surface layer in the northwestern and northern quadrants, which were evident in satellite images during mid-April, may have played a role in the development of a chlorophyll a maxima along the western boundary region as these streamers were swept around the ring in a clockwise direction. The elevation in the temperature of this nutrient-rich shelf water would be expected to promote increased phytoplankton productivity. However, the airborne data alone do not clarify whether these observations are a snapshot in the early development of a maxima that eventually extended further around the ring, or whether nutrient levels were essentially spent before the boundary waters were circulated into the southeast quadrant. In the latter case, the airborne data would represent a longer term, steady-state image of the chlorophyll a distribution.

A general coherence between the chlorophyll a and phycoerythrin profiles is evident in Figures 6 and 7. Gross correlation between these photopigments has also been observed in data acquired on the April 29, 1982 flight over warm core Ring 82B and on a flight conducted over warm core Ring 81G on May 15, 1981. Although some small-scale spatial variability in correlation between the photopigments can be seen in Figures 6 and 7, considerably more variability was observed in the mission flown over warm core Ring 81G (*9, 18*). There, small-scale spatial variability was often abrupt and pronounced, and patches of each photopigment were sometimes completely offset from one another.

The relative proportion of chlorophyll a to phycoerythrin changes on a larger spatial scale along the transects shown in Figures 6 and 7. This aspect becomes more apparent in Figure 10, which is a scatter diagram of chlorophyll a signal plotted against phycoerythrin signal on Flight Line 7. Every fifth point from the data used in Figure 6 was selected for plotting in Figure 10 to reduce the density of points for ease of interpretation. The overall high degree of positive correlation in the entire ensemble yielded a linear correlation coefficient of 0.95. The points from various sections along the transect plot occurred in discrete clusters. These clusters have been delineated by ovals in Figure 10 and labeled with "A"–"E." The locations of these segments are provided in Figure 6a by the corresponding sequence of labels. The linear correlation coefficients (and regression slopes) for the individual segments are also given in Figure 10.

These differences in the relative proportions of chlorophyll a and phycoerythrin concentrations are due to differences in phytoplankton or cyanobacteria assemblages. Currently, no explanation is offered to evaluate the observed differences in terms of species composition.

Conclusions

The results indicate that during mid-April the chlorophyll a and phycoerythrin photopigment concentration and variability in warm core Ring 82B were higher in the ring boundary region than in either the adjacent

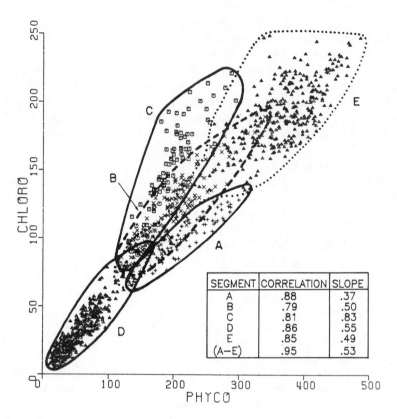

SEGMENT	CORRELATION	SLOPE
A	.88	.37
B	.79	.50
C	.81	.83
D	.86	.55
E	.85	.49
(A–E)	.95	.53

Figure 10. Scatter diagram of phycoerythrin and chlorophyll a fluorescence. The diagram was constructed by using 532-nm excitation. Very high coherency existed at most points along Flight Line 7. Changes in the ratio of chlorophyll a to phycoerythrin can be seen for segments "A"–"E," the locations of which are given in Figure 6a.

shelf- or slope-water masses. The levels of the photopigments were especially low and monotonous in the high-temperature region of the ring core. Contoured projections of the data collected on the five flight lines showed chlorophyll a maxima along the eastern wall of the WCR. However, no similar maxima were found for phycoerythrin. The formation of the elevated region of chlorophyll a concentration is perhaps related to increased productivity from streamers of shelf water being drawn into the WCR in the northwest quadrant. More precise interpretation of these observations is difficult from the single airborne data set alone. General coherence between chlorophyll a and phycoerythrin was found; however, smaller scale variations and district differences in the ratio between these photopigments were also observed.

An explanation for the variation in concentration levels between the chlorophyll a and phycoerythrin photopigments shown in the preceding section will require further analysis using forthcoming data obtained from the participating vessels and some future research to understand the role of phycoerythrin in marine phytoplankton. Chlorophyll a is both widely used and accepted in studies of marine phytoplankton; however, phycoerythrin is not. Several aspects related to the airborne laser-induced phycoerythrin observations need further clarification. In particular, measurements of the photopigment concentration and identification of the species responsible for the phycoerythrin LIF need to be made from a supporting vessel. Progress has been reported in several relevant areas of investigation. These areas include studies of marine phycoerythrin-bearing organisms (*19*), a proposed method for extracting and quantifying the photopigment concentration (*20*), and experiments designed to assess the spectral characteristics of phycoerythrin fluorescence in certain marine and estuarine phytoplankton (*21*).

Acknowledgments

We wish to extend our personal thanks to the many persons associated with the warm core ring experiments and with the AOL project. In particular, we are indebted to Jack L. Bufton and the Instrument Electro-Optics Branch for the loan of the frequency-doubled YAG laser. We also thank Wayne E. Esaias and the Oceanic Processes Branch of NASA Headquarters for their assistance and encouragement in various aspects connected with these experiments.

Literature Cited

1. Kim, H. H. *Appl. Opt.* 1973, *12* (7), 1454.
2. Bristow, M.; Nielsen, D.; Bundy, D.; Furtek, R.; Baker, J. "Airborne Laser Fluorosensing of Surface Water Chlorophyll a," Report No. EPA-600/4-79-048, Environmental Protection Agency, Las Vegas, 1979.
3. Bristow, M.; Nielsen, D.; Bundy D.; Furtek, R. *Appl. Opt.* 1981, *20* (*17*), 2889.
4. Farmer, F. J.; Brown, C. A., Jr.; Jarrett, O., Jr.; Campbell, J. W.; Staton, W. *Proc. 13th Int. Symp. for Remote Sensing of Environ.* 1979, *3*, pp. 1793–805.
5. Jarrett, O., Jr.; Esaias, W. E.; Brown, C. A., Jr.; Pritchard, E. B. *NASA Conf. Proc.* 1980, Document No. 2188.
6. Schink, D.; McCarthy, J.; Joyce, T.; Flierl, G.; Wiebe, P.; Kester, D. *Trans. Am. Geophys. Union* 1982, *63*, 834.
7. Hoge, F. E.; Swift, R. N. *Appl. Opt.* 1981, *20*, 3197.
8. Hoge, F. E.; Swift, R. N. *NASA Conf. Proc.* 1980, Document No. CP2188.
9. Hoge, F. E.; Swift, R. N. *Appl. Opt.* 1983, *22*, 2272–81.
10. Hoge, F. E.; Swift, R. N. *Appl. Opt.* 1980, *19*, 3269.
11. Croswell, W. F.; Fedors, J. C.; Hoge, F. E.; Swift, R. N.; Johnson, J. C. *IEEE Trans. Geosci. Rem. Sens.* 1983, *GE-21*, Nos. 1, 2.

12. Hoge, F. E.; Swift, R. N. *Appl. Opt.* **1981**, *20*, 1191.
13. Hoge, F. E.; Swift, R. N. *Int. J. Rem. Sens.* **1982**, *3 (4)*, 475.
14. Backus, R. H.; Flierl, G. R.; Kester, D. R.; Olson, D. B.; Richardson, P. L.; Vastano, A. C.; Wiebe, P. H.; Wormuth, J. *Science (Washington, D.C.)* **1981**, *212 (4499)*, 1091.
15. Wiebe, P. H.; Hulbert, E. M.; Carpenter, E. J.; Jahn, A. E.; Knapp, G. P., III; Boyd, S. H.; Ortner, P. E.; Cox, J. L. *Deep-Sea Res.* **1976**, *23*, 695.
16. Hoge, F. E.; Swift, R. N.; Frederick, E. B. *Appl. Opt.* **1980**, *19*, 871.
17. Poole, L. R.; Esaias, W. E. *Appl. Opt.* **1982**, *21 (20)*, 3756.
18. Hoge, F. E.; Swift, R. N. *Trans., Am. Geophys. Union* **1982**, *63 (3)*, 60.
19. Moreth, C. M.; Yentsch, C. S. *Limnol. Oceanogr.* **1970**, *15*, 313-17.
20. Stewart, D. E. "A Method for Extraction and Quantitation of Phycoerythrin from Algae"; 1982, NASA Contractor Report No. 165966.
21. Exton, R. J.; Houghton, W. M.,; Esaias, W.; Haas, L. W.; Hayward, D. *Limnol. Oceanogr.* **1983**, *28*, 1225-31.

RECEIVED for review July 5, 1983. ACCEPTED March 7, 1984.

Application of Satellites to Chemical Oceanography

EUGENE D. TRAGANZA

Department of Oceanography, Naval Postgraduate School, Monterey, CA 93943

Of the sensors now on satellites, none has been directly useful to chemical oceanography; however, concentrations of any chemical species that covary with temperature can be inferred from satellite measurements of temperature. This relationship is illustrated with satellite IR imagery off central California where large-scale changes in physical, chemical, and biological properties can occur and are frequently focused along frontal boundaries of upwelling systems. High inverse correlations of temperature with normally nonconservative plant nutrients, nitrate and phosphate, are used to calibrate satellite IR images as chemical maps yielding standard deviations of 14 and 7%, respectively. Chemical maps with satellite-derived maps of chlorophyll reveal a regional relationship between phytoplankton and chemical structure in the California coastal zone and the importance of chemical fronts as sites of chemical exchange and primary production.

Sᴇɴsᴏʀs ᴄᴜʀʀᴇɴᴛʟʏ ᴏɴ sᴀᴛᴇʟʟɪᴛᴇs have little if any conceivable direct chemical oceanographic capability (1). However, chemical properties may be inferred from the other parameters measured. Improvement of IR sensors on satellites has enabled accurate analysis of sea surface temperature and the detection of thermal patterns associated with currents and upwelling at the ocean's surface. Hypothetically, the concentration of any chemical species that covaries with temperature could be related to satellite measurements of temperature.

Temperature and plant nutrients (nitrate, phosphate, and silicate) were found to be (2–4) inversely correlated in the California Current system. Zentarra and Kamykowski (3) found latitudinal relationships along the west coast of North and South America that could conceivably supplement remote-sensing data with a climatological estimate of the median and

0065-2393/85/0209-0373$06.00/0

range of nutrient concentration (3). According to these authors, the least nutrient variability at a temperature may occur in 10° geographic bands at 35° N and 35° S. These areas are the central regions of the California Current and Peru Current "where the water masses are most uniformly distributed" (3). However, the relationship between temperature and plant nutrients can be even closer in new waters that upwell at the coast (4). Studies by Traganza et al. (4–6) between 1978 and 1980 off Point Sur, California 36° N, 121.5° W) show persistent inverse correlations of nitrate and phosphate with temperature in spite of the nonconservative nature of these elements. These studies show that analytical predictions of nitrate and phosphate distributions can be made from satellites in regions of strong upwelling, inferred estimates of nitrate and phosphate can be used to map the "chemical mesoscale" of a coastal upwelling region, and "chemical fronts" are important sites of nutrient exchange and phytoplankton production.

Methods

The data and procedures described here deal with a June 9–11, 1980 cruise (6). Some of these procedures have been superseded. In particular, image processing has changed with the development of new algorithms and software. This change will be noted; however, in principle the procedures are the same.

Capture of IR Data from Satellites. IR data are broadcast continually from the National Oceanic and Atmospheric Administration (NOAA) satellites and are recorded on a selective basis for later transmission to ground stations (7). Satellite data-receiving facilities used in these studies along the U.S. west coast include the NOAA field station at Redwood City, California; the Scripps Satellite Oceanographic Facility (SSOF) at La Jolla, California; and the NOAA field station in Anchorage, Alaska. IR data are available at high (~1.1 km) resolution called local area coverage (LAC) when played back, or high picture resolution transmission (HPRT) when transmitted in real time. These satellites are in view of a particular antenna (horizon to horizon) for about 16 min depending on the altitude. Ground coverage "viewed" by the satellite is a swath of about 2000 km (8). The television IR observational satellite (TIROS-N) series, which began in October 1968, operates in near polar (98° inclination), sun-synchronous orbits. The period is about 102 min, which produces 14.2 orbits per day and permits two overhead passes per day (night and day) at each ground station. Because the number of orbits per day is not an integer, the suborbital tracks do not repeat on a daily basis. Successive views of the same area on earth are made from different angles, and hence the swaths are somewhat, but systematically, displaced on each pass. The local solar time (LST) of each pass is essentially unchanged for any latitude (i.e., sun-synchronous). Orbital information for each orbit, including orbit number, date, and time, is referred to as the ephemeris.

A major improvement introduced with TIROS-N was the advanced, very high-resolution radiometer (AVHRR). The AVHRR is a cross-track scanning radiometer with five spectral channels: 0.55–0.90, 0.725–1.10, 3.55–3.93, 10.5–11.5, and 11.5–12.5 μm. The instantaneous field of view (IFOV) of each sensor is about 1.4 mrad. This IFOV leads to a resolution at the satellite subpoint of ~ 1.1 km (the size of each picture element or pixel increases away from the nadir because of the earth's curvature and the incidence angle). The scanning rate is 360 scans/min. A total of 2048 samples (pixels) is obtained per channel per earth scan. The analog data from the sensors are digitized on board and are transmitted with calibration coefficients that are used to convert the raw sensor counts into calibrated radiance values. These data are stored by the receiving station on magnetic tape.

After extracting an image such as 512 lines × 512 samples, counts are converted into radiance values by applying the calibration coefficients. The image is geometrically corrected, and geographic coordinates are assigned to each pixel. Radiance values are then converted to temperature by inverting the Planck function. To increase the contrast or enhance the image, the temperature range is "stretched" over the available count range, 0–255 for 8-bit data and 0–1023 for 10-bit data, so that the thermal resolution is ~ 0.5 or 0.2 °C, respectively. The temperature as measured at the satellite is corrected to take into account the attenuation and addition of thermal radiance by the atmosphere. Thermal IR is absorbed by water vapor and thus requires a correction of ~ 1–10 °C. Aerosols require a correction of ~ 0.1–1 °C, and carbon dioxide and ozone require a correction of ~ 0.1 °C. With 10-bit data, multichanel algorithms (9, 10) can be used to extract sea surface temperature from satellite data to within 0.5 °C of true surface temperature. However, the truncated 8-bit data available for this study were not precise enough to use these algorithms. Therefore, in situ measurements of sea surface temperature gradients were used to derive regional atmospheric corrections. Following Maul et al. (11), temperature as measured at the satellite was corrected to within 1 °C of true surface temperature.

Underway Analysis of the Sea Surface.   All in situ measurements are made while underway to minimize the problem of comparing data collected almost instantaneously by the satellite with time-integrated data collected by the ship. In the June 1980 study, measurements were made at 9 knots, but the speed decreased as weather and high seas increased. Thermal calibrations of the satellite images were based on ship data collected nearest in time to the satellite overpass.

Nutrients were measured in a Technicon AA-II Autoanalyzer every 2 min on a continuous flow of seawater pumped to the instrument. A 2.54-cm inside diameter (ID) nylon hose with the intake at 2.5 m was led on board to a deck-mounted, air-operated double diaphragm pump (Wilden Pump and Engineering Company, Model M-4, Colton, CA). A debubbler was used in line between the pump, the Autoanalyzer, and a fluorometer (Turner Designs III), which measured fluorescence continuously. Fluorescence was calibrated against discrete chlorophyll measurements. Temperature was sensed by a thermistor located at 2.5 m in the ship's sea chest and recorded continuously. Details of these procedures and calibration of satellite visible imagery as "chlorophyll" distribution were reported previously (4–6).

Temperature Algorithm. IR radiation from the ocean originates in the "skin" or upper fraction of a millimeter of the sea surface. Some IR radiation is absorbed in the atmosphere before it reaches the satellite. Some IR radiation is added by thermal emission of the atmosphere along the propagation path from the sea surface to the satellite radiometer (Figure 1). The algorithm for obtaining sea surface temperature from radiance measured at the satellite, therefore, must include corrections for "atmospheric transmittance" and "path radiance." In theory, absolute measurements of sea surface "skin" temperature can be obtained from solutions of the radiative-transfer equation, $L(T_\lambda) = L_o(T_s)\tau(p_o) + L_*$, where: $L(T_\lambda)$ is the radiance of wavelength measured by the satellite radiometer in watts per steradian per square meter per micron at temperature T; T_λ is the temperature obtained by inverting Planck's function for the measured value of radiance in Kelvins; $L_o (T_s)$ is the total radiance emitted from the surface at temperature T_s in watts per steradian per square meter per micron; $\tau(p_o)$ is the atmospheric transmittance between the pressure levels 0 (top of the atmosphere) and p_o (sea level); and L_* is the path radiance due to the isotropic thermal emission of photons by the atmosphere along the propagation path from the sea surface to the radiometer in watts per steradian per cubic meter.

To derive the parameters τ, the atmospheric transmittance, and L_*, the path radiance, linear regression analyses were performed on radiances from the satellite measurements, L, and radiances derived from shipboard measurements of temperature, L_o, versus elapsed distance, X, along selected parts of the track. These track segments were selected nearest in time to the satellite overpass and where temperature gradients were large enough to minimize the relative errors due to satellite-measured thermal noise and the lower AVHRR temperature resolution of 0.5 °C in the 8-bit data from Redwood City.

The ratio of the slopes of the regression lines m_1/m_2 gives the transmittance τ. If the sea surface temperature gradient is much larger than the moisture gradient, that is, $| \Delta L_o | \gg | \Delta L_* |$, then by the radiative-transfer equation, $\tau = \Delta L/\Delta L_o$, and $\tau = m_1/m_2$ (11).

The intercept difference in the regression line gives the path radiance L_*. Radiance at the satellite and sea surface is obtained from the intercepts, L' and L_o', of the regression lines, for satellite and in situ radiance versus distance along the cruise track, namely, $L = m_1X + L'$, and $L_o = $

$m_2X + L_o'$. These values are substituted in the radiative-transfer equation to obtain path radiance, that is, $L_* = (L' + m_1X) - (L_o' + m_2X)m_1/m_2$, and $L_* = L' - \tau L_o'$. When both transmittance and path radiance are known, sea surface radiances as measured at the satellite are converted to sea surface radiances by the radiative-transfer equation. These radiances in turn are converted to temperatures through the Planck function (6).

Temperature Maps.   The Point Sur upwelling scene was selected from the satellite image, magnified, and "stretched" so each radiometric count was a difference of ~0.5 °C. The scene was reproduced as a digital printout of radiometric counts, and a computer program (12) was used to locate the cruise track on a line-by-pixel basis. Regression lines were computed for satellite radiance and sea surface radiance versus distance along the cruise track. For example, on June 9 these elapsed distances were 4–16 km, 30–60 km, and 64–83 km (Figure 1). From the slopes and intercepts, τ and L_* values were computed from the algorithm to yield a mean transmittance of 0.693, and a mean path radiance of 2.15 W/m$^2 \cdot$Sr$\cdot \mu$m ^{3}. Satellite radiances were corrected to sea surface radiances by using these mean values in the radiative-transfer equation, and then they were converted to sea surface temperature through the Planck function.

Temperature maps were produced by overlaying each satellite image on a navigation chart with a zoom-transfer optical scope. (Algorithms now available at SSOF place the coastline and lines of latitude and longitudes directly on the image.) Each radiometric count was assigned a color to make it possible to easily draw isotherms between pixels at a 0.7 °C temperature interval (the resolution after atmospheric correction). Each isotherm temperature was assigned from the average between pixel temperature values on each side of the isotherm.

Nutrient Maps.   Regression equations (6) for nitrate and phosphate against temperature were used to convert the temperature maps to nutrient maps. Each isopleth is the average value of the pixel nutrient concentrations on each side.

Results

Figures 2 and 3 are IR images of the eastern north Pacific Ocean taken by the NOAA-7 satellite on June 9 and 11, 1980. The development of a cyclonic upwelling system is clearly evident off Point Sur, California. This system formed in a few days and became a tightly curled cyclonic feature in less than 48 h (Figures 2 and 3). These features are typically 50-100 km large and persist for 10 days or more. The strong inverse correlation of nutrients and temperatures along the cruise track is evident in Figures 4 and 5. Figure 6 shows a magnified and enhanced view of the cold surface feature off Point Sur in an early phase in which it was similar in shape to the local bottom topography along the break in slope at 150 m. In the next 36–

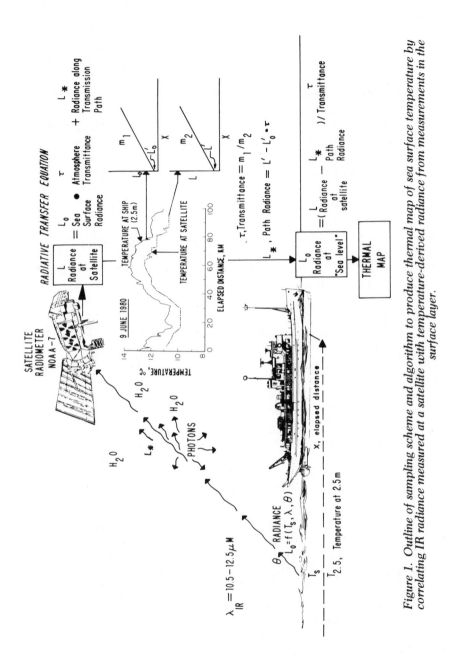

Figure 1. *Outline of sampling scheme and algorithm to produce thermal map of sea surface temperature by correlating IR radiance measured at a satellite with temperature-derived radiance from measurements in the surface layer.*

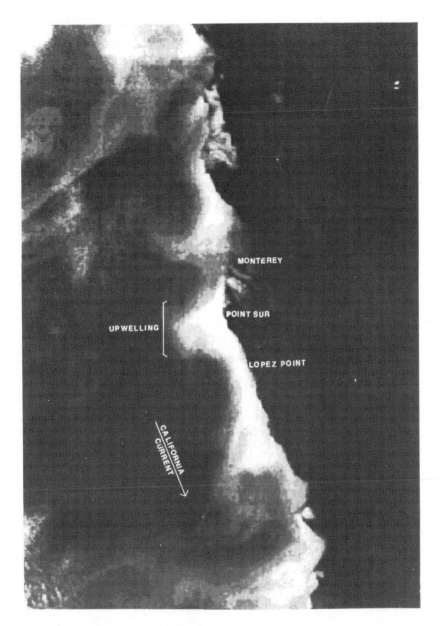

Figure 2. Upwelling system off Point Sur, California is evident in NOAA-7 satellite IR image from June 9, 1980. Gray shades represent ~0.5 °C (lightest = 8 °C; darkest = 15 °C).

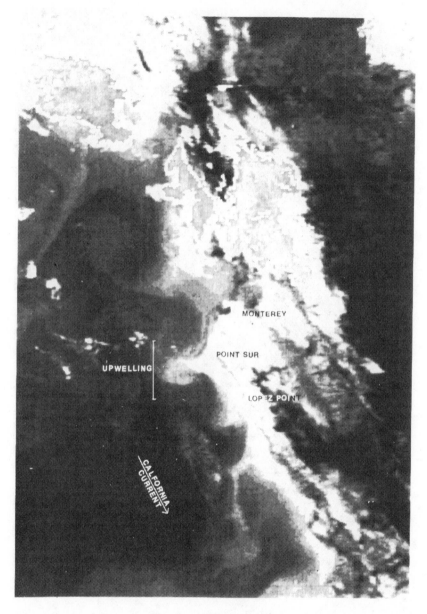

Figure 3. Upwelling system with cyclonic appearance is evident in NOAA-7 satellite IR image from June 11, 1980. Gray shades represent ~0.5 °C (lightest = 7 °C; darkest = 15 °C).

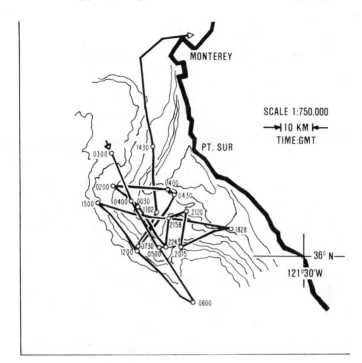

Figure 4. Track of the June 9–11, 1980 cruise and outline of upwelling system as interpreted from satellite IR image from June 9, 1980.

48 h the feature became cyclonic (Figure 3) with its equatorward edge over the axis of the Sur Canyon. Water of 8.7 °C appeared nearshore (Figure 6, Point A), moved seaward (while warming) toward the center of the feature, and was replaced with colder (7.39 °C) water at its original location (6). At the same time, nitrate and phosphate concentrations increased from 18.85 and 2.40 μM to 26.17 and 2.72 μM, respectively. Relationships between nutrients and temperature were obtained from linear regressions (Figure 7) and were used to convert temperature maps (Figure 8) to nutrient maps (Figure 9) of the Point Sur upwelling center.

Biomass was highest (6.25 mg of chlorophyll/m³) in the frontal transition zone on the equatorward side (Figure 8), lower on the poleward side, and very low (< 0.5 mg of chlorophyll/m³) in low-temperature water inside the feature and in the warm (low nutrient) water outside its frontal boundary. During formation, the upwelling system moved faster than the ship so that chlorophyll contours do not exactly represent the area (cross hatched) in which phytoplankton were concentrated. Satellite information showing the actual "bloom" location was provided when a coastal zone color scanner (CZCS) image (Figure 10) from June 12, 1980 was processed

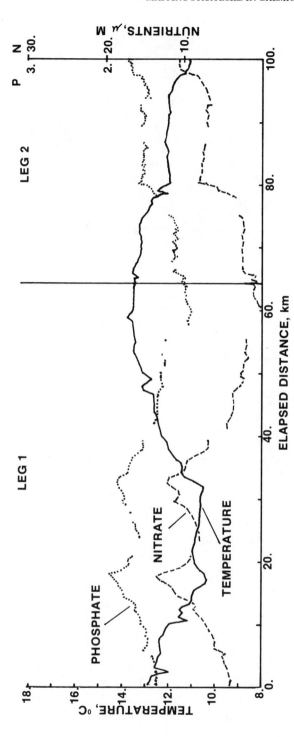

Figure 5a. Nitrate, phosphate, and temperature at 2.5 m versus elapsed distance along the track of the June 9–11, 1980 cruise for Legs 1 and 2.

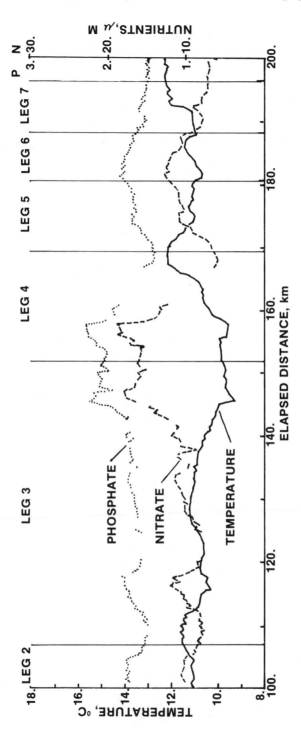

Figure 5b. Same as in Figure 5a for Legs 3–7.

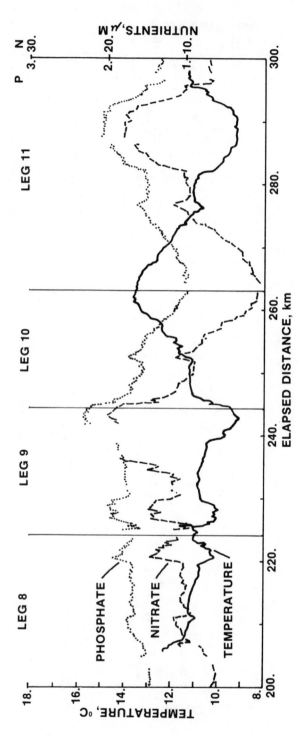

Figure 5c. Same as in Figure 5a for Legs 8–11.

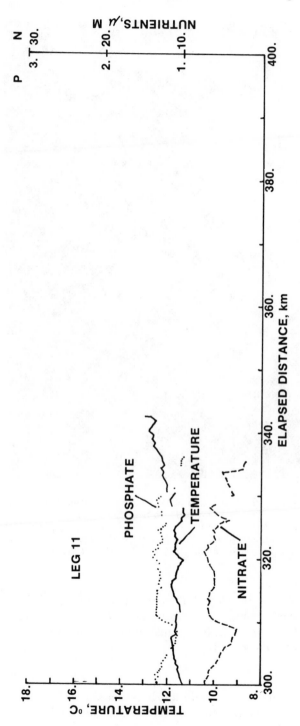

Figure 5d.   Same as in Figure 5a for Leg 11.

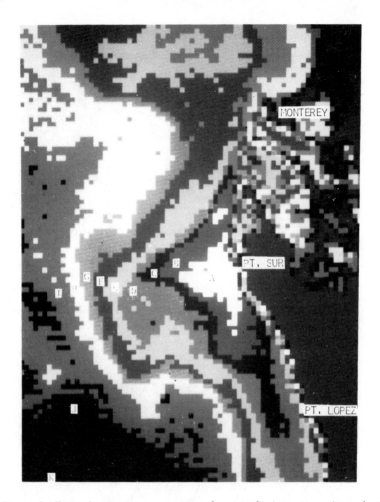

Figure 6. First of a two image sequence showing the formation of a cyclonic upwelling system off Point Sur, Calif., June 9, 1980. Letters associated with gray shades represent average pixel temperatures (pixel = ~1.1 km², the resolution of the image), A–K, 8.57–14.92 °C with a resolution of ~0.7 °C. (Reproduced with permission from Ref. 6. Copyright 1983, Plenum Press.)

by using a previously developed algorithm (*13*). When the IR and color images are viewed together they strongly suggest a regional relationship between the distribution of phytoplankton and the chemical structure of the central California coastal zone.

Discussion

Temperature.   The close correlation between sea surface temperature (2.5 m) and surface temperature as measured at the satellite is evident in

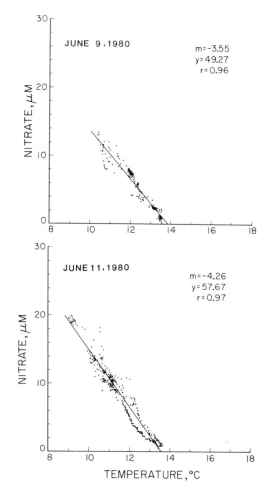

Figure 7. Nitrate vs. temperature with regression line from data collected June 9–11, 1980. (Reproduced with permission from Ref. 6. Copyright 1983, Plenum Press.)

Figure 1. On the basis of this observation, and in the absence of "skin" (< 1 mm) temperature measurements, sea surface radiance was calculated from 2.5-m temperatures and used in the algorithm to correct "satellite temperatures" to sea surface temperatures. This approach worked well in this study because the relative distribution of 2.5-m temperatures was correlated with the relative distribution of skin temperatures. This conlcusion was supported in two ways: (1) the mean difference between surface temperature measured in samples collected in a bucket and the thermistor at 2.5 m was 0.1 °C, and (2) when 2.5-m temperatures were compared with cor-

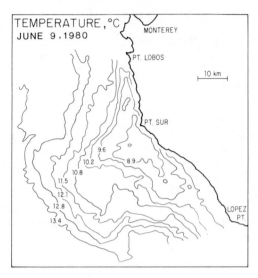

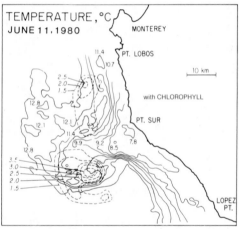

Figure 8. Sea surface temperature maps (°C), inferred from satellite IR images. Contour interval is one radiometric unit ($\simeq 0.7\,°C$). Phytoplankton pigments (hatched area) are shown in June 11 figure with a contour interval of $0.7\ mg/m^3$. (Reproduced with permission from Ref. 6. Copyright 1983, Plenum Press.)

rected "satellite temperatures" from sections of the cruise track not used to formulate the algorithm, they yielded a standard deviation of 0.5 °C.

This agreement is rather good considering it includes errors due to the ocean, the atmosphere, the AVHRR, the satellite navigation procedure, the ship's thermistor, and the ship's navigational ability. The main reasons for this agreement are that the area was cloud-free, and little advection occurred in the interval between ship measurements and satellite measure-

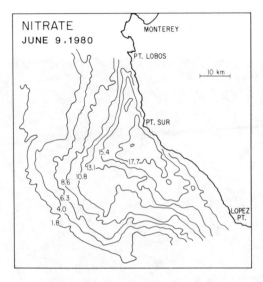

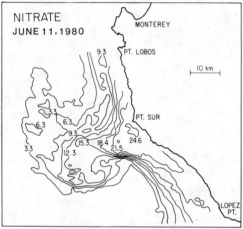

Figure 9. Sea surface nitrate maps generated from correlation with sea sur-face temperature. Contour interval is one radiometric unit (≃ 0.7 °C); iso-pleths are in micromolar values. (Reproduced with permission from Ref. 6. Copyright 1983, Plenum Press).

ments. When advection occurs, if the satellite overpass is not close in time to the ship's measurements, the navigational error between the ship and the satellite could easily swamp all other error.

Nutrients.   With an error of 0.5 °C (standard deviation) in a thermal map and by using the nutrient–temperature regression equations, expected errors on the order of 1.7 and 0.1 μM for nitrate (mean ≃ 12 μM), and phosphate (mean ≃ 1.5 μM), respectively, can be calculated. These are er-

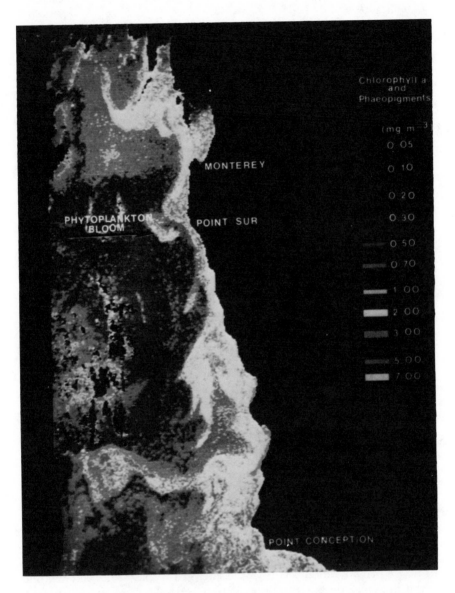

Figure 10. Phytoplankton pigment distribution in gray shades after original color-coded ocean color image. (Reproduced with permission from Ref. 6. Copyright 1983, Plenum Press)

rors of 14 % for nitrate and 7 % for phosphate. When nutrients and temperature are less well correlated, the overall errors will exceed these values (6).

The slope and the degree of covariance of nutrients with temperature may differ in different upwelling systems or in the same system at different times. These differences depend on the initial element concentration in the source water, upwelling velocity and depth (a function of wind strength and duration), and changes that occur by advection, diffusion, mixing, heat transfer, salt transfer, and biological activity. A seasonal variation in upwelling and source water occurs because of climatic change over the year and the influx of different water masses. The June 1980 data show that even with such variability the distribution of nutrients in a strong upwelling system can be inferred from satellites given a minimum set of in situ temperature and nutrient values.

Conclusions

Although existing coastal upwelling theory is unable to predict upwelling distributions or details of upwelling circulation, satellite IR data can be used directly and by inference to show the distribution of temperature and nutrients in an upwelling region. The upwelling systems in the California coastal zone appear to evolve in a sequential series of stages that can be used as a convenient classification.

1. Upwelling, favorable winds occur.
2. "Coastal upwelling bands" follow the coastline and local bathymetry.
3. "Cold wedges" develop and reflect the local bottom topography.
4. Wedges grow into "plumes" at capes and points.
5. Plumes develop as "surface jets" extend seaward tens to hundreds of kilometers, "cyclonic upwelling systems," which form equatorward and poleward of capes and points by interacting with the California Current (possibly eddies—submarine canyons appear to be an influence) or "anticyclonic upwelling systems," which form in a manner similar to cyclonic systems.

As the season progresses a broad band of upwelling develops with many of the structures just outlined embedded or extending from its seaward edge.

Satellites have firmly established that such systems exist (5), and that the formation of upwelling systems is an important process in determining the distribution and production of primary biomass in the California Current system (6). "Nutrient cells" (4) that were apparent in previous data from conventional surveys are explained, as are the sharp thermal and chemical gradients frequently encountered but not identified as the frontal boundaries of upwelling systems. The concentrations of phytoplankton seen along frontal boundaries, when IR and ocean color images are viewed together, are strong evidence that chemical fronts are important sites of chemical exchange and primary production. Whether or not a chemical exchange occurs at fronts is an important question that extends to a variety of fronts in all oceans. Mixing is vigorous at fronts, and instabilities can occur. However, mechanisms for nutrient entrainment across the sloping pycnocline, which characterizes frontal boundaries, have not been demonstrated. Chemical front studies (unpublished data) show localized upwelling and downwelling structure in the sloping pycnocline near upwelling fronts in association with large chlorophyll concentrations and very high phytoplankton-specific growth rates. Sustained upwelling and cross-frontal mixing could explain the persistence of the nutrient–temperature correlation and high primary production observed at these fronts. Experiments are in progress to test this hypothesis by the development and application of an atmospherically coupled biophysical model.

Literature Cited

1. Andersen, N. R., "The Potential Application of Remote Sensing in Chemical Oceanographic Research"; Natl. Sci. Found: Washington, D.C., 1980; 98 pp.
2. Strickland, J. D. H.; Solarzano, L.; Eppley, R. W. *Scripps Inst. Oceanogr. Bull.* 1967, *17*, 1–22.
3. Zentarra, S. J.; Kamykowski, D. *J. Mar. Res.* 1977, *35(2)*, 321–37.
4. Traganza, E. D.; Nestor, D. A.; McDonald, A. K. *J. Geophys. Res.* 1980, *85*, 4101–6.
5. Traganza, E. D.; Conrad, J. C.; Breaker, L. C. In "Coastal Upwelling"; Richards, F. A. Ed.; American Geophysical Union: Washington, D.C., 1981; pp. 228–41.
6. Traganza, E. D.; Silva, V. M.; Austin, D. M.; Hanson, W. E.; Bronsink, S. H., In "Coastal Upwelling: Its Sediment Record"; Suess, E.; Thiede, J., Eds.; Plenum: New York, 1983; pp. 61–83.
7. Schwalb, A. "The TIROS-N/NOAA A-G Satellite Series," National Oceanic and Atmospheric Administration; Washington, D.C., 1978, No. 95.
8. Nagler, R. G., In "Satellite Applications to Marine Technology"; Williams, F. L., Ed.; NASA: Washington D.C., 1977; pp. 157–66.
9. Bernstein, R.L. *J. Geophys. Res.* 1982, 87, 9455–65.
10. McClain, P.E. In "Oceanography from Space"; Gower, J.F.R., Ed.; Plenum, New York, 1981; pp. 73–85.
11. Maul, G.A.; de Witt, P.W.; Yanaway, A.; Baig, S.R. *J. Geophys. Res.* 1978, *83*, 6123–35.
12. Lundell, G., M.S. Thesis, Naval Postgraduate School, Monterey, Calif., 1981.
13. Gordon, H. R.; Clark, D. K.; Mueller, J. L.; Hovis, W. A. *Science* 1980, *210*, 60–63.

RECEIVED for review September 6, 1983. ACCEPTED April 23, 1984.

pH–Temperature Relationships in the Gulf of California

ALBERTO ZIRINO and STEPHEN H. LIEBERMAN

Marine Environment Branch, Code 522, Naval Ocean Systems Center, San Diego, CA 92152

The potential of a glass–reference combination electrode immersed in a pumped stream of seawater was measured while underway in the Gulf of California. The pH at the surface (sampling) temperature was computed from the temperature dependencies of the electrode and buffers. For most of the Gulf, pH and temperature showed a high degree of positive correlation. However, in the warmer waters of the southern Gulf and on spatial scales of less than 20 km, negative correlations were also observed; pH maxima were associated with temperature minima and chlorophyll a maxima. These observations are discussed in terms of oceanographic processes. The positive pH–temperature correlation is probably caused by mixing processes and the cumulative biological depletion of CO_2 from surface waters.

A DIRECT RELATIONSHIP EXISTS BETWEEN THE pH values of the ocean's surface and the potential exchange of CO_2 across the air–sea interface (1). Although the literature contains many discussions about the possible values of the surface pH (1–4), few measurements have been made. Studies of pH and CO_2 relationships across the northern Pacific Ocean have revealed that pH, pCO_2, and temperature correlated closely (5). The pCO_2 (and by inference, pH) was seldom at a value close to that suggested by equilibrium considerations; biological and physical processes strongly affected pCO_2 at the surface.

With a simple underway technique for measuring pH, a strong correlation previously was found between pH, coastal upwelling, and the resulting planktonic production (6–8). However, discussions of these measurements were limited to changes that were greater than 0.1 pH units because, previously, pH was measured electrometrically and the electrometric measurement of pH may involve an absolute error of 0.1 pH units (9); thus, no attempt was made to look at the finer structure.

0065-2393/85/0209-0393$06.00/0

This chapter demonstrates that pH may be measured continuously while underway to 0.01 units, and that pH–temperature relationships can be interpreted in terms of known oceanographic processes. Data from the Gulf of California are used to illustrate these points.

Experimental

All measurements were made aboard the USNS *DeSteiguer* during the VARIFRONT III expedition of the Naval Ocean Systems Center to the Gulf of California. The expedition was carried out during November and December of 1981 in conjunction with scientists from the Centro de Investigaciones Cientificas y Escuela Superior de Ensenada (CICESE). Measurements were made while underway at approximately 10 knots during northerly and southerly transects of the Gulf. The data presented here deal only with the southerly transects shown in Figure 1 as A–B.

pH Measurements. The pH of a flowing stream of seawater was measured in the ship's laboratory. The stream originated from a conduit mounted in the forward transducer well with the intake approximately 1 m under the bow at the keel. Approximately 50 m of Teflon-lined hose was used in the conduit between the intake and the laboratory. Determinations were made with a Corning 476055 combination electrode mounted in a Teflon manifold (8). A single electrode, in conjunction with a pH meter (Corning Model 103), was used for all the measurements. Initially, the electrode was calibrated in the manifold by placing the appropriate buffers in the Teflon reservoir (Figure 2) and recirculating them past the electrode until a constant millivolt reading was obtained. Two buffer solutions, made up to National Bureau of Standards (NBS) specifications (*10*) (available commercially from Beckman Instruments), were used. These solutions were KH_2PO_4, 8.69×10^{-4} M; Na_2HPO_4, 3.043×10^{-2} M (pH = 7.413 at 25 °C); and borax, 0.01 M (pH = 9.180 at 25 °C). A single calibration was used to compute pH over the 3-day measurement period. The manifold temperature, during calibration and during the measurements of seawater, was determined with a thermistor mounted in the manifold alongside the pH electrode. The thermistor output was read with a YSI (Yellow Springs Instrument Company) Model 47 scanning telethermometer. As an electrical check, the buffer solution was placed in contact with the seawater in the Teflon hose by using Pt wire during the standardization. A constant meter reading signified that the seawater and buffer solution were at the same ground potential.

The pH at the in situ temperature T was calculated from the following expression:

$$pH_{(T)} = \{(E_{cell_{(T)}} - E_{7_{(T)}})/S(T/TC)\} + B_{7_{(T)}} \tag{1}$$

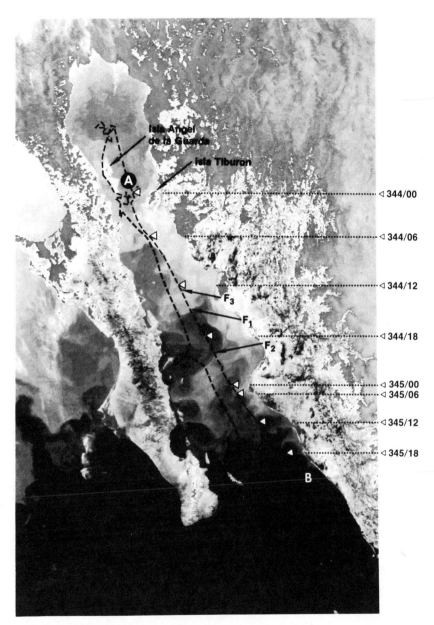

Figure 1. Cruise track of the USNS DeSteiguer *superimposed on an IR image of the Gulf of California. Lighter shades denote colder temperatures.*

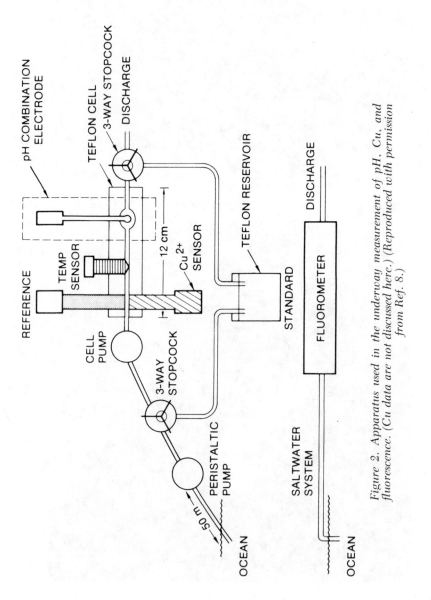

Figure 2. Apparatus used in the underway measurement of pH, Cu, and fluorescence. (Cu data are not discussed here.) (Reproduced with permission from Ref. 8.)

where the ratio T/TC is in Kelvins and B_7 is the NBS determined pH value for the pH 7 buffer at T. $E_{cell(T)}$ is given by

$$E_{cell(T)} = E_{cell(TM)} + 0.27 \text{ mV/deg} (T - TM) \tag{2}$$

Equation 4 defines $E_{7(T)}$, TM is the temperature of the manifold, TC is the temperature during calibration, and 0.27 mV/deg is the whole-cell (inner reference/glass/seawater//outer reference) temperature response of the combination electrode in 34 ppt seawater at 1 atm (*12*).

Similarly, the electrode slope S is given by

$$S = (E_{9(TC)} - E_{7(TC)})/(B_{9(TC)} - B_{7(TC)}) \tag{3}$$

where $E_{7(TC)}$ and $E_{9(TC)}$ are the electrode potentials generated in NBS 7 and 9 buffers at the calibration temperature (TC), respectively, and $B_{7(TC)}$ and $B_{9(TC)}$ are the NBS determined pH values at TC. Finally,

$$E_{7(T)} = E_{7(TC)} + F(T) \tag{4}$$

where $E_{7(T)}$ is the empirically determined temperature-dependent response of the combination electrode in phosphate buffer (pH = 7.413 at 25 °C). Using the method of Flynn (*12*), we found that between 10 and 30 °C the potential generated by this electrode did not differ significantly from $E_{7(TC)}$. In principle, Equations 1–4 should computate the pH, at any temperature T, after calibration with buffers at temperature TC.

Sea Surface Temperature Measurements. Sea surface temperature was measured with a calibrated InterOcean thermistor (0.01 °C) mounted at the conduit under the bow. (Thus, in the figures, recorded temperature slightly precedes recorded pH). Seawater for the determination of chlorophyll a fluorescence was pumped from the sea chest of the USNS *DeSteiguer* and measured separately although in parallel with pH. Fluorescence was measured with a Turner Designs Model 10 fluorometer equipped with a flow-through cell. Chlorophyll a was calculated from fluorescence (*11*) by calibrating the flow-through in vivo fluorescence with discrete acetone-extracted fluorescence measurements of particulate chlorophyll obtained from water samples that passed through the in vivo system. The responses of the electrode, fluorometer, and thermistor were recorded on an inhouse-built data logger. Measurements were recorded every 5 s. Each time (space) series between A and B represents approximately 40,000 individual determinations.

Results

Figure 1 is an IR image of the Gulf of California taken by the National Oceanic and Atmospheric Administration (NOAA) 7 satellite on December

10, 1981, and processed by the National Earth Satellite Service (NESS) at Redwood City, California. In Figure 1, progressively higher temperatures are shown as progressively darker areas. Physically, the Gulf can be divided into two distinct hydrographic regions separated by a shallow sill just south of Angel de la Guarda and Tiburon Islands. The northern, shallower portion is mixed by strong tidal currents with overall lower surface temperatures and slightly higher salinities. Conversely, the hydrographic features of the southern Gulf resemble those of the eastern tropical Pacific Ocean, and show greater stratification, higher surface temperatures, lower surface salinities, and a distinct O_2 minimum from 500 to 800 m. The difference in surface temperatures between the northern and southern portions of the Gulf can be seen in Figure 1, which also shows that the area immediately south of Tiburon Island is a source of cold, upwelled water. This feature, which can be attributed to the combined effect of tidal mixing and shallow topography, is present year-round. Also, from November to April prevailing northerly winds form additional plumes of upwelled water that extend westward of the mainland Mexican coast. Several of these thermal features were crossed by the USNS *DeSteiguer* during its southern transect.

Figure 3 shows a time–space series of $pH(T)$ as computed from Equations 1–4 from the raw data collected during transect A–B. Similarly, the figure also shows plots of chlorophyll a and temperature as functions of cruise time and distance. The x-axis denotes the Julian date and the time of sampling. The straight-line distance from A to B is 850 km; the actual distance traveled was approximately 930 km. Average speed was 18 km/h. Some of the corresponding positions are also indicated in Figure 1 (dotted lines). Frontal areas are denoted by an F.

Temperature in the survey area increased from north to south, going from a low value of 19 °C in the upwelling area to a value of approximately 27 °C at the mouth of the Gulf. All of the major frontal features visible in the IR image appear in the temperature record. Three degree fronts occurred near 344/06 (December 10, 6 A.M.) and at 344/15. Overall, $pH(T)$ follows temperature very closely and increases from a minimum of 8.24 at 110 km south of Tiburon Island (344/06) to approximately 8.45 at the southern portion of the Gulf. On the whole, identifiable major patches with length scales greater than 100 km coincide in temperature and $pH(T)$. Positive correlation between surface temperature and $pH(T)$ is also evident on smaller scales of 20 km or less, but only in the northern, colder portion of the Gulf (to 345/06). In the coldest waters traversed—from 343/23 to 344/06 in Figure 3—correlation is observable on scales of 3 km or less (Figure 4).

On the other hand, in the warmer portion of the Gulf and on space scales of approximately 20 km or less, T and $pH(T)$ are not strongly correlated or are often negatively correlated. Negative correlation is often characterized by a $pH(T)$ maximum appearing in conjunction with a T

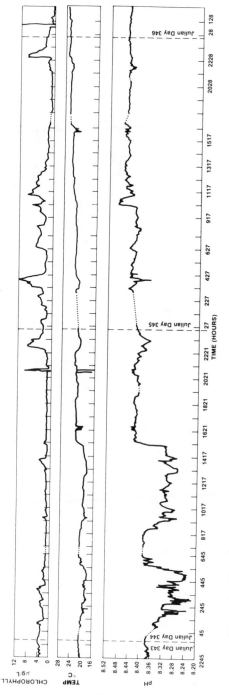

Figure 3. Time (space) series of pH(T) temperature, and chlorophyll a measured in the surface waters of the Gulf of California during December 10–12, 1981.

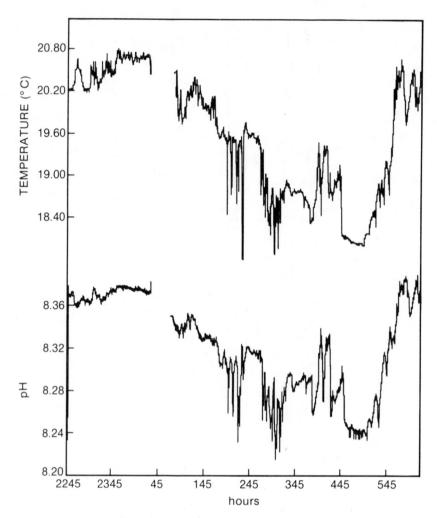

Figure 4. Fine-scale structure of pH(T) and temperature measured on December 10, 1981.

minimum. When pH(T) maxima accompany colder surface waters the pH(T) signal is often matched by a very similar signal in chlorophyll a. Figure 5 shows one such section. Other similar observations are visible in Figure 3 at 344/23, 344/07, and 345/12.

Large-scale correlation does not exist between chlorophyll a and pH(T) or T. Chlorophyll a concentrations, as measured by in vivo fluorescence, average 2–3 μg/L from Point A to the frontal area at 344/06 and decrease to between 1 and 2 μg/L thereafter, to the location at 344/23, where an intrusion of cold water has produced a 1° warm-to-cold front. This upwelled "patch" lies at the beginning of an extensive (330 km) stretch

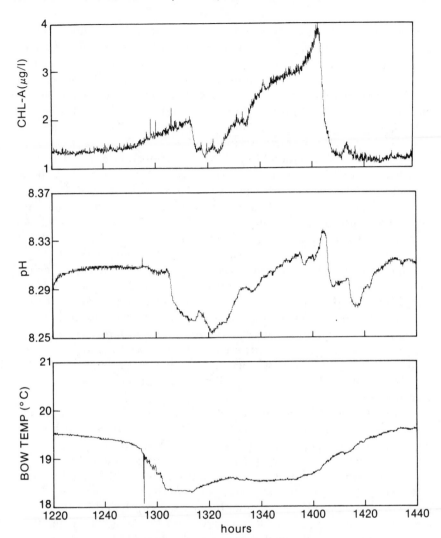

Figure 5. pH(T) and chlorophyll a maxima in association with a temperature minimum (December 10, 1981).

of high chlorophyll a concentration, which averages between 4 and 5 μg/L; individual peak concentrations reach above 9 μg/L. From north to south, the change patches observed in the chlorophyll a concentration is on a considerably greater scale than the 100-km scale changes observable in both temperature and pH(T).

Discussion

Sampling.   The measurement system is able to detect pH changes as small as 0.01 pH units per kilometer (Figure 4). Figure 4 shows changes in pH(T) that have been scaled to resemble changes in T. The high positive correlation between pH(T) and T indicates that the pumping system is adequate for this degree of resolution. Weichart (6) and Simpson (*see* chapter 21) also found that their own pumping sytems were adequate. Additionally, they have both shown that when inorganic phosphate in seawater is measured in parallel with pH, time–space series are obtained in which PO_4^{3-} and pH are highly, albeit inversely, correlated. Thus, little gaseous CO_2 is lost or "smeared" through pumping.

pH(T) vs. T Relationships. The pH–temperature coefficient of seawater (dpH/dT) is not only a property of seawater but a quantity that is dependent on the characteristics of the test (glass) and reference electrodes, as well as on the temperature characteristics of the buffer solutions (10). An average response for the Corning Semimicro (476055) electrode in 34.3 ppt seawater was determined previously (12) in the 10-to-30 °C range to be 0.27 ± 0.03 mV/°C. This value was not sensitive to salinity down to 30.8 ppt, nor to initial pH. When this average value is used to compute pH(T) from Equations 1–4, an average dpH/dT of − 0.012 pH units/°C is obtained (Figure 6). (Actual dmV/dT values increased with decreasing temperature, but the deviations from the mean are small enough to be ignored in this discussion.) This value is in excellent agreement with an average coefficient of − 0.01 [determined by Gieskes (13)] and with Ben Yaakov's computed values of dpH/dT, which were estimated from the temperature dependence of the equilibrium constants of carbonic acid in seawater (14).

From the previous discussion a plot of pH(T) vs. T for the data from transects A–B would be expected to resemble Figure 6. In fact, such a plot possesses a positive slope with dpH(T)/dT = 0.034 with R^2 = 0.88 for a linear fit to the data (Figure 7). Electrode drift during the 3 days of sampling from north to south does not account for this positive trend because other data, taken with the same electrode between December 3 and 5, going from south to north, also plot out linearly with dpH(T)/dT = 0.037 pH units/°C, albeit with less significance (R^2 = 0.55). In addition, the relationship between pH(T) and T demonstrated in Figure 7 is actually a curve whose slope decreases at increasing pH values. This curvature occurs because changes in pH(T) represent actual changes in the hydrogen-ion content of the seawater, and these changes reflect the uptake of CO_2 by primary producers. Our reasoning follows.

Uptake of CO_2 by Marine Organisms. The pH of the surface seawater may be thought to increase because of loss of CO_2 according to the following reaction:

$$HCO_3^- \rightleftharpoons CO_{2(aq)} + OH^- \tag{5}$$

Reaction 5 indicates that the pH of surface seawater may be raised, by loss of CO_2 to photosynthetic organisms or to the atmosphere via carbonic acid,

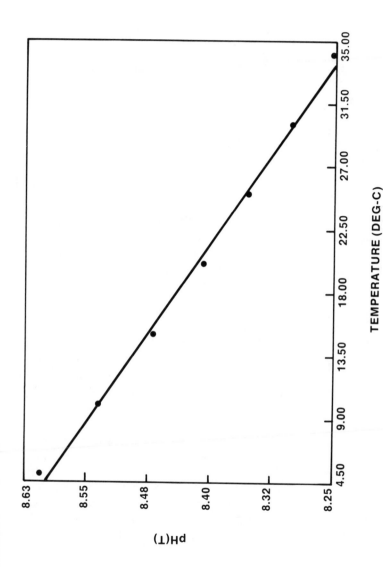

Figure 6. Plot of pH(T) vs. T calculated from Equations 1–4 with E_{cell} = −72.1 mV at 21 °C, $E_{7.413}$ = −15.4 mV at 25 °C, and S = −59.2 mV.

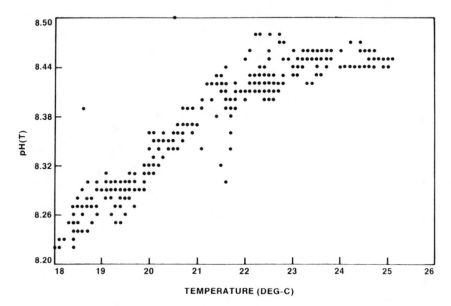

Figure 7. *Plot of pH(T) vs. T for Gulf of California data. pH(T) values are rounded to 0.01, and temperature is rounded to 0.1. Each circle represents a 5-min average, or 60 points.*

without producing a change in the alkalinity (the concentration of weak acid anions titratable to the carbonic acid–bicarbonate equivalence point). Although not strictly true for planktonic photosynthesis (*15*), it is a reasonable first-order approximation because the specific alkalinity (alkalinity/chlorinity ratio × 10^3) is nearly constant for surface waters of the world ocean (*1*). *Prima facie* evidence can be seen in Simpson's data (*see* chapter 21). Of the two processes under consideration, that is, atmospheric venting and photosynthesis, evidence for the latter can be seen in the data (Figure 3, 344/12, 344/22, 345/07, etc.) where pH maxima coincide with chlorophyll a maxima. Largely, for the Gulf area studied, no correlation exists between chlorophyll a and pH(*T*) in Peruvian, and Baja and central Californian upwelling areas as observed previously (*see* chapter 21) (*7,8*). In the latter cases, much higher concentrations of chlorophyll a were measured; thus, recent in situ growth could be correlated directly to pH. At the lower chlorophyll a levels found during this study, however, pH may more clearly reflect cumulative past growth rather than in situ standing stock. Also, variations in the fluorescence yield with physiological state of the organism (*16*), as well as variations in the carbon-to-chlorophyll ratio (*17*), make the correlation between plant growth and pH less obvious at low chlorophyll concentrations.

The overall increase of pH(*T*) with temperature can also not be attributed to loss of CO_2 to the atmosphere. Surface salinities collected on the cruise track indicate that the salinity did not vary greatly from 35.3 ppt. Similarly, the specific alkalinity of Gulf surface waters did not significantly depart from 0.120 (*18*). By using these data and an atmospheric CO_2 value of 330 ppm, equilibrium pH values can be calculated for a portion of the survey area (*19*). These values (Table I) indicate that, under equilibrium conditions, the surface pH value of the Gulf should not vary significantly from 8.23, which is a value close to the low pH values measured at the beginning of the survey. Thus, even within the uncertainties associated with absolute pH measurements, the stepwise increase of pH is probably not due to atmospheric exchange. Finally, the rate of CO_2 uptake associated with planktonic production was measured in Stuart Channel, British Columbia (*20*). In this study, it was observed that CO_2 was removed from the water at a rate approximately 20 times greater than the rate of CO_2 invasion from the atmosphere.

Why then does pH(*T*) vary directly with *T*? Although we can only theorize at present, we can do so within a self-consistent model, the elements of which all appear in the Gulf data. In light-poor, subsurface waters, low pH is associated with low temperatures because respiration processes dominate and CO_2 is accumulated in the water along with micronutrients, in accordance with the Redfield ratio (*21–23*). During upwelling, this water is brought near the surface where it is mixed with surface water by local winds. Low pH and the high positive correlation between pH(*T*) and *T* (Figure 4) reflect the mixing. Positive correlation is maintained as photosynthetic processes become dominant because removal of CO_2 is accompanied by a warming of the water (Figure 7). Because the exchange of CO_2 across the air–sea interface is very slow (*24*), increasing pH will accompany increasing temperature. This correlation will continue as long as the water stays at the surface until either convergence occurs or a

Table I. Equilibrium pH Computed from Temperature and Salinity

Time (day)	T (°C)	S(ppt)	K_1' ($\times 10^6$)	K_2' ($\times 10^{10}$)	K_B' ($\times 10^9$)	α' ($\times 10^2$)	pH
343/2300	20.5	35.49	1.018	7.372	1.943	3.187	8.23
344/0015	20.7	35.53	1.021	7.414	1.953	3.169	8.23
344/0500	18.1	35.21	0.976	6.917	1.836	3.414	8.22
344/0630	20.1	35.26	1.009	7.260	1.919	3.226	8.22
344/1230	19.5	35.24	0.992	7.156	1.894	3.281	8.22
344/1400	18.7	35.17	0.985	7.009	1.860	3.357	8.22
344/1715	22.3	35.21	1.042	7.615	2.005	3.041	8.23
344/1805	22.4	35.23	1.044	7.636	2.010	3.032	8.23
344/2247	21.1	35.15	1.023	7.403	1.955	3.141	8.23
345/0015	21.8	35.23	1.034	7.536	1.986	3.081	8.23

NOTE: The $_pCO_2$ = 3.3 × 10^{-4} atm and specific alkalinity = 0.120.
SOURCE: Reference 1.

wind event (storm) lowers the pH by mixing in more subsurface water. Thus, the coefficient $d\mathrm{pH}(T)/dT$ may be some measure of the net fixation of carbon occurring in the water as it resides at the surface.

Despite these arguments, sharp, positive increases in pH associated with lower temperatures such as those observed here and in the Peruvian upwelling zone (8) are difficult to explain because upwelled waters rich in nutrients for growth are also rich in CO_2. If the Redfield ratio is to be maintained, then a near-equilibrium pH must also be maintained. Brewer (25) has used this argument to estimate past CO_2 partial pressures in the atmosphere. Thus, an additional mechanism for CO_2 removal is required, and the rapid recycling of excreted nitrogen is suggested. Figure 8 shows a 35-km patch from the southern end of the Gulf characterized by a pH maximum in association with a temperature minimum. This negative correlation is characteristic of several other "growth events" already mentioned and suggests that CO_2 has been removed in excess of the CO_2–nutrient ratio of the intruding cold water. Such a deviation from the Redfield ratio could be caused by zooplankters excreting NH_3 into the water without a simultaneous conversion of organic matter into CO_2. Indeed, much nitrogen recycling is evident in surface waters (26). Thus the "hole" in the chlorophyll a profile suggests grazing by zooplankton. Many of the concepts just discussed were used in a much more quantitative manner to explain O_2 supersaturation in surface waters off the northwest African coast (27). In this area a distinct NH_3 subsurface layer was formed when upwelling processes were weak.

Conclusion

By using simple equipment, underway measurements of pH significant to better than 0.01 pH units can be made. In addition, the pH–temperature relationships obtained from underway measurements yield data that can be used to interpret oceanographic processes. In the Gulf of California, the pH of freshly upwelled waters shows a significant positive correlation with the in situ temperature on scales from 3 km to hundreds of kilometers. Older surface waters may show a negative correlation on scales of about 20 km. A negative correlation is usually associated with chlorophyll maxima that resemble the pH maxima. Also, the pH at the in situ temperature increases with temperature beyond that which is accountable to physicochemical processes.

The positive pH–temperature correlation is probably due to the cumulative biological depletion of CO_2 from surface waters. Confirmation of this hypothesis awaits the development of underway techniques for determining alkalinity (28) and total CO_2.

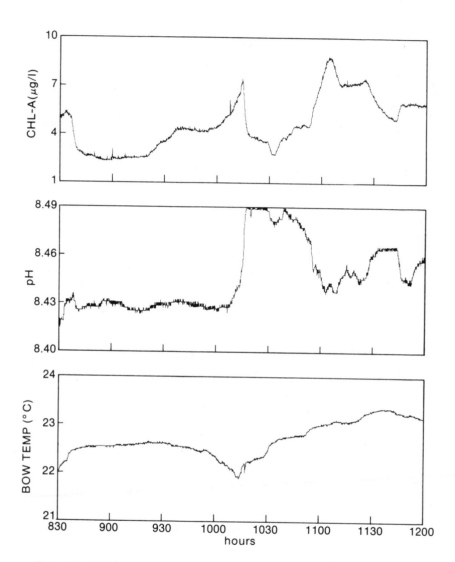

Figure 8. pH(T) maximum associated with temperature and chlorophyll a minima. The "hole" in chlorophyll a suggests grazing (December 11, 1981).

Acknowledgment

We are grateful to T. T. Packard and to our many colleagues at the Scripps Institution of Oceanography for their valuable suggestions. We also wish to thank Saul Alvarez-Borrego, Gilberto Gaxiola, and the students at the Centro de Investigaciones Cientificas y Escuela Superior de Ensenada for making *Varifront III* possible. This work was funded under the NOSC IR/IED program.

Literature Cited

1. Skirrow, G. In "Chemical Oceanography"; Riley, J. P.; Skirrow, G., Eds.; Academic: London, 1975; vol. 2, pp. 1–192.
2. Whitfield, M. *Nature* 1974, *247*, 523–25.
3. Whitfield, M. *Nature* 1975, *254*, 274–75.
4. Dyrssen, D. In "Oceanic Sound Scattering Prediction"; Andersen, N. R.; Zahuranec, B. J., Eds.; Plenum: New York, 1977; pp. 65–84.
5. Gordon, L. I.; Park, P. K.; Hager, S. W.; Parsons, T. R. *J. Oceanogr. Soc. Jpn.* 1971, *27*, 81–90.
6. Weichart, G.; *Meteor. Forschungsergeb.* 1974, *14*, 33–70.
7. Simpson, J. J.; Zirino, A. *Deep-Sea Res.* 1980, *27*, 773–83.
8. Zirino, A.; Clavell, C.; Seligman, P. F.; Barber, R. T. *Mar. Chem.* 1983, *12*, 25–42.
9. Culberson, C. H. In *Marine Electrochemistry*; Whitfield, M.; Jagner, D., Eds.; Wiley: Chichester, England, 1981; pp. 187–261.
10. Bates, R. "Determination of pH—Theory and Practice"; Wiley: New York, 1964; pp. 75–77.
11. Lorenzen, C. J. *Deep-Sea Res.* 1966, *19*, 209–32.
12. Flynn, K. M., M. S. Thesis, San Diego State University, San Diego, CA, 1981.
13. Gieskes, J. M. T. M. *Limnol. Oceanogr.* 1969, *14*, 679–85.
14. Ben Yaakov, S. *Limnol. Oceanogr.* 1970, *15*, 326–28.
15. Brewer, P. G.; Goldman, J. C. *Limnol. Oceanogr.* 1976, *21*, 108–17.
16. Kiefer, D. A., *Mar. Biol.*, 1973, *22*, 263–69.
17. Banse, K., *Mar. Biol.*, 1977, *41*, 199–212.
18. Gaxiola-Castro, G.; Alvarez-Borrego, S.; Schwartzlose, R. A. *Cienc. Mar.* 1978, *5*, 25–40.
19. Culberson, C. H. *Limnol. Oceanogr.* 1980, *25*, 150–52.
20. Johnson, K. S., Pytkowicz, R. M., Wong, C. S. *Limnol. Oceanogr.* 1979, *24*, 474–82.
21. Redfield, A. C.; Ketchum, B. H.; Richards, F. A. In "The Sea"; Hill, N. M., Ed.; Wiley: New York, 1963; vol. II, pp. 26–77.
22. Culberson, C. H.; Pytrowicz, R. M. *J. Oceanogr. Soc. Jpn.* 1970, *26(2)*, 95–100.
23. Alvarez-Borrego, S.; Guthrie, D.; Culberson, C. H.; Park, P. K. *Limnol. Oceanogr.* 1975, *20*, 795–805.
24. Broecker, W. S.; Li, Y.; Peng., T. In "Impingement of Man on the Oceans"; Hood, D. W., Ed.; Wiley: New York, 1971; pp. 287–324.
25. Brewer, P. G. *Geophys. Res. Lett.* 1978, *5*, 997–1000.
26. Harrison, W. G. In "Primary Productivity in the Sea"; Falkowski, P. G., Ed.; Plenum: New York, 1980; pp. 433–60.
27. Minas, H. J.; Packard, T. T.; Minas, M.; Coste, B. *J. Mar. Res.* 1982, *40*, 615–41.
28. Keir, R. S.; Kounaves, S. P.; Zirino, A. *Anal. Chim. Acta* 1977, *91*, 181–87.

Received for review October 10, 1983. Accepted May 9, 1984.

21

Air–Sea Exchange of Carbon Dioxide and Oxygen Induced by Phytoplankton
Methods and Interpretation

JAMES J. SIMPSON

Marine Life Research Group, Scripps Institution of Oceanography, La Jolla, CA 92093

This chapter discusses analytical methods that were used to continuously sample physical, chemical, and biological variability in the oceans while underway in R. V. New Horizon *in July 1979. Some statistical methods useful in the analysis and interpretation of these data also are given. Data indicate that the in situ concentrations of CO_2 and O_2 can depart radically from their expected oceanic equilibrium values with respect to their atmospheric components in active upwelling areas. The departure from equilibrium is caused by both physical and biological processes. Dominant length scales are associated with the biological processes. The global distribution of primary productivity coupled with these and other small- and mesoscale measurements suggests that coastal and polar regions may be in nonequilibrium with respect to O_2–CO_2 exchange between ocean and atmosphere during parts of the year. The equatorial regions may be in nonequilibrium throughout the year. Consequently, the use of direct oceanic CO_2 transport models based on total CO_2, alkalinity, and apparent O_2 utilization relationships are in question.*

APPLICATIONS OF CONTINUOUS, UNDERWAY MEASUREMENT TO MARINE SCIENCE include measurements of chemical variability in the upper ocean, especially near coastal fronts and streamers (*1, 2*); determination of the in situ concentrations of trace metals (*3*) and of dissolved gases near the air–sea interface (*4, 5*); assessment of the types, concentrations, and variations of planktonic communities (*6*); investigation of the mixing and lateral dispersion of pollutants in the sea (*7*); and the acquisition of spatially adequate ground truth for remote sensing of fisheries stocks and oceanic prop-

0065-2393/85/0209-0409$11.50/0
© 1985 American Chemical Society

erties (8). Although the utility of this approach to chemical oceanographic sampling has been emphasized by the National Research Council (9), the literature has not addressed, in detail, the analytical techniques used in data acquisition, and the corresponding data processing techniques required for the efficient and effective analysis of the large data bases (e.g., 10^5–10^6 ordered sets) that underway sampling systems can produce. This chapter addresses these latter concerns by describing an underway physical–chemical–biological data acquisition system and some statistical methods of analysis used in a study of the effects of phytoplankton on the air–sea exchange of O_2 and CO_2 in the coastal regions of an eastern boundary current.

Analytical Methods

Simultaneous measurements of physical, chemical, and biological water properties were made from R. V. *New Horizon* while underway during July 1979. Water for the measurements was drawn from a 3-m depth through the bow intake. A fast-response thermistor was inserted in the flow at the intake to measure in situ temperature. Then, the water was pumped to the laboratory for the remaining analyses. A flow rate of 20–25 L/min was maintained to the laboratory. A debubbler was inserted in the flow (Figure 1) because all of the analytic methods are affected by the presence of bubbles in the sample flow. *Continuous data* were obtained by directly and continuously pumping the sample flow from the debubbler into the various analytical devices diagramed in Figure 1. *Discrete data* were obtained by drawing individual samples from the same flow at specified times for discrete sample analysis.

   Continuous Data.   Salinity was measured continuously with a Bissett–Berman thermosalinograph. Oxygen was measured with a Beckman 147737 polarographic electrode, which consists of a silver anode and a gold cathode enclosed within a single plastic housing. The electrolyte agent is a cellulose-based potassium chloride gel that is held in place by a Teflon membrane. (This electrode is discussed in Reference 10.) The polarographic electrode was sealed in a poly(vinyl chloride) block, and a flow rate of about 2 L/min was maintained past it. This flow rate is equivalent to that past a similar electrode on a conductivity, temperature, and depth (CTD)/O_2 system lowered through the water at 70 m/min. This flow rate ensures that the output current of the electrode corresponds to approximately 100% of the final electrode response. The pH of seawater was measured with a K599360 Cole–Parmer peripheric combination electrode (11). This electrode is different from the one used by Simpson and Zirino (4); however, the electrode used here and the electrode used by Simpson and Zirino (4) respond similarly in the oceanic environment. Details of the cell geometry and flow rate past the electrode used in this measurement also are similar to those given by Simpson and Zirino (4). A multichannel, continuous-flow autoanalyzer system was used to measure the in situ concentrations of nitrate, phosphate, and silicate. A Turner Designs fluorometer equipped with a continuous-flow cuvette was used to measure in vivo chlorophyll a fluorescence. A diagram of the flow, measurement, and recording system is shown in Figure 1.

Data from the pH electrode were measured in the voltage domain throughout the experiment. A precision digital voltmeter was used for about 90% of the data. This device allowed pH to be resolved to within ± 0.005 pH units. During one brief period of the experiment (∼ 10% of the data), however, the resolution was limited to about ± 0.01–0.02 pH units because of equipment failure. This smaller subset of data, however, does not seriously affect the results or the conclusions presented because the observed pH range was about 7.8 to 8.4. Data from the oxygen electrode were measured in the frequency domain using a frequency shift keying (FSK) system. These two different measuring methods, one in the voltage domain and one in the frequency domain, effectively eliminated any potential cross talk between the pH and oxygen electrodes.

Data were sampled from all instruments (Figure 1) at 20-s intervals and recorded digitally. At a ship's speed of about 18 km/h, this sampling interval corresponds to about 100 m in space. From these data, sequential 1-min (∼ 300-m) averages were constructed because both pumping and debubbling of the sample act as low-pass spatial filters and necessarily smooth out some of the fine-scale in situ chemical, physical, and biological structure. Both these processes, coupled with ship's speed, ultimately set a lower limit on the spatial scales of variability that any continuous underway sampling system can realistically resolve.

Discrete Data.   The underway data were calibrated with discrete samples drawn from the flow about every 15 min. On occasion, especially in sharp frontal regions, calibration samples were drawn more frequently. A detailed description of each calibration procedure is given here.

Traditional bucket samples generally are inadequate for the calibration of in situ temperature because preferential near-surface absorption of solar radiation occurs in the upper few meters of the water column (12). Hence, uncertainty in the depth of the bucket sample can introduce significant errors in the in situ temperature measurement. If the flow rate to the ship's laboratory is sufficiently high (≥ 15–20 L/min), then discrete temperature measurements can be made with a laboratory-grade thermometer inserted in the flow prior to the debubbler. The combination of discrete temperature measurements with a precise thermal-bath calibration of the in situ temperature sensor prior to use at sea produces high-quality underway temperature data to within ± 0.03 °C.

Salinity samples were drawn from the flow and allowed to equilibrate to the laboratory temperature for 24 h prior to analysis with a Guildline Autosal. Throughout the cruise the salinometer was calibrated against the

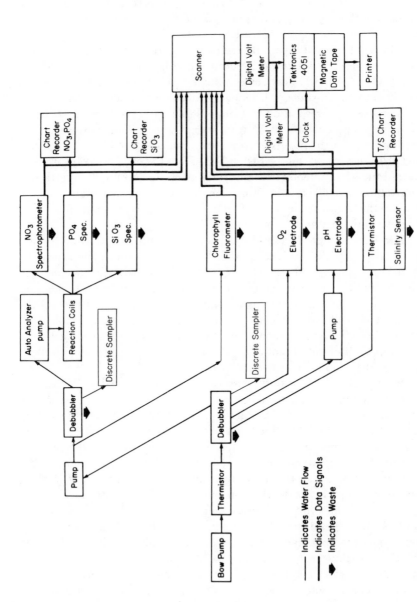

Figure 1. Flow, measurement, and data acquisition system used to take continuous underway measurements during the California Coastal Fronts Program.

international standard seawater. Using these procedures we can determine salinity to within ± 0.003 ppt. (Salinometers have been compared (*13*) to determine the effects of instrument bias on the measurements.)

Dissolved oxygen samples were drawn from the flow and titrated in calibrated 125-mL iodine flasks with a 1-mL microburet (*14*). Preweighed potassium iodate crystals were used to prepare 0.01 N potassium iodate solutions that served as standardizations for the method throughout the cruise. Approximately 1500 discrete oxygen titrations were performed during the experiment to calibrate the polarographic oxygen electrode.

Discrete analyses for nitrate, nitrite, phosphate, and silicate were performed on samples drawn from the flow with a modified Hitachi autoanalyzer system; the full-scale ranges for these measurements are 0–30, 0–10, 0–2.4, and 0–45 μmol/L, respectively. The precision and accuracy of these measurements are 1% of full-scale values. Both Sagami standard nutrient solutions and standard solutions prepared from preweighed standards were used to standardize the nutrient measurements. Beer's law curves were run throughout the experiment to establish the output response of the spectrophotometers. The response functions were linear for the concentrations measured. A complete description of the analytical methods used for nutrients is given in References 15 and 16.

A 100-mL sample was drawn from the flow and glass-fiber filtered to determine the concentrations of chlorophyll a and phaeophytin pigment, which were extracted from the filter in 90% acetone in the dark at 2 °C for 24 h. The chlorophyll a and phaeophytin pigment contents of this extract were determined fluorometrically with a Turner fluorometer calibrated with a pure extract of chlorophyll a (*17*). Several methods currently are used to convert in vivo chlorophyll a fluorescence to chlorophyll a (in micrograms per liter) from extracted chlorophyll a values. For example, a single linear regression model was used previously (*18, 19*) to relate in vivo chlorophyll a fluorescence to extracted chlorophyll a. Data from a study by Mackas et al. (*20*) required two separate calibration curves based on different geographic, hydrographic, and taxonomic criteria found during their underway survey. In studies by Steele and Henderson (*21*), discrete-point calibrations appear to be used; studies by Cox et al. (*6*) used a running average method. The running average method is useful in the coastal waters of the California Current System (CCS) because the CCS behaves largely as a faunal ecotone (*22*): that is, a region where highly variable horizontal advection, in addition to local processes, regulates the floral and faunal composition, biomass, and production. Other studies (*23*) show that small-scale thermal fronts also act as zoogeographic boundaries in marine and freshwater systems. For biological reasons different underway studies may require different calibration procedures for in vivo chlorophyll a fluorescence (*24, 25*). Ultimately, the appropriate calibration procedure is determined by the nature of the phytoplankton found during the underway survey.

The discrete titration alkalinity and the titration total-CO_2 data used in this study were obtained with the automated, potentiometric acid-titration method [e.g., geochemical ocean sections study (GEOSECS) titrators] (26). The method is based on the technique developed by Gran (27), Dyrssen (28), and Edmond (29). This method can determine values of alkalinity and total CO_2 from a single sample. The titrators were calibrated with a sodium borate decahydrate ($Na_2B_4O_7 \cdot 10H_2O$) solution (30) and Na_2CO_3 standard solutions prepared gravimetrically (26). The precision of the method was estimated (13) on the basis of repeated measurements and is $\pm 0.1\%$ for titration alkalinity and $\pm 0.5\%$ for titration total CO_2. The accuracy of the method was estimated (13) on the basis of differences between standard solutions and measured values. The average deviation from the standard solution for titration alkalinity is -0.12% and $+0.7\%$ for the titration total CO_2 (13). A complete discussion of the internal consistency of the carbonate chemistry data taken with the GEOSECS titrators is discussed elsewhere (31). Discrete pH values were calculated from the titration alkalinity and titration total CO_2 (31, 32) and subsequently were used to calibrate the continuous pH measurements just discussed.

Computational Methods

The methods used to compute the physical and chemical properties derived from the continuous and discrete data are discussed here.

Continuous Data.   The density (σ_t) was computed from the in situ temperature (°C) and salinity (ppt) by using the equation of state (33).

Percent saturation of dissolved O_2 was defined as

$$\% O_2 = \frac{[O_2]}{\alpha_{O_2}} \times 100 \tag{1}$$

where $[O_2]$ is the measured concentration of dissolved O_2 and α_{O_2} is the solubility of dissolved O_2 in seawater determined from the equations of Weiss (34).

Chlorinity (parts per thousand) was determined from salinity by using the constant ratio of 1.80655 given by Wilson (35). The calculated titration alkalinity (CTA) was calculated from the chlorinity (36) by using a specific alkalinity of 0.126. This method has been used for the calculation of alkalinity in upwelling areas off northwest Africa (37) and Peru (4). The use of a constant specific alkalinity to calculate the distribution of the partial pressure of CO_2 (pCO_2) in upwelled waters does not introduce a significant error in pCO_2 because changes in pH caused by changes in the specific alkalinity of surface waters due to the upwelled subsurface waters generally are negligible ($\sim 1\%$) compared to changes produced by in situ biological processes (4). The carbonate alkalinity (A) was calculated from the relation (36)

$$A = CTA - \left\{ \frac{K'_B \Sigma B}{a_H + K'_B} \right\} \tag{2}$$

where ΣB is the total borate of seawater, K'_B is the first apparent dissociation constant for boric acid in seawater, and a_H is the hydrogen ion activity. For most ocean water, the total borate (moles per liter) can be adequately estimated from the chlorinity (parts per thousand) by using, $\Sigma B = 2.2 \times 10^{-5} \times$ chlorinity (38).

The continuous distributions of pCO_2 and of the total inorganic CO_2 (ΣCO_2) were determined from pH (continuous measurement) and alkalinity (i.e., CTA) measurements (11, 36). The pCO_2 in seawater was calculated according to the equation given in Reference 36:

$$pCO_2 = A \left\{ \frac{a_H}{\alpha K'_1 (1 + 2K'_2/a_H)} \right\} \tag{3}$$

where α is the solubility of CO_2 in seawater, and K'_1 and K'_2 are the first and second apparent dissociation constants for carbonic acid in seawater, respectively. In Equation 3, the solubility of CO_2 in seawater was computed from the empirical relation given by Weiss (39). The deviation of pCO_2 from atmospheric equilibrium was defined as $\Delta pCO_2 = pCO_2 - 330$ ppm. The commonly used atmospheric equilibrium value for pCO_2 is 330 ppm (4, 36). The total inorganic CO_2 (ΣCO_2) was computed from the relation given in Reference 36:

$$\Sigma CO_2 = A \left\{ \frac{1 + K'_2/a_H + a_H/K'_1}{1 + 2K'_2/a_H} \right\} \tag{4}$$

The Lyman dissociation constants (40) were used in Equations 2–4 because they were determined with an electrode system similar to that used here and they are also the constants used previously (36) in the derivation of Equations 2–4.

Discrete Data.   Generally, the computational methods used for the discrete data were the same as those used for the continuous data. The discrete computations of the components of the carbonate system, however, were an exception. In this case, the observables were titration alkalinity (TA) and the total CO_2 (TCO_2) rather than pH and calculated alkalinity (CTA). Both TA and TCO_2 were measured with the GEOSECS titrators (*see* previous discussion). The equations and constants used to calculate pCO_2 from TA and TCO_2 are given by (*32*)

$$TA = A(Z) + \frac{K_B(Z)TB}{K_B(Z) + a_H} \qquad (5)$$

$$[A(Z)/TCO_2][a_H/K_1(Z)]^2 + \{[A(Z)/TCO_2] - 1\}[a_H/K_1(Z)]$$
$$+ [K_2(Z)/K_1(Z)]\{[A(Z)/TCO_2] - 2\} = 0 \qquad (6)$$

$$pCO_2 = TCO_2\{2 - [A(0)/TCO_2]\}/\alpha_s\{2 + [K_1(0)/a_H]\} \qquad (7)$$

and

$$(CO_3^{2-}) = [A(Z) - TCO_2]/\{1 + [a_H^2/K_1(Z)K_2(Z)]\} \qquad (8)$$

where TA is the titration alkalinity (per unit mass); TCO_2 is the total CO_2 concentration (per unit mass); TB is the total borate concentration (moles per kilogram); $A(0)$ and $A(Z)$ are the carbonate alkalinity at depths of 0 and Z km, respectively; a_H is the hydrogen ion activity; $K_1(0)$ and $K_1(Z)$ are the first apparent dissociation constants for carbonic acid in seawater at depths of 0 and Z km, respectively; $K_2(0)$ and $K_2(Z)$ are the second apparent dissociation constants in seawater at depths of 0 and Z km, respectively; $K_B(0)$ and $K_B(Z)$ are the apparent dissociation constants for boric acid in seawater at depths of 0 and Z km, respectively; and α_s is the solubility of CO_2 in seawater at 1 atm (*39*). The values of $K_1(0)$ and $K_2(0)$ used in Equations 5–8 are those of Mehrbach et al. (*41*), and the value of $K_B(0)$ is from Lyman (*40*) as expressed by Li et al. (*42*). The choice of dissociation constants used in this case is based on internal consistency checks with the Atlantic GEOSECS data for titration alkalinity, titration TCO_2, and pCO_2 (*43*).

Comparison of Methods.   Here, a comparison is made between the results obtained from the different analytical and computational methods used to determine the parameters of the carbonate system. Results obtained from Equations 2–4 (i.e., those given in Reference 36) are expressed in moles per liter, whereas results obtained from Equations 5–8 (i.e., those given in Reference 32) are expressed in moles per kilogram. For ocean surface waters, the difference between the two is small. Prior to a given intercomparison, however, the relevant results were converted to a common scale by using the observed density of seawater, σ_t. Similarly, the concentration of dissolved oxygen is usually measured in milliliters per liter. In stoichiometric applications, however, the concentration of dissolved oxygen should be reported in moles per kilogram. Where necessary, the ideal gas law for oxygen and the measured density were used.

For the continuous data, the alkalinity (CTA) was calculated from the continuous underway measurement of salinity by assuming a constant specific alkalinity (36). For the discrete data, the titration alkalinity (TA) was measured directly with the GEOSECS titrators. The CTA is plotted as a function of the measured TA in Figure 2. The mean value of TA was 2274.40 ± 11.60, and the mean value of CTA was 2294.43 ± 9.69. On the average, TA is 20.03 units (or about − 0.88%) less than CTA. The sign of this difference is consistent with that reported by Bainbridge (13) who showed that titration alkalinity measured with the GEOSECS titrators typically deviates from the standard solution by − 0.12%. The standard deviation for CTA, however, is about 10% less than that for TA. A linear regression analysis on the data (Figure 2) yielded a slope m = 1.009 with a correlation coefficient r = 0.999. The measured slope differs slightly from the ideal slope of m = 1 because of the mean offset of 20.3 alkalinity units between CTA and TA. Therefore, alkalinity can be calculated from salinity to within 1% of the measured TA. A more detailed discussion of the horizontal and vertical structure of TA in coastal upwelling areas, based on direct measurements with the GEOSECS titrators, is found elsewhere (44).

Figure 3 shows a typical time series of pCO_2, calculated from Equation 3 with pH and CTA as observables. Also shown in this figure are 10

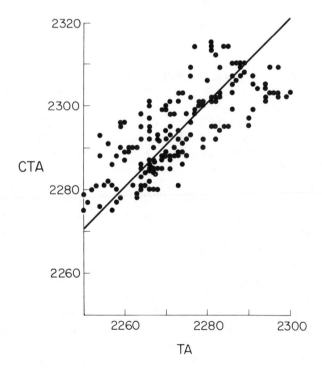

Figure 2. Calculated alkalinity (CTA) shown as a function of measured titration alkalinity (TA). Linear regression curve (slope = 1.009) also is shown.

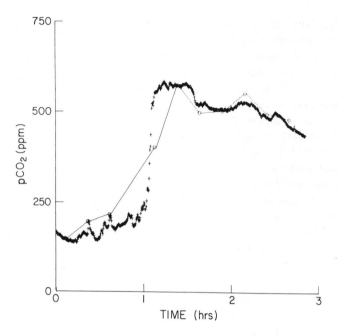

Figure 3. Continuous pCO_2 *data overlaid with 10 discrete values of* pCO_2.

discrete values of pCO_2 calculated from Equations 5–8 with TA and TCO_2 as observables. The line segments in Figure 3 connect the discrete values of pCO_2. Again, these two methods agree well.

Figures 4a and 4b show typical space–time series of temperature, salinity, σ_t, dissolved oxygen, pH, chlorophyll a, nitrate, phosphate, and silicate taken with the underway system (Figure 1). The underway data and the calibration samples agree (*see also* References 2 and 5). This agreement confirms the utility of underway sampling in chemical and biological studies of coastal processes. Also shown in Figure 4b are typical time series of discretely measured TA and titration TCO_2. These data, like the data in Figure 2, also show that changes in TA are controlled largely by changes in salinity, even in the biologically productive areas of the CCS. Changes in TCO_2, however, are controlled largely by biological processes.

Previously, the oxygen electrode proved troublesome in many oceanographic applications. Several factors have combined to optimize the performance of the oxygen electrode in this application, compared to the usual CTD/O_2 application. For example, the pressure and temperature effects on the permeability of the Teflon membrane used in the oxygen electrode can be modeled (10) with separate exponential functions (e.g., $P_m = P_0 e^{-kp}$ where P_m is the membrane permeability at a given absolute pressure p, P_0 is the membrane permeability at zero pressure, and k is a con-

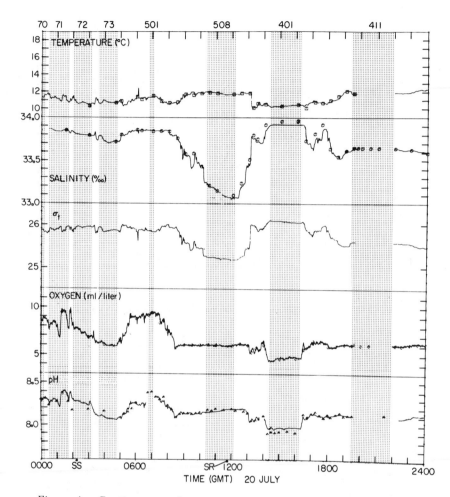

Figure 4a. Continuous underway measurements of temperature, salinity, dissolved O_2, and pH. Density, computed from temperature and salinity, also is shown. Calibration data are shown with a symbol. Numbers at the top of the figure refer to CTD station numbers and the stippled areas show the time on station.

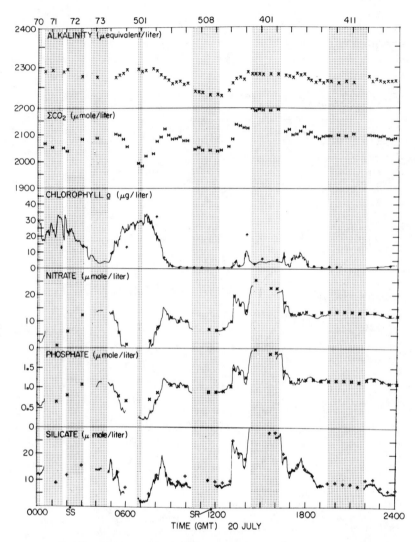

Figure 4b. Continuous underway measurements of nitrate, phosphate, silicate, and chlorophyll a. Calibration data and discrete determinations of TA and titration total CO_2 (taken with the GEOSECS titrators) are shown with a symbol.

stant for the membrane material). In this application, the pressure-induced change in membrane permeability is negligible. Temperature-induced changes in membrane permeability also are small because the range of surface temperatures measured is small compared to that measured during a typical CTD cast. In addition, for a given temperature, the effect of salinity on the electrode measurement of dissolved oxygen is approximately linear. A salinity-induced error of about 0.5% in the electrode determination of dissolved O_2 occurs for salinities between 32 and 36 ppt (*10*). The range of surface salinities measured in this experiment is small ($\sim$ 0.8 ppt); hence, the salinity-induced error in the electrode measurement of O_2 is correspondingly small. Finally, the absence of pressure effects on this measurement eliminates hysteresis-induced temperature and pressure lags in the electrode output current. In CTD/O_2 applications, these effects generally are significant and require corrections to the data with model lag functions.

The agreement (Figure 4a) between the discrete and continuous measurement of dissolved O_2 conservatively corresponds to a maximum error in the percent saturation of dissolved O_2 of only $\pm$ 1–2%. The agreement (Figure 3) between discrete and continuous pCO_2 values conservatively corresponds to a typical error of $\pm$ 15–20 ppm. The observed percent saturation of dissolved O_2 ranges from 70 to 185%. The range of pCO_2 values is 150–700 ppm. Also, for a given temperature, salinity, and pH an error of 1% in TA results in an equivalent error in pCO_2 (*see* Equation 3). Hence, the uncertainties in the measurements of O_2, alkalinity, and pCO_2 have an insignificant effect on the results and conclusions presented in this chapter. Further, the consistency between the continuous and discrete data (*see* Figure 4) demonstrates that the results and conclusions presented herein are not critically dependent on a single measurement. Rather, each measurement within the ensemble of measurements independently supports the results presented and the conclusions reached.

Statistical Procedures

The observational areas and the oceanographic sampling plan used in this experiment are described here. The choice of the appropriate statistical methods to use in data analysis generally is dependent on the nature of the data and how the data were taken.

A region of the southern California Current off Point Conception and a region of the northern California Current between Points Reyes and Arena were sampled continuously while underway from R. V. *New Horizon* by following the zigzag patterns shown in Figure 5. Cross-shelf transects in the northern region are labeled numerically; those in the southern region are labeled alphabetically. Point Arena frequently is a center of active coastal upwelling, and although upwelling does occur off Point Conception, it often is less intense than that off Point Arena (*6*).

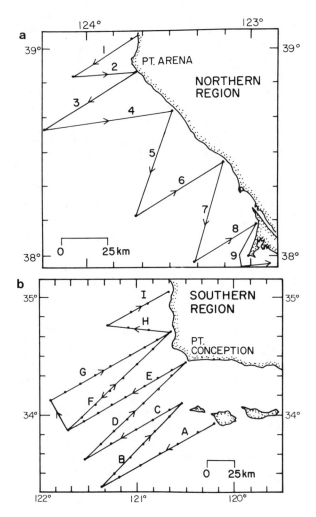

Figure 5. *Ship's tracks during the July 1979 California Coastal Fronts exper-*
iment. Cross-shelf transects are identified with numbers in the northern re-
gion (a) and letters in the southern region (b).

Typical cross-shelf space–time series of underway data are shown in
Figure 6. Line E is from the southern region, and the direction of travel is
from the coast to the offshore (*see* Figures 5 and 10). Line 4 is from the
northern region, and the direction of travel is from the offshore to the coast
(*see* Figures 5 and 11). Such space–time series, $y(\mathbf{r}, t)$, are considered space
series, $y(\mathbf{r})$, because the spatial (i.e., cross-shelf) variability in all signals
generally is much greater than the temporal variability. This assumption
has been validated (2) for short-time scales (~ 1–2 days) with direct obser-

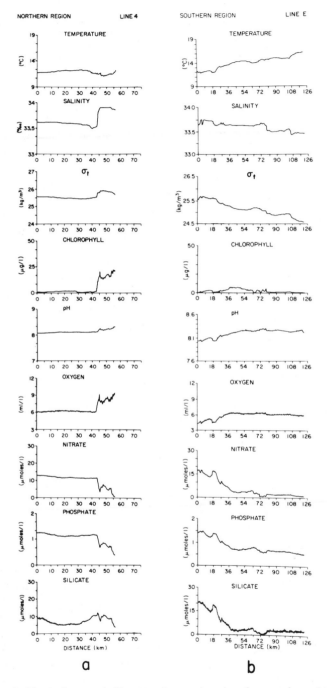

Figure 6. Typical cross-shelf space–time series of underway data. Line E is from the southern region; Line 4 is from the northern region.

vations in the CCS. Generally, each line consists of three separate domains: a region of enhanced property values, a region of diminished property values, and a region of rapid transition (i.e., frontal zone) between the two. The frontal zone may be sharp (Figure 6a) or more gradual (Figure 6b). The intensity of the frontal zone varies from property to property. The chemical and biological fronts, however, are in general more pronounced than the physical fronts. Spatial trends also appear in the data. Superimposed on these trends are small-amplitude fluctuations (small compared to the amplitude of the mean trend) in signal strength. The data (Figures 4 and 6) show pH and chlorophyll a values nearly as high as those observed in the Peruvian coastal upwelling area (4). The data also show that the nutrients are in phase and that pH, O_2, and chlorophyll a, although in phase with each other, are out of phase with nutrients. These qualitative relationships indicate the importance of phytoplankton photosynthesis in the chemistry of the coastal waters of the CCS.

Data generally are classified as either deterministic or random. *Deterministic* means that the process under study can be described by an explicit mathematical relationship. *Random* means that the phenomenon under study cannot be described by an explicit mathematical function because each observation of the phenomenon is unique. A single representation of a random phenomenon is called a sample function. If, as in all experiments, the sample function is of finite length, then it is called a sample record. The set of all possible sample functions, $\{ y(\mathbf{r}) \}$, which the random process might produce, defines the random or stochastic process. The mean value and autocorrelation function for a random process are defined by

$$\mu_y(\mathbf{r}) = \lim_{N \to \infty} \frac{1}{N} \sum_{k=1}^{N} y_k(\mathbf{r}) \tag{9}$$

and

$$R_y(\mathbf{r}, \mathbf{r} + \delta) = \lim_{N \to \infty} \frac{1}{N} \sum_{k=1}^{N} y_k(\mathbf{r}) y_k(\mathbf{r} + \delta) \tag{10}$$

A random process is weakly stationary if its mean value and autocorrelation function are independent of $\mathbf{r}$. Thus, for a weakly stationary random process, the mean value is a constant [$\mu_y(\mathbf{r}) = \mu_y$] and the autocorrelation function depends only on the spatial lag δ [e.g., $R_y(\mathbf{r}, \mathbf{r} + \delta) = R_y(\delta)$]. A random process is strongly stationary if the infinite collection of higher order statistical moments and joint moments are space invariant. Most geophysical phenomena are not strongly stationary. However, the random process under study must be at least weakly stationary, otherwise the results of the space- or time-series analysis can be suspect. An extensive treatment of these statistical concepts is available (45, 46). A detailed review of the misapplication of some techniques of space- and time-series analysis to biological processes is given in Reference 47.

In Figure 6a, the dominant feature is a sharp discontinuity in spatial structure that can be modeled as a step function. In Figure 6b, the dominant feature is a region of enhanced spatial gradient in properties that can be modeled as a ramp function. These two classes of structure are representative of all the data taken along the individual transects shown in Figure 5. Each class of structure exhibits a form of frontal boundary in the data. The presence of frontal boundaries in the data complicates the statistical analysis for two reasons. First, frontal boundaries render the space series nonstationary because the value of μ_y is critically dependent upon the particular segments of data used to calculate μ_y. Second, such space series are not likely to represent a uniform random process; fronts act as physical and ecological barriers (*see* calibration procedure for discrete chlorophyll a data in previous section) (*23*). Moreover, observations (*48*) of remotely sensed ocean color and phytoplankton abundance made with the coastal zone color scanner (CZCS) and IR thermal imagery taken with the advanced very high-resolution radiometer (AVHRR), coupled with in situ observations show very sharp, large-scale frontal boundaries in the California current. Again, different dynamical and biological processes were observed to occur on different sides of the front. Hence, in general, the same random processes may not operate on both sides of a frontal boundary.

Step-function space series were initially divided into three segments: inshore data, frontal zone data, and offshore data. Frontal zone data were not used in the subsequent statistical analyses for the reasons just cited. The inshore and offshore data segments were tested for the presence of trends and then for stationarity with the procedures given by Bendat and Piersol (*46*). In some instances, the inshore and offshore data segments were further subdivided in an attempt to satisfy the conditions of weakly stationary data.

Ramp-function space series were initially divided into two segments: ramp data and nonramp data. The ramp data show the presence of a linear trend. This trend was removed, and the residual data series was retained. Virtually all the ramp data in this study are representative of inshore (i.e., coastal) data. The offshore (i.e., nonramp) data also were tested for trends. The detrended segments were tested for stationarity with the procedures just cited.

A *trend in the data* is defined (*46*) as any frequency component whose period is longer than the sample record. Failure to remove these trends often produces large distortions in correlation functions (and spectra), especially at low frequencies. A least-squares detrending procedure (*46*) was used to remove linear trends in the data. This method differs from the endpoint technique (*49, 50*). The end-point technique was not used here because it is recommended for use on data in which a large deterministic component, whose period is about equal to the length of the record, is superimposed on a random process. Typical examples of data segments with trends, and the trends for the southern region, are shown in Figures 7a, 8a, and 9a. Additional examples (from the northern region) are found elsewhere (2).

Not all trends are linear. The nature of the trend is determined by the data itself. Nonlinear trends can be important and frequently are more difficult to remove from the data. For example, second- and fifth-order polynomials were used previously (51) to remove nonlinear trends in radiometrically sensed sea surface temperature data. Nonlinear trend removal, however, was not used in this study because the linear trend analysis proved adequate. A method suitable for the removal of nonlinear trends from underway data can be found in Reference 47.

Figures 7b, 8b, and 9b show typical examples of detrended data segments from the southern region. Unlike the data in Figure 6 (i.e., entire data records along a given line), the data in Figures 7–9 are for much smaller length segments extracted from a given line, which emphasize small-scale features (i.e., patches) not adequately resolved in Figure 6. Hence, the origin of a given data segment shown in Figures 7–9 is keyed to the location of a patch and does not necessarily correspond to the origin of the parent line (Figure 5) from which it was taken. The actual locations of the data segments used in Figures 7–9 are shown in the surface distribution of chlorophyll a for the southern region (Figure 10), and the full transects (i.e., Lines E, F, and I) from which the segments were extracted are shown in the distribution of the percent saturation of dissolved oxygen. A brief discussion of each data segment is given to aid in the interpretation of these segments.

The data in Figure 7 are from Line E of the southern region, and the origin of this data segment is at the coast. Maximum nutrient concentrations, minimum pH, and minimum dissolved oxygen and chlorophyll concentrations occur at the beginning of this data segment (Figure 5) because the coastal origin of Line E coincides with the upwelling center off Point Conception. This coincidence is evident from the spatial distribution of sampled properties in the southern region, overlaid with the cruise track for Line E (Figure 10). The data in Figure 8 are from Line I of the southern region. The origin of this data segment is offshore in water with a chlorophyll concentration between 2 and 4 $\mu g/L$ (compare Figures 8 and 10) and does not correspond to the origin of Line I (Figure 5) where the chlorophyll concentration is less than 2 $\mu g/L$ (see Figure 10). The direction of travel is toward the coast. The large, irregularly shaped phytoplankton patch (Figure 10) with chlorophyll concentrations between 6 and 10 $\mu g/L$ corresponds to the large peak in chlorophyll shown in Figure 8. The data in Figure 9 are from Line F of the southern region. Here, the origin of the data segment is at the coast. The chlorophyll structure (Figure 9) has two peaks;

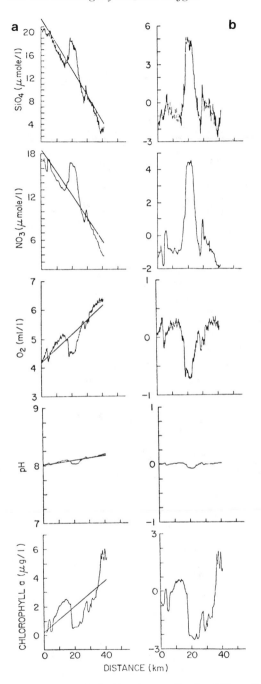

Figure 7. Spatial segments of silicate, nitrate, dissolved O₂, pH, and chlorophyll a from Line E of the southern region (Figure 6). Key: left, data and mean spatial trend; and right, the detrended series.

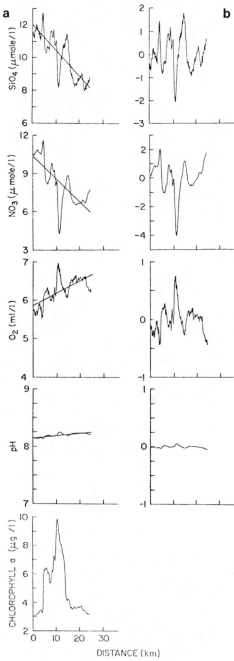

Figure 8. Spatial segments of silicate, nitrate, dissolved O_2, pH, and chloro-phyll a from Line I of the southern region (Figure 6). Key: left, data and mean spatial; and right, the detrended series. (Detrending of chlorophyll a was not possible in this segment.)

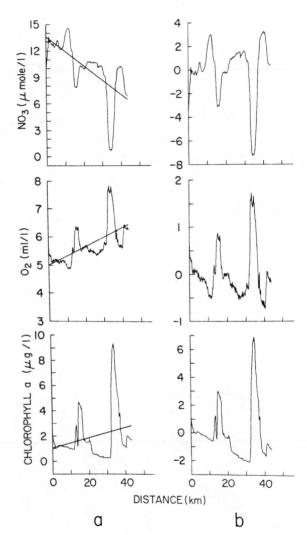

Figure 9. *Spatial segments of nitrate, dissolved O₂, and chlorophyll a from Line F of the southern region (Figure 6). Key is the same as in Figure 8.*

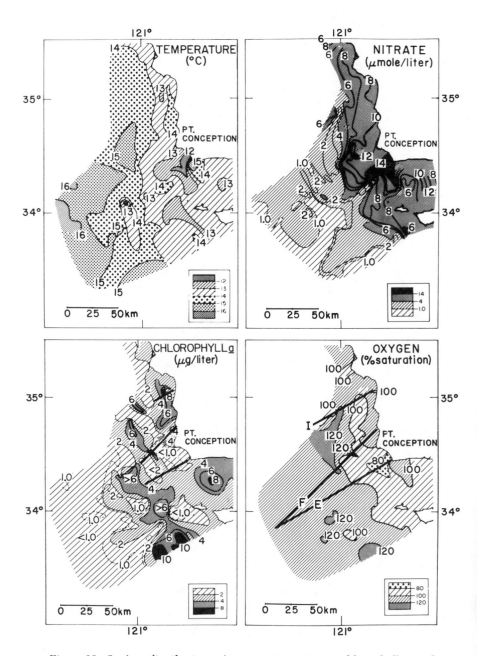

Figure 10. Surface distributions of temperature, nitrate, chlorophyll a, and percent saturation of dissolved O_2 for the southern region (Figure 6).

the width of the first peak is about twice that of the second peak. These peaks in chlorophyll (Figure 9) correspond to the two small patches (Figure 10) that Line F intersects.

A comparison between underway data segments (Figures 7–9) and smoothed spatial distributions (Figure 10) is often helpful, but it must be done carefully. Frequently, small-scale features observed in underway segments may not appear clearly (or at all) in one or more of the corresponding surface maps because the surface maps are made by smoothing and interpolating the original data. Hence, the effective spatial sampling interval (ΔL) appropriate for the surface maps often is considerably larger than that associated with the underway data. If ΔL is the spatial separation between samples, then the Nyquist maximum cutoff wavenumber is $k_{max} = 1/(2\Delta L) = 1/(\lambda_{min})$ where λ_{min} is the smallest resolvable length-scale in the data. Consequently, many features observed in underway data will be missed entirely with traditional discrete bottle sampling and/or smoothed out in the data processing that produces contour maps. A more detailed discussion of the Nyquist criteria is given in Reference 46.

Typically, these detrended data segments (Figures 7b, 8b, and 9b) are nonstationary for at least one of three reasons: they have a spatially varying mean value, they have a spatially varying mean-square value, and they have a spatially varying wave-number structure. Hence, these detrended segments are unsuitable for either correlation or spectral analyses. The dominant feature in these nonstationary segments is a discontinuity in structure that can be modeled as a binary flip-flop. The width of these pulses is typically in the range 1–10 km. Overall, the smallest component of natural variance in the data is associated with space scales between 0.5 and 1.0 km. The sampling procedures used here preclude a meaningful discussion of that portion of the natural variance associated with space scales less than 0.3 km.

No linear detrended chlorophyll a segment is shown in Figure 8b because Figure 8a indicates that no linear trend exists in this particular segment of chlorophyll a data. Strong linear trends, however, are seen in the associated structures of silicate, nitrate, dissolved O_2, and pH (e.g., Figure 8b). Moreover, although linear trends are present in the chlorophll a data shown in Figures 7b and 9b, they are not as pronounced as the corresponding linear trends in the associated chemical structures. This apparent discrepancy can be understood simply in terms of photosynthesis and the limitations of the in vivo chlorophyll a fluorescence technique. Photosynthesis can be modeled by a generalized chemical reaction of the form (52)

$$106CO_2 + 16NO_3^- + H_2PO_4^- + 17H^+ + 122H_2O$$
$$\rightleftharpoons C_{106}H_{263}O_{110}N_{16}P + 138O_2 \quad (11)$$

Thus, the strong linear trends in chemical variables (i.e., NO_3, O_2, and pH) are consistent with strong spatial gradients in primary productivity because the reactants and products of photosynthetic activity should show simple linear relationships given approximately by the ratios of the stoichiometric coefficients in Reaction 11. Unfortunately, the in vivo chlorophyll a fluorescence technique does not measure the phytoplankton biomass indexed to carbon that appears as a product in Reaction 11, but rather as an in vivo fluorescence yield. Many physical and physiological factors can affect the magnitude of the fluorescence yield. Such effects could easily reduce or eliminate any linear trends in chlorophyll a data. Similarly, processes independent of chemistry and physiology could also remove trends, for example, losses to grazing by zooplankton.

All the data taken during the experiment, of which Figures 6–9 are representative examples, support the conclusion that most of the variance in an arbitrary cross-shelf space series can be modeled as superposition of step functions, ramp functions, binary flip-flops, and smaller-scale fluctuations. A biological and physical interpretation of these length scales is given in the discussion section.

Results

Near-surface distributions of temperature, chlorophyll a, percent saturation of dissolved O_2, and nitrate for the northern region (Figure 5a) are shown in Figure 11. The cold plume of nutrient-rich water, which extends seaward from Point Arena, confirms that Point Arena is a center of local upwelling. The temperature distribution shows that the subsurface source waters reach the surface only within a narrow band (~ 10 km) adjacent to the coast. These data are consistent with the results obtained earlier (53) that indicated that upwelling is confined to within a narrow coastal zone whose offshore length scale is given by the baroclinic radius of deformation $R = hN/f$, where h is the water depth, N is the mean Brunt–Väisälä frequency, and f is the Coriolis parameter. Typically, R is about 10–15 km in most upwelling regions of the world. The temperature distribution also shows that the water adjacent to Point Reyes is more typical of warm oceanic water than of freshly upwelled water. Although the nitrate distribution shows some of the gross features of the temperature distribution (e.g., upwelling source at Point Arena and mean southward advection parallel to the coast), it also shows significant alteration by in situ biological processes. For example, the concentration of nitrate is greatly reduced from its source value along a narrow band of inshore water parallel to the coast and in a larger region near Point Reyes. The regions of reduced nitrate concentration are spatially coherent with regions of high supersaturation of dissolved O_2 and with regions of high chlorophyll a concentration, (which imply high phytoplankton biomass).

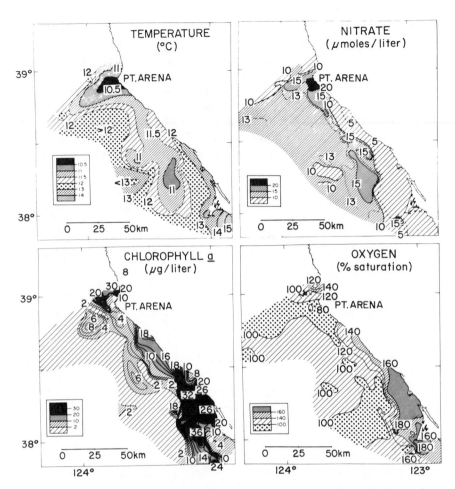

Figure 11. Surface distributions of temperature, nitrate, chlorophyll a, and percent saturation of dissolved O_2 for the northern region (Figure 6).

Corresponding surface distributions for the southern region (Figure 5b) are shown in Figure 10. Generally, the surface temperature range is warmer and the nitrate concentration range is lower than those in the northern region. The concentrations of chlorophyll a and the percent saturation of dissolved O_2, although much higher than the typical open-ocean surface values (54), are significantly lower than the corresponding concentrations measured in the northern region.

All the data are consistent with the interpretation that upwelling was more intense off Point Arena than off Point Conception during the experiment. The observed differences in the frequency and intensity of upwelling events off Point Arena and off Point Conception result from differences in the local wind field, local bathymetry, coastline topography, variation of the Coriolis parameter with latitude, and the interaction of the local flow field with the mean flow of the CCS (2).

Significant biological differences between the two regions also existed during the experiment. For example, the spatial patterns of grazing-related parameters (e.g., activity of the enzyme laminarinase in zooplankton) and their relationship to phytoplankton distributions have been examined (6). In both the northern (Point Arena–Point Reyes) and southern (Point Conception) regions, zooplankton abundance was highest in regions of high chlorophyll concentration, but this relationship was most pronounced where zonal upwelling produced alongshore bands of high phytoplankton and grazer abundance (i.e., the northern region). In the northern region, a significant negative correlation previously was shown (6) between grazer abundance and laminarinase activities. This correlation was not shown in the southern region where zonal patterns were interrupted by offshore extensions in the surface distributions of temperature and chlorophyll (Figure 10). These offshore extensions resulted in offshore patches of high chlorophyll concentration. The correspondence of high chlorophyll and high zooplankton biomass in zonal features (northern region) was interpreted previously (6) as a behavioral response (i.e., localization) by grazers to the upwelling circulation regime that optimizes residence time in the phytoplankton-dense inshore region (see Figure 11). Complex circulation in areas where zonal upwelling is interrupted by unique topographical features, such as Point Conception, however, may counteract behavioral localization by grazers and reduce their potential to persist in regions of high phytoplankton standing crop.

Figure 12 shows plots of the mean trends in the water properties sampled along the inshore ends of the cross-shelf transects (Figure 5b) for the southern region; analogous plots of the offshore trends are shown in Figure 13. Lines I and H in Figure 5b were not included in Figures 12 and 13 because they were considerably shorter than the other southern region lines. Hence, no offshore-type water was sampled along Lines I and H. Similar plots for the northern region are available (2). These mean trends were computed with the statistical procedures discussed earlier. These trends are a measure of the mean cross-shelf spatial gradients (denoted by ∇) in water properties sampled along the different line segments shown in Figure 5b. Generally, the magnitudes of these spatial gradients are greatest for the inshore segments. Spatial gradient structure is more similar among the data for the inshore segments than for the offshore segments.

The data (Figure 12) show that the mean spatial gradients of nitrate, phosphate, and silicate are spatially coherent and in phase with each other. These gradients, however, are out of phase by about $180°$ with the corre-

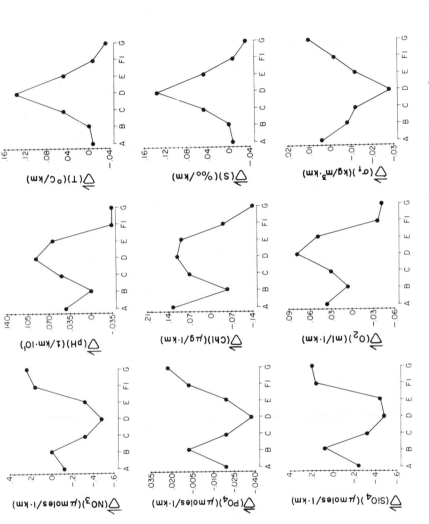

Figure 12. Mean spatial gradients (∇) of the various water properties sampled along inshore segments. The letters correspond to the cross-shelf transects in Figure 5b.

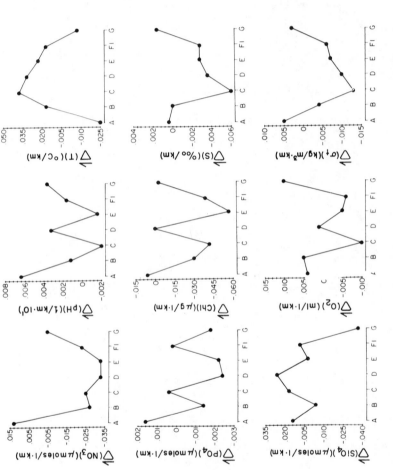

Figure 13. Mean spatial gradients (∇) of the various water properties sampled along offshore segments. The letters correspond to the cross-shelf transects in Figure 5b.

sponding spatial gradients of pH, chlorophyll a, and dissolved O_2. The coherent and in phase relationships among the inshore gradients of pH, dissolved O_2, and chlorophyll a show the importance of phytoplankton photosynthesis as a mechanism for enhanced O_2 production and CO_2 consumption in the inshore regions of the CCS. Coherent structure in the spatial gradients of offshore properties (Figure 13) is greatest among pH, dissolved O_2, and chlorophyll a. Thus, even in the offshore regions of the CCS, the importance of phytoplankton photosynthesis in the dynamics of the air–sea exchange of O_2 and CO_2 cannot be ignored. The offshore spatial gradients of nitrate, phosphate, and silicate are less coherent among themselves than when compared to their inshore counterparts and compared to the offshore spatial gradients in pH, dissolved O_2, and chlorophyll a. Also, the phase relationships among offshore spatial gradients in nitrate, phosphate, and silicate are more random compared to the regular pattern shown by the inshore spatial gradients in nutrients. The diminished levels of coherence and the more random phases in the offshore spatial gradients compared to the inshore spatial gradients partially occur because the magnitude of the spatial gradients typically decreases with increasing distance from the coast. This result is consistent with larger scale structure reported by Kenyon (55). The limiting case, of course, is the surface water of the great central gyres. Here, the magnitude of the mean spatial gradients in surface properties is near zero.

For brevity, ∇ is introduced to denote the mean spatial gradient of a water property. Thus, the notation $\nabla(O_2)$ defines the mean spatial gradient in dissolved O_2 in milliliters per liter per kilometer. Figure 14 shows plots of $\nabla(PO_4)$ versus $\nabla(NO_3)$ and of $\nabla(O_2)$ versus $\nabla(pH)$ for both inshore and offshore segments from the southern region. Analogous plots of $\nabla(NO_3)$ versus $\nabla(O_2)$ and of $\nabla(NO_3)$ versus $\nabla(pH)$ are shown in Figure 15. The inshore linear correlation coefficients (R) for $\nabla(O_2)$ on $\nabla(pH)$, $\nabla(PO_4)$ on $\nabla(NO_3)$, $\nabla(NO_3)$ on $\nabla(O_2)$, and $\nabla(NO_3)$ on $\nabla(pH)$ are 0.992, 0.961, -0.973, and -0.957, respectively. These mean correlation coefficients agree with the enhanced CO_2 and NO_3 consumption and O_2 production by photosynthesis in the inshore regions of the CCS. In addition, freshly upwelled waters have PO_4/NO_3 ratios in close agreement with the model ratio of Redfield et al. (52). The corresponding offshore mean correlation coefficients for the four plots in Figures 14 and 15 are 0.72, 0.62, 0.49, and 0.74, respectively. These lower correlation coefficients indicate that, in the CCS, nitrogen is the biologically limiting nutrient (56), and hence, that the PO_4/NO_3 ratio frequently departs from its expected Redfield ratio in the more nutrient-depleted surface waters of the offshore CCS. The warmer offshore surface waters of the CCS have pH values that approach the midlatitude equilibrium value 8.2 (36). Thus, air–sea exchange processes are probably more important in the offshore region. Under these circumstances, different exchange rates for O_2 and CO_2 also would contribute to the diminished correlations (57).

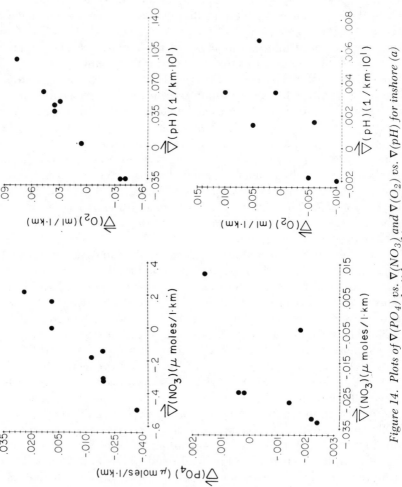

Figure 14. Plots of $\nabla(PO_4)$ vs. $\nabla(NO_3)$ and $\nabla(O_2)$ vs. $\nabla(pH)$ for inshore (a) and offshore (b) segments.

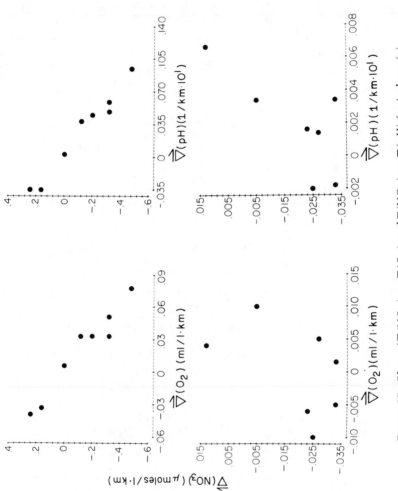

Figure 15. Plots of $\nabla(NO_3)$ vs. $\nabla(O_2)$ and $\nabla(NO_3)$ vs. $\nabla(pH)$ for inshore (a) and offshore (b) segments.

Discussion

Percent saturation of dissolved O_2 is shown as a function of the departure from atmospheric equilibrium of pCO_2 in Figure 16. Data from the northern and southern regions were sorted according to the two criteria: inshore and offshore. Inshore means inshore of the upwelling frontal boundary. The pCO_2-distribution values have a range ± 200 ppm of the atmospheric equilibrium value (~ 330 ppm), and the O_2 distributions have values as high as 185% supersaturated and as low as 70% undersaturated with respect to the in situ equilibrium. If the in situ concentrations of O_2 and CO_2 were in equilibrium, then the data would be clustered near the equilibrium coordinate (0, 100) in Figure 16. The ocean–atmosphere system in this re-

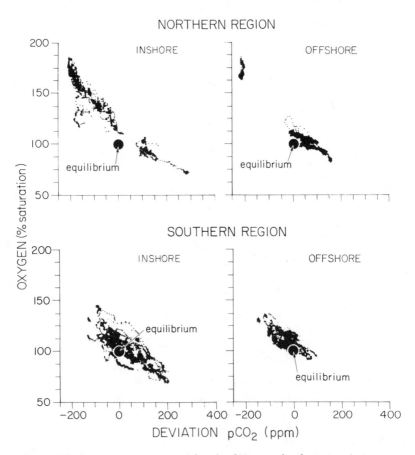

Figure 16. Percent saturation of dissolved O_2 vs. the deviation from atmospheric equilibrium of pCO_2 for both the northern (top) and southern (bottom) regions. Data are subdivided into inshore and offshore regions. The large dot indicates the equilibrium coordinate (0, 100).

gion of the CCS was not in equilibrium with respect to the air–sea exchange of O_2 and CO_2 during this experiment. The day–night differences in $\Delta p CO_2$ and ΔO_2 given by Simpson (2) are small compared to the inshore offshore differences shown in Figure 16. The offshore deviations from equilibrium are small compared to the inshore deviations, except in the region where water of inshore origin had advected offshore from the upwelling source prior to in situ chemical alteration by biological processes. Although the offshore deviations from equilibrium are small compared to the inshore deviations, they are large compared to the deviations from equilibrium observed in near-surface, open-ocean water (58).

The data in Figure 16 show that percent saturation of dissolved O_2 and the deviation of $p CO_2$ from atmospheric equilibrium are useful chemical variables for tracking the oceanic equilibrium of the in situ concentrations of CO_2 and O_2 with respect to their atmospheric counterparts. Photosynthetic alteration of in situ chemical species, however, is better determined from the stoichiometry of the photosynthetic equation (i.e., Reaction 11). Hence, the total inorganic CO_2 (micromoles per kilogram) is plotted as a function of the concentration of dissolved O_2 (micromoles per kilogram) for both the inshore and offshore areas of the southern region in Figure 17. [Analogous data for the northern region were given by Simpson (2)]. Approximately 10,000 ΣCO_2–O_2 pairs are shown in Figure 17. The mean slope (i.e., $\Sigma CO_2/O_2$) for the inshore area is -0.82, and the mean slope for the offshore area is -0.83. The negative sign shows that ΣCO_2 is consumed as O_2 is produced, a result consistent with the chemical model of photosynthesis (Reaction 11). Further, these mean slopes compare favorably with the stoichiometric ratio of 0.78 obtained directly from Reaction 11. Thus, the departure from equilibrium of the in situ concentrations of CO_2 and O_2 with respect to their atmospheric counterparts (Figure 16) results from phytoplankton photosynthesis.

In upwelling areas, such as those off Peru and California, cold, nutrient-rich subsurface waters supersaturated with respect to CO_2 and depleted of dissolved O_2 are brought to the surface (59). Thus, the physical process of upwelling can explain only part of the deviation from equilibrium shown in Figure 16. Rapid equilibration of these freshly upwelled waters with the atmosphere would tend to reduce $p CO_2$ to about 330 ppm and increase the in situ O_2 concentration to nearly 100% saturation. The small day–night differences given by Simpson (2) do not support a very rapid air–sea exchange of O_2 as some previous investigations (57) have suggested. Moreover, the equilibration times for both O_2 and CO_2 are not well known (36, 60), and the data suggest that they may be considerably longer than the typical upwelling event time scale (approaching days). Further, equilibration of midlatitude waters with the atmosphere produces a very narrow in situ pH distribution centered about pH $\simeq$ 8.2 (36). The degree of supersaturation of dissolved O_2 ($\sim$ 185%) shown in Figure 16 cannot be

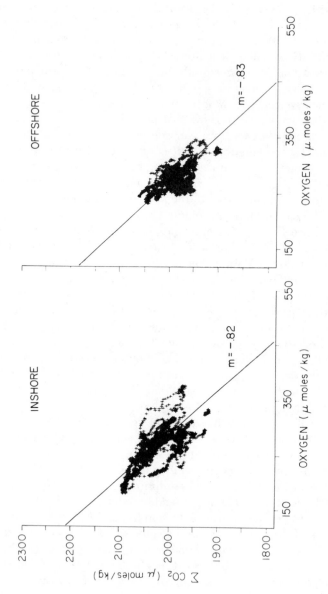

Figure 17. ΣCO_2 vs. O_2 for both the inshore and offshore areas of the southern region. Mean slopes through the data also are given.

explained by summer heating of the near-surface waters because the range of surface temperature (Figures 10 and 11) is too small to significantly affect the solubility of O_2 in seawater (34). Mixing of O_2-saturated waters of different temperatures and salinities also is inadequate to explain the level of O_2 supersaturation (*see* Reference 61). Again, only an in situ biological process (photosynthesis) can explain those regions of the data (Figure 16) undersaturated with respect to CO_2 and supersaturated with respect to O_2. The coherence and phase relationships among the mean spatial gradients in water properties shown in Figures 12–15, as well as the fluctuations about these spatial gradients (Figures 7b, 8b, and 9b), also support this conclusion. Respiration reduces the in situ concentration of dissolved O_2. Simpson (2) indicated that the effects of respiration, however, are small compared to those of photosynthesis in the biologically productive surface waters of the CCS, at least on time scales of several days. Another study (61) suggested that this condition is true for even longer time scales and in other areas of the ocean.

The data show that a clear distinction is necessary between the air–sea equilibrium concentrations of CO_2 and O_2 and equilibrium conditions associated with the in situ physical and biological processes that produce these concentrations. Analysis of the activity of the enzyme laminarinase in zooplankton samples and of phytoplankton biomass (6) showed that the approach to equilibrium conditions between zooplankton grazing rates and phytoplankton production is a function of both zooplankton localization (e.g., behavioral aggregation) and the circulation pattern associated with the coastal upwelling event. Localization occurs because of vertical migratory responses within an upwelling circulation regime (62). A complex interplay between these two processes determines whether biological equilibrium (balance of grazing rate and phytoplankton production rate) is ever approached. The physical process of upwelling is inherently a nonequilibrium process because it produces abnormally high (or low) concentrations of chemical species in the surface waters of the euphotic zone. Thus, observed nonequilibrium concentrations of O_2 and CO_2 can be interpreted as indicators of in situ nonequilibrium physical and biological processes.

The length-scale decomposition of underway data (Figures 7–9) and the spatial gradient analyses (Figures 12–15) suggest that there is a set of dominant length scales over which the major, biologically induced departures from in situ equilibrium conditions for the air–sea exchange of CO_2 and O_2 are partitioned. The most dominant length scale in this experiment appears to be mesoscale (typically, $20 \text{ km} \leq R_m \leq 50 \text{ km}$) and is associated with the mean spatial gradients in water properties along a coastal transect. The length of this scale is determined (to first order) by the position of the upwelling frontal boundary relative to the coast. The position and strength of the local upwelling source and its interaction with the mean flow field, in turn, determine the position of the upwelling front. Superimposed on this spatially varying DC component of the signal are three types of AC signals. The dominant AC component in the signal is the flip-flop type of structure (Figures 7b, 8b, and 9b) whose typical width is 1 km $\leq R_p$ $\leq$ 10 km. These structures are of the same scale as the patches discussed by Bainbridge (63). Most of the spatial variability observed is associated with the length scales R_m and R_p. Signal fluctuations associated with length scales smaller than 1 km are either coherent or random. These small-scale coherent fluctuations can be associated with smaller biological patch scales (0.3 km $\leq R_{sp} \leq 1$ km) (64). The random AC fluctuations within this same range of length scales probably are not associated with biological processes but rather with the physical processes of air–sea exchange. The possibility of coherent, biologically induced fluctuations on even smaller space scale [e.g., the centimeter scales of Cassie (65)] exists. The limitations of known underway sampling technology, however, preclude a discussion of such very small (meter scale) and fine (centimeter scale) structure. The current data set, however, suggests that the importance of biological processes on CO_2 and O_2 air–sea exchange decreases and the importance of physical processes increases with decreasing length scales in a coastal upwelling area. A more detailed discussion of sampling, statistics, and patch size is given elsewhere (66).

   This chapter has considered the importance of in situ biological processes in determining the local air–sea exchange of CO_2 and O_2 in two different regions of the CCS. Shown in Figure 18 is the worldwide distribution of phytoplankton primary productivity. These data (Figure 18) suggest that in situ biological processes within the euphotic zone have significance for the global CO_2–O_2 balance. Eastern boundary currents (e.g., Benguela current, Peru current, and California current), the Equatorial and Antarctic circumpolar currents, large areas of the Kuroshio, and the polar regions are a few of the oceanic areas with enhanced primary productivity. Generally, the large-scale spatial gradient in primary productivity (Figure 18) decreases with increasing distance from the coast. Lowest near-surface primary productivity occurs in the central gyres. The mesoscale spatial gradients (Figures 12 and 13) reported in this chapter are consistent with the large-scale gradient in Figure 18. This large-scale pattern of primary productivity (with implied CO_2 consumption and O_2 production) is consistent with small-scale and mesoscale observations reported here, observations of pH off Peru (4), observations of dissolved O_2 in the coastal waters off Oregon and Washington (68), and observations of O_2 and phytoplankton biomass off northwest Africa (69). The global distribution of primary productivity suggests that over the central gyres of the oceans the air–sea exchange of CO_2 and O_2 generally will tend to approach equilibrium conditions because near-surface biological production is low, the time scales of the dominant physical processes that affect the circulation are

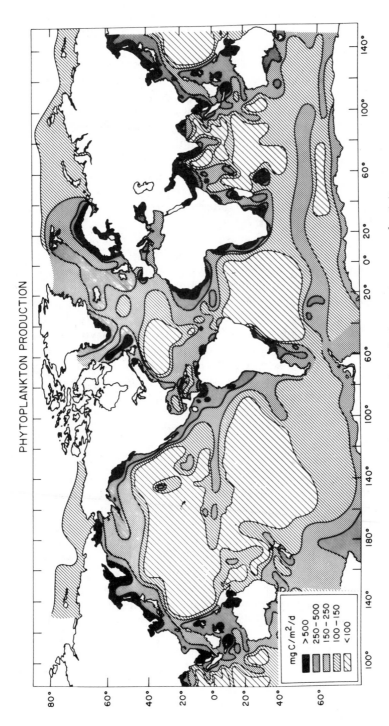

PHYTOPLANKTON PRODUCTION

mg C/m²/d

> 500
250–500
150–250
100–150
< 100

Figure 18. Global distribution of phytoplankton primary productivity. (Adapted from Ref. 67. Copyright 1981. the Food and Agriculture Organization of the United Nations.)

long, and "event"-type processes are infrequent. In most coastal domains, however, both the biological and physical processes that affect in situ CO_2 and O_2 concentrations are likely to produce large departures from equilibrium. The time scales over which these departures from equilibrium can be maintained or regenerated by physical processes (e.g., upwelling and near-surface stratification) and biological processes (e.g., photosynthesis) against the equilibrium restoring forces of air–sea exchange are uncertain. Coastal boundary currents and other biologically productive areas of the world's oceans are frequently out of equilibrium with the atmosphere. In general, the equatorial region may be in disequilibrium because upwelling is a semipermanent feature of equatorial circulation. The data in Figure 18 suggest that the global air–sea boundary condition on CO_2–O_2 exchange is complex. The effect of such a boundary condition on numerical models of CO_2 exchange and climate is discussed elsewhere (2).

Water-mass formation occurs at the surface in many of the highly productive areas of the world's oceans shown in Figure 18. The results in Figure 17 and those given in Reference 2 have shown that the near-surface ΣCO_2 in these areas is affected significantly by phytoplankton photosynthesis in the euphotic zone. These analyses support the conclusion (70) that the large uncertainties in the direct estimates of oceanic CO_2 transport based on ΣCO_2, titration alkalinity (TA), and apparent oxygen utilization* relationships (71, 72) partially result from an incomplete knowledge of the effects of biological processes on ΣCO_2, TA, and O_2 in biologically productive areas of the world's oceans. Perhaps the greatest flaw in the Brewer model (71), however, is the equilibrium assumption. A detailed analysis of this assumption is available (2).

Conclusions

The data taken from R. V. *New Horizon* indicate that, in active upwelling areas of the California Current, the in situ concentrations of CO_2 and O_2 can depart radically from equilibrium values. The departure from equilibrium results from a combination of physical (e.g., coastal upwelling, near-surface stratification, and advection) and biological (e.g., phytoplankton photosynthesis) processes. The continuous nature of the data permitted the determination of the dominant length scales associated with the biologically induced departures from the in situ equilibrium concentrations of O_2 and CO_2. The global distribution of primary productivity suggests that, in general, coastal and polar regions may not be in equilibrium with respect to O_2–CO_2 exchange between ocean and atmosphere during parts of the year. The equatorial region may be in nonequilibrium throughout the year. Water-mass formation occurs at the surface in many of these highly productive areas of the world's oceans. The high primary productivity and

*See Ref. 68 for a complete discussion of apparent oxygen utilization.

the subsequent departures from surface equilibrium conditions in these areas may seriously hamper the use of simple, direct oceanic CO_2 transport models as tools in assessing the effects of increased anthropogenic CO_2 levels on the global CO_2 and climate balance.

Acknowledgments

This work was supported by the State of California through the Marine Life Research Group (MLRG) of the Scripps Institution of Oceanography, the Calspace Institute, and the Foundation for Ocean Research. The Physical and Chemical Oceanographic Data Facility (PACODF) took the data. Steven Tighe, Bruno Bittner, and Ed Soler assisted with the computations. The figures were prepared by the MLRG Illustrations Group under the direction of Fred Crowe.

Special thanks to Nancy Hulbirt, René Wagemakers, and Guy Tapper; Ruth Ebey and Sharon McBride for typing the manuscript; Captain L. Davis and the crew of R. V. *New Horizon* for their cooperation and skill that were essential to the success of this work; and L. Haury for carefully reviewing the manuscript and making comments that were most helpful in preparing the final draft. Likewise, the constructive comments of the two reviewers greatly improved the manuscript. I am deeply indebted to J. L. Reid, Chairman of the Marine Life Research Group, for his continued support and encouragement. I also wish to thank A. Zirino, who first introduced me to the exciting problems in marine chemistry.

List of Symbols

General

t	in situ temperature (°C)
σ_t	sigma-t (kg/m^3)
$[O_2]$	measured concentration of dissolved O_2 (mL/L)
α_{O_2}	solubility of dissolved O_2 in seawater (mL/L) at 1 atm (*34*)
$\% O_2$	percent saturation of dissolved O_2

Notation for Carbonate System Calculation (Continuous Data)

CTA	calculated titration alkalinity (units per liter)
A	carbonate alkalinity (units per liter)
ΣB	total borate (mol/L)
K_B'	first apparent dissociation constant for boric acid in seawater (*40*)
pCO_2	partial pressure of CO_2 in seawater (atm); then converted to ppm for ΔpCO_2 calculation
ΔpCO_2	deviation of pCO_2 from atmospheric equilibrium (ppm)
α	solubility of CO_2 in seawater (mol/L·atm) (*39*)

K_1' first apparent dissociation constant for carbonic acid in seawater (40)

K_2' second apparent dissociation constant for carbonic acid in seawater (40)

a_H hydrogen ion activity

ΣCO_2 total inorganic CO_2 (mol/L)

Notation for Carbonate System Calculations (Discrete Data)

TA measured titration alkalinity (units per kilogram)

TCO_2 measured total CO_2 concentration (mol/kg)

$A(0)$, $A(Z)$ carbonate alkalinity at depth 0 and Z km, respectively

$K_B(0)$, $K_B(Z)$ apparent dissociation constants for boric acid at depths 0 and Z km, respectively $(40, 42)$

TB total borate concentration (mol/kg)

$K_1(0)$, $K_1(Z)$ first apparent dissociation constant for carbonic acid at depths 0 and Z km, respectively (41)

$K_2(0)$, $K_2(Z)$ second apparent dissociation constant for carbonic acid at depths 0 and Z km, respectively (41)

α_s solubility of CO_2 in seawater at 1-atm pressure (39)

Literature Cited

1. Traganza, E. D.; Conrad, J. C.; Breaker, L. C. *Coastal Upwelling-Coastal and Estuarine Sci.* 1981, *1*, 228–41.
2. Simpson, J. J. *Deep-Sea Res.* submitted.
3. Taylor, R. J. in "Proc. Nat. Res. Council Symp. Water Sampling While Underway"; National Research Council: Washington, D.C., 1980; p. 193.
4. Simpson, J. J.; Zirino, A. *Deep-Sea Res.* 1980, *27*, 733–44.
5. Simpson, J. J. in "Gas Transfer at Water Surfaces"; Brutsaert, W.; Jirka, G. H., Eds.; D. Reidel Publ. Co.: 1984; pp. 505–14.
6. Cox, J. L.; Haury, L. R.; Simpson, J. J. *J. Mar. Res.* 1982, *40*, 1127–53.
7. Orr, M. H.; Winget, C. L. in "Proc. Nat. Res. Council Symp. Water Sampling While Underway"; National Research Council: Washington, D.C., 1980; p. 147.
8. Lasker, R.; Peláez, J.; Laurs, R. M. *Remote Sens. Environ.* 1981, *11*, 439–53.
9. "Proc. Nat. Res. Council Symp. Water Sampling While Underway"; National Research Council: Washington, D.C., 1980; 293 pp.
10. Greene, M. W.; Gafford, R. D.; Rohrbaugh, D. G. "Marine Technology Society Annual Report"; National Academy Press: Washington, D.C., 1970, vol. 2.
11. Park, P. K. *J. Oceanogr. Soc. Jpn.* 1968, *3*, 1–7.
12. Paulson, C. A.; Simpson, J. J. *J. Phys. Oceanogr.* 1977, *7*, 952–56.
13. "GEOSECS Atlantic Expedition. Volume 1—Hydrographic Data"; Bainbridge, A. E., Ed.; U.S. Government Printing Office: Washington, D.C., 1981; 121 pp.
14. Carpenter, J. H. *Limnol. Oceanogr.* 1965, *10*, 141–43.
15. Hager, S. W.; Atlas, E. L.; Gordon, L. I.; Mantyla, A. W.; Park, P. K. *Limnol. Oceanogr.* 1972, *17*, 931–37.
16. Atlas, E. L.; Hager, S. W.; Gordon, L. I.; Park, P. K. "A Practical Manual for Use of the Technicon AutoAnalyzer in Seawater Nutrient Analyses"; Ref. No. 71–72, Oregon State Univ.: Corvallis, Oreg., 1971; 49 pp.
17. Strickland, J. D. H.; Parsons, T. R. *Bull. Fish. Res. Board Can.* 1968, *167*.

18. Savidge, G. *Estuarine Coastal Mar. Sci.* 1976, *4*, 617–25.
19. Lekan, J. F.; Wilson, R. E. *Estuarine Coastal Mar. Sci.* 1978, *6*, 239–51.
20. Mackus, D. I.; Louttit, G. C.; Austin, M. J. *Can. J. Fish. Aquat. Sci.* 1980, *37*, 1476–87.
21. Steele, J. H.; Henderson, E. W. *Deep-Sea Res.* 1979, *26A*, 955–63.
22. McGowan, J. A. in "Ocean Sound Scattering Prediction"; Anderson, N. R.; Zahuranec, B. J., Eds.; Plenum: New York, 1977; pp. 423–44.
23. Brandt, S. B.; Wadley, V. A. *Ecol. Soc. Aust.* 1980, *11*, 13–26.
24. Marra, J.; Heinemann, K. *Limnol. Oceanogr.* 1982, *27*, 1141–53.
25. Abbott, M. R.; Richerson, P. J.; Powell, T. M. *Limnol. Oceanogr.* 1982, *27*, 218–24.
26. Bos, D. L.; Williams, R. T. in "Proc. Workshop on Oceanic CO₂ Standardization"; U.S. Dept. of Energy, Rept. #CONF-7911173 1979; 119 pp.
27. Gran, G. *Analyst* 1952, *77*, 661–71.
28. Dyrssen, D. *Acta Chem. Scand.* 1965, *19*, 1265.
29. Edmond, J. M. *Deep-Sea Res.* 1970, *17*, 737–50.
30. Vogel, A. E. "A Text Book of Quantitative Inorganic Analysis, Theory and Practice"; Longmans, Green and Co.: London, 1957; 238 pp.
31. Takahashi, T. in "GEOSECS Atlantic Expedition. Volume 1—Hydrographic Data"; Bainbridge A. E., Ed.; U.S. Government Printing Office: Washington, D.C., 1981.
32. Broecker, W. S.; Takahashi, T. *Deep-Sea Res.* 1978, *25*, 69–95.
33. Cox, R. A.; McCartney, M. J.; Culkin, F. *Deep-Sea Res.* 1970, *17*, 679–89.
34. Weiss, R. F. *Deep-Sea Res.* 1970, *17*, 721–35.
35. Wilson, T. R. S. in "Chemical Oceanography"; Riley, J. P.; Skirrow, G., Eds.; Academic: New York, 1975; vol. I, pp. 365–413.
36. Skirrow, G. in "Chemical Oceanography"; Riley, J. P.; Skirrow, G., Eds.; Academic: New York, 1975; vol. II, pp. 1–181.
37. Weichart, G. *"Meteor" Forschungsergeb. A14*, 33–70.
38. Culkin, F. in "Chemical Oceanography"; Riley, J. P.; Skirrow, G., Eds.; Academic: New York, 1975; vol. I, pp. 121–61.
39. Weiss, R. F. *Mar. Chem.* 1974, *2*, 203–15.
40. Lyman, J. Ph.D. thesis, University of California, Los Angeles, 1956.
41. Mehrbach, C.; Culberson, C. H.; Hawley, J. E.; Pytkowicz, R. M. *Limnol. Oceanogr.* 1973, *18*, 897–907.
42. Li, Y. H.; Takahashi, T.; Broecker, W. S. *J. Geophys. Res.* 1969, *74*, 5307–23.
43. Takahashi, T.; Kaiteris, P.; Broecker, W. S.; Bainbridge, A. E. *Earth Planet Sci. Lett.* 1976, *32*, 450–67.
44. Simpson, J. J. *Deep-Sea Res.*, in press.
45. Jenkins, G. M.; Watt, D. G. "Spectral Analysis and Its Application"; Holden Day: San Francisco, 1968; 523 pp.
46. Bendat, L. S.; Piersol, A. G. "Random Data: Analysis and Measurement Procedures"; Wiley: New York, 1971; 407 pp.
47. Dengler, A. T., Jr., Ph.D. thesis, University of California, San Diego, 1981.
48. Peláez-Hudlet, J. Ph.D. thesis, University of California, San Diego, 1983.
49. Fasham, M. J. R.; Pugh, P. R. *Deep-Sea Res.* 1976, *23*, 527–38.
50. Frankignoul, C. *J. Geophys. Res.* 1974, *79*, 3459–62.
51. Paulson, C. A.; Simpson, J. J. *J. Geophys. Res.* 1981, *86*, 11,044–54.
52. Redfield, A. C.; Ketchum, B. H.; Richards, F. A. in "The Sea"; Hill, M. H., Ed.; Interscience: New York, 1960; vol. 2, pp. 26–77.
53. Yoshida, K. *Rec. Oceanogr. Wks. Jpn.* 1955, *2*, 8–20.
54. Venrick, E. L. *Deep-Sea Res.* 1979, *26*, 1153–78.
55. Kenyon, K. E. *Deep-Sea Res.* 1983, *30*, 349–70.
56. Thomas, W. H. *Bull. Inter-Amer. Trop. Tuna Comm.* 1977, *17*, 173–212.
57. Kanwisher, J. *Deep-Sea Res.* 1963, *10*, 195–207.
58. Keeling, C. D. *J. Geophys. Res.* 1968, *73*, 4543–53.
59. Smith, R. L. in "Oceanography and Marine Biology, An Annual Review"; Barnes, Harold, Ed; 1968; vol. VI.
60. Broecker, W. S.; Li, Y.-H.; Peng, T.-H. in "Impingement of Man on the Oceans"; Hood, P. W., Ed.; Wiley: New York, 1971; pp. 287–324.

61. Shulenberger, E.; Reid, J. L. *Deep-Sea Res.* **1981**, *28*, 901–19.
62. Peterson, W. T.; Miller, C. B.; Hutchinson, A. *Deep-Sea Res.* **1979**, *26*, 467–94.
63. Bainbridge, R. *Biol. Rev.* **1957**, *32*, 91–115.
64. Bernhard, M.; Rampi, L. *Botanica Gotheburgersia* **1965**, *3*, 13–24.
65. Cassie, R. M. *New Zealand J. Sci. Technol.* **1959**, *2*, 398–409.
66. Simpson, J. J.; Tighe, S. *Biol. Oceanogr.*, in press.
67. "Atlas of the Living Resources of the Sea"; Food and Agriculture Organization of the United Nations: Rome, 1981; pp. 13–19.
68. Stefansson, U.; Richards, F. A. *Deep-Sea Res.* **1964**, *11*, 355–80.
69. Minas, H. J.; Packard, T. T.; Minas, M.; Coste, B. *J. Mar. Res.* **1982**, *40*, 615–41.
70. Schiller, A. M.; Gieskes, J. M. *J. Geophys. Res.*, **1980**, *83*, 2719–27.
71. Brewer, P. G. *Geophys. Res. Lett.*, **1978**, *3*, 997–1000.
72. Chen, C. T.; Pytkowicz, R. M. *Nature*, **1979**, *281*, 361–65.

RECEIVED for review July 5, 1983. ACCEPTED March 14, 1984.

INDEXES

AUTHOR INDEX

SUBJECT INDEX

A

453

Copyediting and indexing by Deborah Corson and Robin Giroux
Production by Meg Marshall and Karen McCeney
Jacket design by Pamela Lewis
Managing Editor: Janet S. Dodd

Typeset by Action Comp Co., Baltimore, Md.
and Hot Type Ltd., Washington, D.C.
Printed and bound by Maple Press Co., York, Pa.